MOLECULAR OPTICAL ACTIVITY
AND THE CHIRAL DISCRIMINATIONS

Molecular optical activity and the chiral discriminations

STEPHEN F. MASON

Professor of Chemistry
University of London
King's College

CAMBRIDGE UNIVERSITY PRESS

Cambridge

London New York New Rochelle

Melbourne Sydney

Published by the Press Syndicate of the University of Cambridge
The Pitt Building, Trumpington Street, Cambridge CB2 1RP
32 East 57th Street, New York, NY 10022, USA
296 Beaconsfield Parade, Middle Park, Melbourne 3206, Australia.

First published 1982

Printed in Great Britain at the University Press, Cambridge

Library of Congress catalogue card number: 82–1125

British Library cataloguing in publication data

Mason, Stephen F.
Molecular optical activity and the chiral
discriminations.
1. Olefins 2. Circular dichroism
I. Title
547′ . 412 QD305.H7
ISBN 0 521 24702 0

CONTENTS

	Preface	vii
1	INTRODUCTION	1
1.1	Provenance and discovery	1
1.2	The origin of optical rotation	2
1.3	Optical activity and molecular dissymmetry	6
1.4	The foundation of classical stereochemistry	7
1.5	Relative stereochemical configuration	9
2	OPTICAL ROTATORY DISPERSION AND CIRCULAR DICHROISM	18
2.1	Optical activity and molecular spectroscopy	18
2.2	The quantum theory of optical activity	21
2.3	The evaluation of transition strengths	23
2.4	Chiroptical instrumentation	26
3	GENERAL MECHANISMS FOR OPTICAL ACTIVITY	34
3.1	The independent-systems procedure	34
3.2	The one-electron theory	34
3.3	The dynamic polarization mechanism	38
3.4	The two-group electric-dipole mechanisms	42
3.5	Relations between the general models for optical activity	47
4	THE SYMMETRIC CHROMOPHORE IN A CHIRAL MOLECULAR ENVIRONMENT	51
4.1	The symmetric and the dissymmetric chromophore	51
4.2	The carbonyl chromophore	51
4.3	The octant rule for the carbonyl $n \rightarrow \pi^*$ Cotton effect	52
4.4	The octant rule and the one-electron mechanism	54
4.5	The octant rule and the dynamic polarization mechanism	55
4.6	The olefin chromophore	59
5	INHERENT DISSYMMETRY	64
5.1	Helical four-centre systems	64
5.2	Dipole-length and dipole-velocity rotational strengths	66
5.3	The helicene series	69
6	DIMERS AND POLYMERS	72
6.1	The biaryl series	72
6.2	Dimeric alkaloids	80
6.3	Dihedral coordination compounds	83

6.4	Biopolymers	88
6.5	Microscopic and macroscopic helices	97
7	CHIRAL TRANSITION-METAL COMPLEXES	103
7.1	The circular dichroism spectroscopy of tris-chelate complexes	103
7.2	Conformational and configurational optical activity	108
7.3	Crystal-field theory of d-electron optical activity	116
7.4	Dynamic-polarization theory of d-electron optical activity	120
7.5	Chiral tetrahedral coordination	125
7.6	Chiral trigonal bipyramid coordination	132
8	VIBRATIONAL OPTICAL ACTIVITY	138
8.1	The fixed partial charge model	138
8.2	Dynamic coupling models	139
8.3	Raman optical activity	143
9	THE CHARACTERISATION OF CHIRAL STRUCTURES	148
9.1	Pasteur on dissymmetry in nature	148
9.2	Enantiomer–racemate phase equilibria	152
9.3	Phase-equilibria methods for relative configuration	162
9.4	Enantiomer and racemate crystal structures	164
9.5	Absolute stereochemical configuration	175
10	ENANTIOMERIC DISCRIMINATION	186
10.1	Optical resolution procedures	186
10.2	Chiral photodiscrimination	196
10.3	Optical activation	202
10.4	Chiral biodiscrimination	219
11	CHIRAL ENERGY DISCRIMINATION	227
11.1	Magnitudes of the energy differentials	227
11.2	Intermolecular discrimination energies	228
11.3	Origins of molecular chirality	238
	Suggestions for further reading	248
	Bibliography and references	250
	Author index	266
	Subject index	271

PREFACE

Michael Faraday (1791-1867) regarded chemistry as part of a single physical science, with optics and electromagnetism. Already, in Faraday's later years, organic and inorganic chemistry had divided, and physical chemistry separated off as an individual discipline before the end of the nineteenth century. These traditional divisions have fossilised to a degree, and no longer correspond wholly to the contemporary state of chemical science. The search for new substances, and for more efficient ways of synthesising known materials, endures as the central core of the science, forming the major division of the subject. The second and more internal quest, that of evolving new unifying principles and improved practical and theoretical analytical procedures, became progressively dependent upon developments in atomic and molecular physics from the beginning of the twentieth century, forming the subject of chemical physics. The now-extended division of chemical physics, with spectroscopic and diffraction procedures and wave-particle theory, restores something of Faraday's earlier vision.

During the period when there was a distinct and largely autonomous chemical methodology, as formulated by Laurent and practised by Pasteur, Kekulé, Le Bel, van't Hoff, and Werner, chiral substances played a central role in the development of classical stereochemistry. Faraday's perception, 'that polarized light was a most subtle and delicate investigator of molecular condition', became amply illustrated in the later development of both organic and inorganic structural theory.

The present book is mainly concerned with the chemical physics of the unique properties of chiral molecules, their optical activity and their discriminatory interactions. The general principles of enantiomer separation and asymmetric synthesis are discussed, together with those underlying chiral bioapplications. In addition, the successive stages in the development of structural theory which were dependent upon the study of optically-active materials are briefly surveyed. The book is intended for graduate students and other research workers in the chemical and allied fields, being based upon courses of graduate lectures given over the past decade.

In assembling material for the book, I am indebted to numerous former and current coworkers and colleagues, especially to Dr Alex Drake and Dr Reiko Kuroda, for their constructive and stimulating contributions. Particular thanks are due to Dr George Ryback, of the Shell Biosciences Laboratory, for the statistical analysis of the entries in *The Pesticide Manual,* cited in § 10.4, and to Professor Jean Jacques and Dr André Collet, of the Collège de France, for the opportunity to consult a manuscript draft of their recent book, with Dr Samuel H. Wilen, of the City University of New York, on *Enantiomers, Racemates and Resolutions* (Wiley, New York, 1981). Dr Collet is additionally thanked, together with Dr Laurence Barron of Glasgow University, for constructive comments and criticism of the manuscript.

Finally, I must record my appreciation of the skilled secretarial assistance of Mrs T.L. Chin, for her patience and precision in typing the manuscript.

London   Stephen Mason

1 INTRODUCTION

1.1 Provenance and discovery

A few years before Le Bel and van't Hoff joined his laboratory, Wurtz (1869) produced a history of chemical theories from the time of Lavoisier, opening with a claim, deprecated by his English translator, Watts, that 'chemistry is a French science'. Restricted to the early study of optically-active materials, the remarkable claim of Wurtz has some substance. With the exception of the notable contribution of J. van't Hoff (1852-1911), the main advances in the field of optical activity and molecular dissymmetry during the nineteenth century were due to J. -B. Biot (1774-1862), A. Fresnel (1788-1827), L. Pasteur (1822-1895), J.A. Le Bel (1847-1930), A.A. Cotton (1869-1951), and other French scientists. Although the tradition later became attenuated, the first of the modern photoelectric circular dichroism spectrometers originated in France (Grosjean and Legrand, 1960).

While the prime mover of the chemical revolution, Lavoisier (1743-1794), was destroyed by the political revolution in France, the Convention set up, from 1794, a number of scientific and technical institutions, the École Polytechnique being the most notable. The teachers and former students of these schools discovered a number of novel optical effects and established the transverse wave theory of light, while effectively founding the subject of physico-chemical optics. Malus (1809) discovered that light reflected from transparent media, glass or water, gives one instead of two images when transmitted through a doubly-refracting crystal, such as calcite. According to the then-current Newtonian theory, light consists of a beam of asymmetric particles, which are normally randomly orientated but become ordered in a doubly-refracting crystal to give two rays. Malus supposed that the light particles are dipolar, like a magnet, and that a reflecting surface orientated the dipoles, to give what he termed 'polarized light'.

Arago (1811) modified the experiment of Malus by inserting a quartz plate, cut perpendicular to the optic axis, between the polarizer, composed of a pile of glass plates, and the calcite analyser, observing a spectrum of coloured images as the polarizer or the analyser was rotated. His colleague Biot (1812) showed that the effect arises from the rotation of the plane of polarization through increasing angles from the red to the violet region of the visible spectrum, and that there are two forms of quartz, one dextrorotatory and the other laevorotatory. Subsequently, Biot (1817) established an approximate inverse square law between the angle of the optical rotation, α, and the wavelength of the radiation, λ,

$$\alpha = K/\lambda^2 \tag{1.1}$$

where K is a constant characteristic of the optically-active substance, crystalline or fluid.

From the general prevalence of optical activity among the natural products in the liquid or vapour phase, Biot concluded that optical rotation is essentially a molecular property, characterised by what he described as the 'molecular rotatory power', or the specific rotation as the property was subsequently termed. Biot (1838) defined the specific rotation, $[\alpha]$, for a given wavelength, λ, and temperature, T, by the relation,

$$[\alpha]_{\lambda}^{T} = \alpha/lc \tag{1.2}$$

where α is the observed rotation in degrees, l the pathlength in decimetres, and c is the concentration of the optically-active substance in g cm^{-3}. Biot adopted the convention of expressing the pathlength in decimetres for fluids, as opposed to millimetres for crystals, 'in order that the significant figures may not be preceded by two zeros to no purpose', and in like manner the molar rotation was defined (Biot, 1842) as the product of the specific rotation and the molecular mass, M, 'for brevity divided by 100',

$$[M]_{\lambda}^{T} = (M/100) [\alpha]_{\lambda}^{T} \tag{1.3}$$

Biot was the author of the chemical convention that positive optical activity corresponds to a clockwise rotation of the plane of polarization to an observer viewing the source of radiation through the polarizer, sample tube, and analyser. Herschel (1822) proposed the contrary convention, which was adopted by physicists until recent years, that the observer view the optical train from the radiation source, with a clockwise rotation taken as positive. Thus a Herschel dextrorotation corresponds to a Biot laevorotation.

1.2   The origin of optical rotation

Optical activity raised serious problems for the traditional particle theory of light, then generally ascribed to Newton. As a supporter of the

Newtonian theory of light, Biot abandoned the study of optical activity from 1818, when the transverse-wave theory began to gain ground, and resumed his optical rotation investigations only in the 1830s, after the wave theory had become firmly established. The longitudinal-wave theory of light, proposed by Huygens (1690), was revived by Young (1802) in order to account for diffraction effects. These effects were investigated by Fresnel (1816) who showed that two parallel-polarized light rays from a single source give an interference pattern, whereas two rays with a mutually perpendicular polarization, produced with a doubly-refracting crystal from the same source, do not interfere with one another. Writing to Arago in 1817, Young suggested that Fresnel's experiments required light waves to have transverse components. Fresnel (1819, 1821) developed the transverse wave theory of light, accounting for most of the optical effects then known, and testing the novel expectations of the theory, such as the luminous area expected at the centre of the shadow of a small circular object.

Subsequently, Fresnel (1824) applied the transverse wave model of light to optical activity. The application led him to the discovery of a new form of polarization, left- and right-circularly polarized light. Fresnel ascribed optical rotation to the circular birefringence of optically-active media. A plane-polarized transverse wave form may be regarded as the resultant of a left- and a right-circularly polarized component with equal amplitudes. For circularly polarized light of frequency, ω, in radians per second, propagated in the z-direction through a medium of refractive index, n, the amplitudes of the radiation field are,

$$a = a_0 \{ \text{i} \cos[\omega(t - nz/c)] \mp \text{j} \sin[\omega(t - nz/c)] \} \tag{1.4}$$

where t is the time and c is the velocity of light in vacuo. The upper sign on the right-hand side of (eqn 1.4) refers to right- and the lower to left-circular radiation, and i and j are unit vectors parallel to the x- and the y-axis, respectively, perpendicular to the direction of propagation. At a given time, e.g. $t = 0$, the envelope of the amplitude vectors has the spatial form of a right-handed helix around the z-direction for right-circularly polarized (RCP) light, or a left-handed helix for left-circularly polarized (LCP) radiation (fig. 1.1). At a given point, e.g. $z = 0$, the amplitude vectors appear, to an observer looking in the $-z$-direction at the light source, to be rotating in time clockwise for RCP light and anticlockwise for LCP light (fig. 1.2).

Propagated through an optically-active medium the two circular components of a plane-polarized transverse wave travel with different velocities. If the incident radiation is x-polarized, with the phase relation (eqn 1.4) between the two circular components, and the refractive index of the medium is larger for LCP than RCP light, $(n_L > n_R)$, the former component is relatively delayed in traversing the medium, giving rise to a clockwise, (+)- or dextrorotation of the plane-polarized resultant transverse wave towards the y-axis (fig. 1.2). For incident x-polarized radiation the linear amplitude, a_y, is zero, but both of the linear amplitudes are non-zero in the emergent plane-polarized transverse wave, lying in the ratio,

$$a_y/a_x = \tan[\omega l(n_L - n_R)/2c] \tag{1.5}$$

Fig. 1.1. The spatial form at a fixed time of the transverse-wave amplitude-vectors for, (a) right-handed circularly-polarized light, (b) left-handed circularly polarized light, and (c) the plane-polarized resultant of a superposition of (a) and (b). The horizontal arrow denotes the direction of propagation, and the arrows perpendicular to this direction represent the instantaneous spatial distribution of the radiation field.

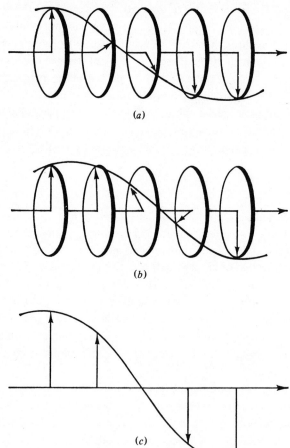

(a)

(b)

(c)

The ratio of the linear amplitudes represents the tangent of the angle, ϕ, in radians, through which the plane of polarization has been rotated in traversing the medium, giving Fresnel's equation,

$$\phi = (n_L - n_R)l\pi/\lambda \tag{1.6}$$

where the pathlength, l, is measured in the same units as the wavelength, λ.

Fresnel designed three devices for the production of circularly polarized light; a triprism of $(+)$- and $(-)$-rotatory crystalline quartz, the quarter-wave plate, and the Fresnel rhomb (Lowry, 1935). The triprism produces rays of LCP and RCP light, with an angular separation dependent upon the circular birefringence of quartz and the dimensions of the prisms, from a collimated incident beam of unpolarized or plane-polarized light. The other two devices are based upon the view that RCP or LCP light is made up of two orthogonal plane-polarized transverse wave components with equal amplitudes and a mutual phase retardation of $\pm\,\lambda/4$. The z-propagation of plane-polarized radiation through a linearly-birefringent plate, with the plane orientated at $\pm\,\pi/4$ to the principal directions of the plate birefringence ($n_x - n_y$), produce LCP or RCP radiation if the plate thickness, l, satisfies the quarter-wave condition,

$$l(n_x - n_y) = \lambda/4 \tag{1.7}$$

The quarter-wave plate is effective only for the wavelength, λ, whereas the Fresnel rhomb, in which the appropriate phase retardation is achieved by internal reflection in a transparent medium, is approximately achromatic over a wide wavelength range, e.g. the visible region for a glass rhomb. With the devices for

Fig. 1.2. The time-dependency of the amplitude vectors of left- and right-circularly polarized radiation, and the plane-polarized resultant at, (*a*) the point of incidence and (*b*) the point of emergence, from a dextrorotatory optically-active medium.

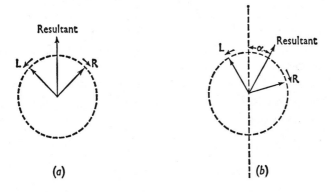

(*a*)   (*b*)

the production of RCP and LCP light, Fresnel's equation (eqn 1.6) was found to be supported by direct measurements of the circular birefringence, e.g. $(n_L - n_R) = 7.1 \times 10^{-6}$ for dextrorotatory crystalline quartz at the sodium D-line, 589 nm, where the optical rotation is $+21.7°$ mm^{-1}.

From his conception of the spatial helical morphology of RCP and LCP radiation, Fresnel (1824) proposed by analogy that the stereochemical form of the molecules making up crystalline quartz is essentially helicoidal. Herschel (1822) had already noted that dextrorotatory and laevorotatory quartz crystals, while similarly hexagonal in their main features, are morphologically distinguished by minor facets which relate the two crystal types by object to mirror-image symmetry, like the optical rotation sign.

1.3   Optical activity and molecular dissymmetry

The optical and crystallographic developments of the 1820s were only slowly assimilated into the chemical theory of molecular structure. Thomas Graham in his textbook *Elements of Chemistry* (1842) informs his students that the problem of optical isomerism 'defeated every attempt at explanation'. The sodium ammonium salts of (+)-tartaric acid and of the corresponding racemic acid, Graham continued, 'not only coincide in the proportion of their water and other constituents, and in the composition of their acids, but also in their external form, having been observed by Mitscherlich to be isomorphous'.

The chemical significance of crystallography was generally appreciated even before Mitscherlich's law of isomorphism (1819), particularly in France, where the concept prevailed throughout the nineteenth century that the molecule and the corresponding crystal in morphological form are 'images of each other'. It was a concept which had led Ampère in 1814 to a rival version of Avogadro's hypothesis (1811), based upon the view that all molecules, including those of the elementary gases, are necessarily three dimensional and thus are at least tetra-atomic. For Ampère, one-dimensional diatomic and two-dimensional triatomic molecules were inadmissible, as there could be no corresponding macroscopic crystal form. Later the concept led Le Bel in 1892 to the conclusion that the valencies of carbon are not exactly tetrahedral, since carbon tetrabromide and carbon tetraiodide at ambient temperature form anisotropic crystals belonging to the monoclinic system. According to Le Bel, if the valencies of carbon were strictly tetrahedral, all CX_4 derivatives would form isotropic crystals, belonging to the cubic system. In fact, carbon tetrabromide undergoes a transition at 46°C to a plastic cubic phase in which the molecules rotate in the crystal lattice.

More significantly the concept encouraged Louis Pasteur as a student in 1848 to examine in detail the supposed isomorphism of the (+)-tartrate and the inactive racemate salts. At ambient temperature the inactive sodium ammonium racemate gave hemihedral crystals of two enantiomorphous types, separable by hand-sorting, one type having the morphology of the corresponding (+)-tartrate crystals and the other exhibiting mirror-image facets. In solution each crystal type was optically active, the specific rotations having an equal magnitude but opposed signs. The conclusion of Pasteur, that the molecular structures of (+)- and (-)-tartaric acid have mirror-image morphologies, like the corresponding crystal forms, was taken up only 25 years later, notably by Le Bel. Meanwhile, on repeating the work of Pasteur in Italy, at a temperature above the transition point (27°C), Scacchi (1865) obtained only one crystal type, with a holohedral form, that of sodium ammonium racemate with both (+)- and (-)-tartrate in a single centrosymmetric lattice.

Before the time of Pasteur, enantiomeric substances were termed 'physical isomers', since they appeared to differ in no respect other than the property of rotating the plane of polarized light in a clockwise or anticlockwise sense. Pasteur added to this property three types of chemical or physical chiral recognition and discrimination, summed up in his review lectures on molecular dissymmetry of 1860. Pasteur coined the term 'dissymmetric' to describe structural forms which are not superposable by translations and rotations alone upon the corresponding mirror-image structure. The equivalent term 'chiral' (handed) was introduced by Kelvin in his Baltimore lectures of 1884 (Kelvin, 1904). The first type of chiral discrimination identified by Pasteur is the preferential co-crystallisation of an enantiomer with its optical antipode, shown by the majority of racemates, which crystallise as such from solution or from the melt. Only some 250 cases have been identified, of the many thousand studied, where a racemate forms a conglomerate of two enantiomorphous crystal types, each type containing a single optical isomer. These are the relatively rare cases of 'spontaneous resolution by crystallisation'. The second of Pasteur's discoveries, diastereomeric discrimination, covers the different chemical and physical properties of salts and other combinations of the two optical isomers in a racemate with a single enantiomer of another chiral substance (+)-A(+)-B and (-)-A(+)-B. The unequal solubilities of such diastereomers affords a general procedure for the optical resolution of racemates. The third discrimination, reflecting Pasteur's growing preoccupation with microbiology, appeared in the biochemical selectivity of micro-organisms between the two enantiomers in a racemic mixture, exemplified by the metabolism of specifically the (+)-isomer of racemic tartaric acid by *Penicillium* moulds.

1.4 The foundation of classical stereochemistry

Pasteur did not concern himself greatly with the internal development of the chemical sciences after 1860, although one of his teachers, Laurent, had laid the foundations of structural chemistry in his *Chemical Method*, (1853, English edition 1855). The book was regarded by its English translator, Odling, to be as significant as the *Elements of Chemistry* by Lavoisier (1789). Laurent argued that, while it is not possible to deduce molecular structures from chemical reactions, it is feasible to deduce possible chemical reactions and products from a hypothetical structure of the starting material. Laboratory experiments would then support or falsify the proposed hypothetical structure. The chemical method of Laurent was taken up, notably, by Kekulé who proposed the regular hexagonal structure for benzene in 1866 and worked out the possible number and types of structural isomers afforded by mono-, di-, and poly-substitution reactions, for both the homo- and hetero-substitution case. The proposed hexagonal benzene structure was amply supported by the subsequent experimental work of his students, particularly that of Körner, and Kekulé was tempted to extend his two-dimensional structural chemistry to aliphatic compounds and inorganic substances, depicting perchloric acid as a chain of four oxygen atoms, terminated by a hydrogen atom at one end and a chlorine atom at the other.

The optical isomerism of aliphatic compounds was anomalous in the structural theory of Kekulé, as Wislicenus pointed out in 1869 when he isolated more isomers of lactic acid than could be accommodated by flatland molecular models. The anomaly was resolved in 1874 independently by van't Hoff, who extended the principles and procedures of Kekulé and Laurent, and Le Bel, who gave a structural form to the general morphological analogies drawn by Fresnel and Pasteur between circular radiation and a chiral molecule and between the latter and a hemihedral crystal. From the model that the valencies of carbon are directed towards the vertices of a regular tetrahedron van't Hoff derived a number of detailed expectations. If a molecule contains n inequivalent asymmetric carbon atoms, there are 2^n optically-active isomers, grouped into enantiomeric pairs. If, however, the n asymmetric carbon atoms are equivalent, as in tartaric acid, there are $(1/2)2^{n/2}$ inactive *meso*-isomers and $(1/2)2^n$ optically-active isomers, again grouped into enantiomeric pairs. In the cumulene series, $RR'C{=}(C)_n{=}CRR'$, and the analogous spiran series, the isomerism is optical for n odd, and geometric, *cis* and *trans*, for n even, including the olefins ($n = 0$). The investigation of these and other expectations of the tetrahedral model for the carbon valencies gave organic stereochemistry its definitive three-dimensional form.

Inorganic structural chemistry remained under the domination of the unproductive organic analogies deriving from Kekulé until the work of Werner. The complex hexammines, such as $[Co(NH_3)_6]^{3+}$, were regarded as linear chains of ammonia groups terminated by the metal atom, like the chains of methylene groups terminated by a halogen in the alkyl halides. Adopting the procedure of the organic stereochemists, going back to the chemical method of Laurent, Werner proposed the metal-centred octahedral structure for the hexammine complexes, and deduced the number and type of isomeric products expected from ligand-replacement reactions, exhanging the ammonia for halide ligands, or for the chelating ligand, ethylenediamine. Disubstitution with a halide affords two geometric isomers, *cis* and *trans*, for octahedral coordination, as opposed to three geometric isomers, one being resolvable into optical isomers, for the trigonal prism coordination polyhedron. Replacement of the six ammonia ligands by three ethylenediamine chelate rings gives two optical isomers for octahedral coordination, in contrast to the two achiral geometric isomers expected for regular trigonal prismatic coordination. By investigating these and analogous expectations, Werner established, between 1893 and 1914, the octahedral stereochemistry of six-coordinate metal complexes and the square-planar structure of a number of their four-coordinate analogues.

1.5   Relative stereochemical configuration

The use of three-dimensional molecular models to interpret the structure of natural products and synthetic organic compounds raised the problem of absolute and relative stereochemical configuration. While it was not possible to specify for a given chiral molecule which of the two structural models, related by mirror-symmetry, is the (+)- or the (−)-isomer, it appeared feasible to relate the stereochemical configuration of a series of enantiomers by chemical interconversions. Emil Fischer in 1891 introduced a convention, which was systematised by Rosanoff (1906), that the stereoisomers chemically related to (+)-glyceraldehyde have the D-configuration, while those connected by chemical interconversions with (−)-glyceraldehyde have the L-configuration. An absolute character was imposed on the convention by assigning to D-(+)-glyceraldehyde an arbitrary absolute stereochemcial configuration. In Fischer's convention, the valency tetrahedron of the asymmetric carbon atom in compounds of the R—CHX—R′ type is so orientated that the tetrahedral edge of the main chain, R—C—R′, lies to the rear and is directed vertically with the most highly oxidised group, R or R′, at the top. The tetrahedral edge connecting the H atom to the X group then lies to the front with a horizontal orientation. If the group X lies on the right-hand side, with the H atom to the left, the structure has the Fischer D-configuration, as in D-(+)-glyceraldehyde and in D-(−)-lactic acid. The converse relation of the X group to the H atom assigns the structure to the L-series, as in the case of the majority of the natural α-amino acids (fig. 1.3).

The application of Fischer's convention was somewhat limited by the uncertain prevalence of Walden's inversions. Walden found (1896*a*, 1897) that the regeneration of a chiral substance in a closed cycle of reactions gives, in some cases, a product which is the optical antipode of the starting material, so that an inversion of stereochemical configuration must occur in one, or an odd number, of the stages in the cycle (fig. 1.4). The majority of the conversions relating configuration according to the Fischer convention did not involve reaction at the asymmetric centre or centres and were free from the ambiguity posed by the Walden inversion. In 1951, when Bijvoet, Peerdeman and van Bommel determined the absolute configuration of (+)-tartaric acid from the anomalous scattering observed in an X-ray diffraction study of the sodium rubidium salt, it was found that Fischer had chosen the correct configuration for D-(+)-glyceraldehyde, which served as the basis of his convention.

The Fischer–Rosanoff convention gives rise, however, to some ambiguities of nomenclature. Natural (+)-tartaric acid belongs to the L-series if the lower asymmetric carbon atom is taken as the reference centre, but to the D-series if the assignment is based upon the upper asymmetric centre. The original convention for the carbohydrate series took the lowermost asymmetric carbon as the primary reference centre for specifying a D- or L-configuration, whereas the uppermost asymmetric carbon in the Fischer projection structure was adopted as the reference centre in the amino-acid series.

In order to avoid such ambiguities, and to accommodate the wider range of chiral molecular types which had become available over the intervening period,

Fig. 1.3. The tetrahedral form (*a*) and the Fischer projection (*b*) of D-(+)-glyceraldehyde; and the absolute stereochemical configuration (*c*) with the Fischer projection (*d*) of (+)-tartaric acid.

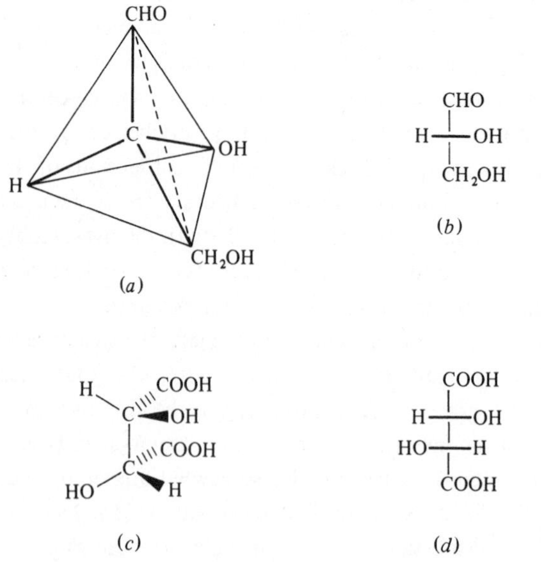

Cahn, Ingold and Prelog developed an alternative convention for the description of absolute stereochemical configuration, the sequence-rule method, given a definitive form in 1966. According to the sequence-rule procedure, the atoms directly bonded to a tetracovalent chiral centre, X, in a molecule, Xabcd, are arranged in a priority order of increasing atomic number, or of mass number in the case of isotopic chirality. A missing substituent, represented by the lone-pair of an asymmetric sulphur atom, for example, has an image of zero atomic number for sequence priority. If two or more of the atoms of the groups, a, b, c, or d, directly bonded to the chiral centre in Xabcd are identical, the next-nearest neighbouring atom to X in the group determines, by its atomic or mass number, the priority of that group. Should the next nearest neighbours be equivalent, successive neighbouring atoms outwards from the chiral centre are considered until the priority of the groups is determined.

With an established priority order, a > b > c > d, in terms of decreasing atomic or mass number, the group of lowest priority, d, in a tetrahedral model is directed away from the observer, and the clockwise or anticlockwise aspect of the priority sequence, a > b > c, around the tetrahedral face presented to the observer is ascertained. If the observed order presents a clockwise sequence, the configuration is specified as (R), from *rectus* (right), or as (S) for a counter-clockwise sequence, from *sinister* (left) (fig. 1.5). Thus (+)-tartaric acid has, unambiguously, the (2R,3R)-configuration. Many of the simpler D-molecules have the (R)-configuration, as in the case of (+)-glyceraldehyde, and most of the L-α-amino acids have the (S)-configuration, but the change to a priority based upon atomic number, rather than oxidation state, requires L-cysteine to be assigned the (R)-configuration.

In chiral biaryls, allenes, spirans, and other molecules without a particular asymmetric centre, a chiral axis is defined by the direction of the internuclear biaryl bond, or that of the cumulated double bonds in allenes. A projection of the molecular structure on a plane perpendicular to the axis of chirality gives a diagram analogous to a Fischer projection formula with two vertically-related

Fig. 1.4. The Walden inversion of the enantiomorphous chlorosuccinic acids through (+)- and (−)-malic acid.

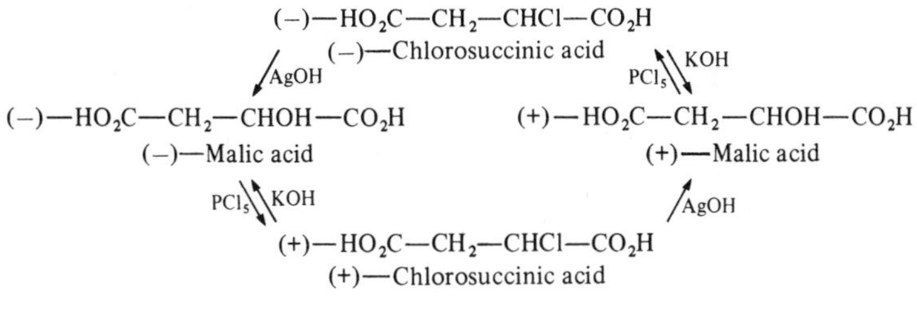

$$(-)-HO_2C-CH_2-CHCl-CO_2H$$
$$(-)-\text{Chlorosuccinic acid}$$

$$/AgOH \qquad\qquad PCl_5 \backslash\backslash\ KOH$$

$$(-)-HO_2C-CH_2-CHOH-CO_2H \qquad (+)-HO_2C-CH_2-CHOH-CO_2H$$
$$(-)-\text{Malic acid} \qquad\qquad (+)-\text{Malic acid}$$

$$PCl_5\backslash\backslash\ KOH \qquad\qquad /AgOH$$

$$(+)-HO_2C-CH_2-CHCl-CO_2H$$
$$(+)-\text{Chlorosuccinic acid}$$

and two horizontally-related groups, representing the nearer and the further substituents, respectively, in the original three-dimensional model or, equally, the converse representation (fig. 1.5). The additional sequence rule, that near groups have priority over far groups, irrespective of the atomic or mass number of those groups in a comparison of the two sets, allows an assignment of the (R)- or the (S)-configuration to the structure from the clockwise or anticlockwise sequence of the priority order of the four groups in the projection diagram. The original sequence rule, based upon atomic or mass number, is employed to establish the priority of the individual groups within a particular set, the near or the distant set (fig. 1.6).

In the case of planar chirality, such as that of the ring-substituted para-cyclophanes, the enantiomer necessarily contains an atom or atoms bonded, directly or indirectly, to the plane but disposed above or below the plane. The atom of highest atomic or mass number so bonded is termed the pilot atom, (p), and it determines the particular side of the plane from which the structure is viewed (fig. 1.6). The atom directly bonded to the plane and to the pilot atom has precedence, (a), and the succession of bonds through the in-plane path follows the atomic or mass number ordering of the atoms bonded to establish the clockwise (R) or anticlockwise (S) configurational sequence in the particular plane of chirality (fig. 1.6).

The (R,S) configurational nomenclature is extended to chiral six-coordinate metal complexes by taking Werner's octahedral numbering scheme as a standard (fig. 1.7). For the wholly unidentate complexes, such as, (+)-*all-cis*-$[Co(H_2O)_2(NH_3)_2(CN)_2]^+$, the sequence rule based upon atomic number applies to the

Fig. 1.5. The (R)-configuration (*a*) and the (S)-configuration (*b*) of Xabcd with the atoms bonded to the asymmetric centre X lying in the atomic or mass number order, a > b > c > d.

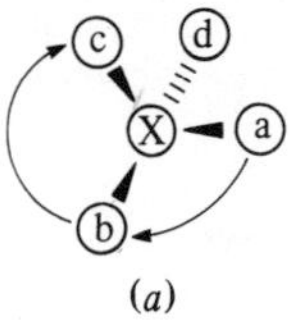

(*a*)

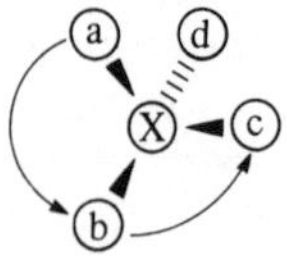

(*b*)

ligand atom directly coordinated to the metal ion. The highest ligand priority is assigned to the 1-position, and the successive positions in serial order are allocated to the ligand atoms of decreasing precedence, down to that of lowest priority in the 6-position. If the sequence of position numbers, 1,2,3, in serial order appears in a clockwise right-handed sequence to an observer on the near side, remote from the 4,5,6-positions, the configuration is designated (R) or, if the aspect of the 1,2,3-positions presented is anticlockwise and left-handed, the configuration is specified as (S) (fig. 1.7). The (R,S) nomenclature is applicable to chelating ligands through the sub-rules that ligands of higher denticity have precedence over those of lower and, for equivalent chelate rings, the second terminal coordinating atom in a given ring takes over the next serial number after that of the first coordinated atom of the ring in Werner's position-numbering scheme. The first determination of the absolute configuration of a chiral metal coordination compound by the anomalous X-ray diffraction method, due to Saito, Nakatsu, Shiro and Kuroya (1954), was that of the tris-ethylenediamine complex of cobalt(III), $(+)$-$[Co(en)_3]^{3+}$, which has the (R)-configuration in the (R,S)-nomenclature.

An ancillary nomenclature for molecules with a helical form, such as those of the helicene series, or the polypeptides forming the α-helix structure, is contained in the helicity rule. If the overall helical morphology of the molecule is right-handed, the configuration is described as P(plus) or, if left-handed, as M(minus). The helicity rule may be extended to propeller or screw-shaped

Fig. 1.6. The (S)-configuration of $(+)$-penta-2,3-diene (a), from the orthoaxial projection (b) with the sequence order, $[(Me > H)near] > [(Me > H)distant]$; and the (S)-configuration of a 4-substituted-$[2,2]$-paracyclophane (c) from the sequence order, $a > b > c$, viewed from the pilot atom position, p, for the precedence $R > H$, e.g. for R = deuterium, in (S)-$(-)$-4d-$[2,2]$ paracyclophane.

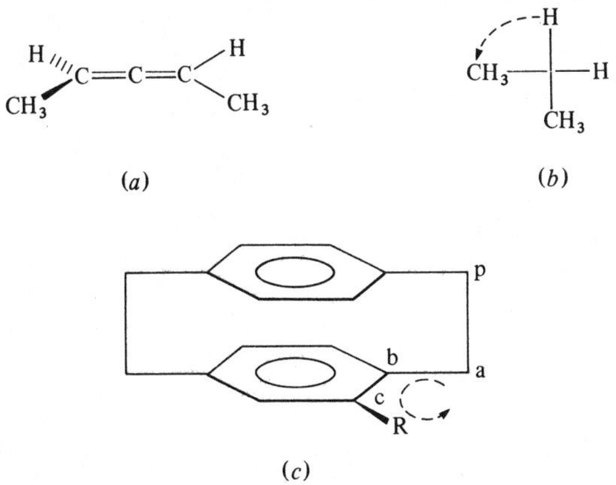

(a)

(b)

(c)

Fig. 1.7. Werner's octahedral numbering-scheme (*a*), viewed in a projection parallel to a triangular octahedral face (*b*), giving the (R)-configuration of the hexa-unidentate complex, *all-cis*-[M(a)$_2$(b)$_2$(c)$_2$] for the precedence order a > b > c, e.g. (R)-(+)-*all-cis*-[Co(H$_2$O)$_2$(NH$_3$)$_2$(CN)$_2$]$^+$, where, a = O, b = N, and c = C.

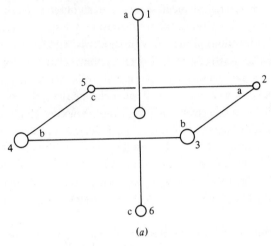

(*a*)

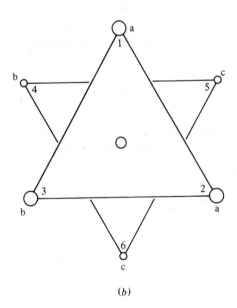

(*b*)

molecules, such as the chiral biaryls and tris-chelated coordinated compounds, in which the dual morphological helicity of chiral structures is particularly evident. ·The isomer, (+)-[Co(en)$_3$]$^{3+}$, has D$_3$ symmetry with a principal threefold rotation axis (C$_3$) and three twofold rotation axes (C$_2$) perpendicular to the principal axis. Viewed along the C$_3$ axis, the isomer, (+)-[Co(en)$_3$]$^{3+}$ has the form of a left-handed three-bladed propeller, but along the direction of a C$_2$ axis the structure presents the aspect of a right-handed propeller form (fig. 1.8). The full specification of configuration as M(C$_3$)P(C$_2$) contains a redundancy, however, as the two helicities are complementary and the one implies the other. The

Fig. 1.8. The (R)- or Λ-configuration of the (+)-[Co(en)$_3$]$^{3+}$ complex ion and the M(C$_3$) helicity of the enantiomer viewed along the direction of the principal C$_3$ axis (*a*), with the complementary P(C$_2$) helicity presented along the direction of a twofold rotation axis, (*b*).

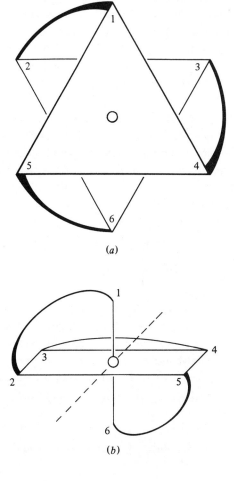

(*a*)

(*b*)

dual complementary helicities of an enantiomer, implying optical activity with opposite sign, connect with the classical and quantum-mechanical sum-rule, which specifies that the optical rotatory strength of all the transitions of a chiral molecule sum to zero over the electromagnetic spectrum as a whole.

The ambiguity in the helical description of bis- and multi-chelate metal complexes is avoided in a IUPAC convention (1970) whereby, in a six-coordinate complex, the octahedral edge spanned by one of the chelate rings serves as a reference axis. Relative to the reference axis, the octahedral edge spanned by a second chelate ring forms a segment of a left-handed (Λ) or a right-handed (Δ) helix, if the two rings do not share an octahedral vertex, nor possess a common mean plane. In a coordination compound containing more than one pair of chelate rings, the overall configuration is defined as Λ if the number of Λ ring-pairs exceeds the number of pairs with a Δ-chirality, or as Δ if the converse inequality holds. Each of the three pairs of chelate rings in the isomer, $(+)\text{-}[Co(en)_3]^{3+}$, display the Λ mutual relation, giving the Λ-configuration overall. The convention is extended to the description of the conformation of a chelate ring as λ or δ, depending upon the left- or the right-handed mutual helical relation of the line connecting the coordinated atoms of the ligand and the line joining the atoms bonded to those ligators, e.g. the $N\cdots N$ direction and the C—C bond of the ethylenediamine ligand.

The principal nomenclature systems are artificial in the sense that two analogous enantiomers with the same configurational label do not necessarily have the same absolute stereochemical form, nor need be related chemically, by interconversions, or physically, by the sign of their optical rotation properties. Physically-based descriptions of stereochemical configuration have been proposed, generally based upon the helix model for optical activity. Crum Brown (1890) and Guye (1890) defined an asymmetry product for a tetra-covalent asymmetric centre, X, in a molecule, Xabcd, as the function,

$$(a - b)\,(b - c)\,(c - d)\,(a - c)\,(a - d)\,(b - d) \tag{1.8}$$

where the quantities a,b,c,d were identified with the respective masses of the four groups bonded to the asymmetric centre but, in principle, might be some other physical property of the group or remain an unspecified algebraic quantity. Walden (1895) showed that the relative masses of the four groups were not significant determinants of the optical activity. Boys (1934) identified the quantities a,b,c,d of function 1.8 with the group volumes or radii and pointed out that the order of decreasing size of the groups defined a left- or right-handed helical configuration for the enantiomer. More recently, Brewster (1967) proposed that the sequence of the polarizabilities of the four groups and their relative disposition around the asymmetric centre, X, defines a left- or right-handed helical ensemble which governs the optical activity. Functions analogous to eqn 1.8, with the quantities a,b,c,d referring to the group refractivities, are found to be consistent with the sodium D-line molecular rotations of the enantiomers investigated.

In algebraic form, eqn 1.8 becomes the chirality function for tetrahedral co-ordination, in the sense employed by Ruch (1972). The chirality functions are based upon the theory of the permutation groups and may be written for any given achiral structural type, tetrahedral, octahedral, trigonal or tetragonal bipyramid or prismatic coordination. For a given structural type, the chirality function represents any pseudoscalar property of the molecule, and it is non-vanishing only for a dissymmetrically substituted derivative. The molecular rotations of a series of such derivatives, based on a common achiral structural type, allow the evaulation of the quantities in eqn 1.8, or its analogues for co-ordination polyhedra other than the tetrahedron, and the quantities thus obtained afford expectation values of the molecular rotations of further members of the series.

2 OPTICAL ROTATORY DISPERSION AND CIRCULAR DICHROISM

2.1 Optical activity and molecular spectroscopy

The development of a molecular model for optical isomerism in the dissymmetrically-substituted tetrahedral carbon atom stimulated the investigation of the physical basis of optical activity in the light of the electromagnetic reformulation of the transverse-wave radiation theory, due to Clerk Maxwell (1865). Following the work of Le Bel and van't Hoff, Boltzmann argued in 1874 that optical activity is an effect dependent upon the ratio of a molecular dimension, d, to the wavelength of the radiation, so that the optical rotatory power of a medium must be diminished in the infrared region and vanish at infinite wavelength. According to the standard dipole approximation, the molecule is point-like in size, relative to the much larger wavelength of light, and it experiences, as a function of time, only a one-dimensional oscillating electrical field, the electric-dipole component of the radiation field. Abandoning the dipole approximation, Boltzmann showed that the dispersion of the optical rotation with respect to wavelength is expected to follow a relation of the form $[(B/\lambda^2) + (C/\lambda^4) + \ldots]$, rather than Biot's simple inverse-square law, or relations with a wavelength-independent term.

In developing the general electromagnetic theory of optical phenomena, Drude showed in 1893 that Boltzmann's relation represents the initial terms of an expansion, in ascending powers of $1/\lambda^2$, of a general optical rotatory dispersion (ORD) equation,

$$[M]_\lambda = \sum_m K_m [\lambda^2 - \lambda_m^2]^{-1} \tag{2.1}$$

where λ_m is the wavelength of the radiation interacting resonantly with a charged particle, or particles, oscillating at a characteristic frequency of c/λ_m in the molecule, and K_m is the molecular rotation constant for that characteristic vibration. Drude's analysis of the ORD of the quartz crystal appeared to support Boltzmann's conclusion. The optical activity of quartz derived largely from an ultraviolet interaction at $0.103\,\mu m$, the corresponding constant, K_m, being dominant, with no detectable contribution from the infrared vibrations at 8.84 or 20.75 μm, for which the corresponding rotation constants were effectively zero.

In the electromagnetic theory, as in the earlier transverse-wave theory of quasi-mechanical vibrations in an all-pervading luminiferous ether, the refractive index of a medium is connected with the corresponding absorption index, as the real and the imaginary parts of a general complex refractivity. Linear birefringence implies linear dichroism and, equally, optical activity or circular birefringence implies circular dichroism in the absorption frequency region. Although observed as early as 1845 in amethyst quartz crystals, circular dichroism (CD) was first characterised as a molecular property by Cotton in 1895 from spectroscopic absorption measurements over the visible-wavelength region of aqueous solutions of copper(II) and chromium(III) (+)-tartrate, employing LCP and RCP radiation. Using plane-polarized radiation, Cotton observed the reversal of the sign of the optical rotation on scanning through an absorption band with respect to wavelength, and found the emergent radiation to be elliptically polarized, on account of the differential absorption of the LCP and the RCP components of the incident radiation. For a positive CD absorption, measured as the differential extinction coefficient, $\Delta\epsilon = (\epsilon_L - \epsilon_R)$, at λ_m, the optical rotation changes from positive at longer wavelengths, $\lambda > \lambda_m$, to negative at shorter wavelengths, $\lambda < \lambda_m$, with the converse sign change if the CD absorption is negative (fig. 2.1).

Drude's equation (eqn 2.1) accounted for the sign reversal in the ORD at λ_m, the sign of the corresponding molecular rotation constant, K_m, reflecting that of the CD absorption at λ_m. Drude assimilated CD absorption along with ORD into a general molecular model for optical activity, which offered a prospect, however remote at the period, of determining the absolute stereochemical configuration of an enantiomer from its observed optical rotatory properties. According to the Drude model, a charged particle in a chiral molecule is constrained stereochemically to a helical oscillation on interacting with a radiation field, either resonantly at one of the λ_m where energy is taken from or given up to the radiation field, or by a non-resonant polarization in the transparent wavelength regions. For a right-handed helical oscillation of the charge in the enantiomer, the interaction is stronger for LCP than RCP radiation, resulting in a positive CD absorption and dextrorotation at longer wavelengths. Conversely, a left-handed helical charge displacement in the molecule results in a negative CD absorption at λ_m and laevorotation at longer wavelengths.

Drude's theory of optical activity did not command a wide following, since it was essentially a one-particle model. In the generally-received view, deriving from Boltzmann, two or more spatially-separated charged particles, or polarizable groups, in a molecular structure were required for optical activity, each charge interacting with the other as well as with the radiation field. Born (1918) found that a minimum of four different isotropically-polarizable groups in a three-dimensional array, typically tetrahedral, are essential for optical activity in an electric-dipole radiation field. Subsequently Kuhn (1930) showed that two anisotropically-polarizable groups in a molecule sufficed, provided that the two groups were spatially separated in the molecular structure and were not related to one another by any secondary symmetry element, S_p, namely, a rotation by $(2\pi/p)$ followed by reflection in a plane perpendicular to the rotation axis.

Fig. 2.1. The Cotton effect in an absorption-frequency region, expressed as the circular dichroism ($\Delta\epsilon = \epsilon_L - \epsilon_R$) (full line) and the 'anomalous' dispersion of the molar optical rotation $[M]$ (broken line). A positive Cotton effect is given by the isomer dextrorotatory at longer wavelengths (a), and a negative Cotton effect by the laevorotatory enantiomer (b).

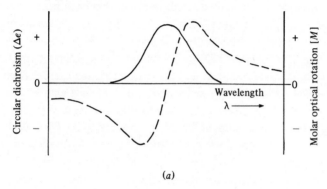

(a)

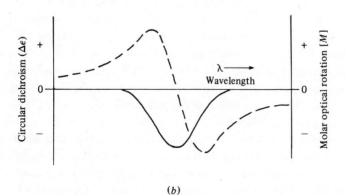

(b)

The equations of Born and Kuhn for the molecular rotatory constants, equivalent to the K_m of the Drude equation (eqn 2.1), contained the Boltzmann ratio (d/λ), expressing the vanishing of the optical rotation in the long-wavelength limit. The ratio represents an approximate measure of the accessibility of an ORD or CD absorption Cotton effect, governed experimentally by the magnitude of Kuhn's dissymmetry factor, $g = \Delta\epsilon/\epsilon$. Kuhn gave the optimum value of the factor as $g_m = (2\pi d/\lambda_m)$, implying that, with a given instrumental sensitivity, the Cotton effects of an enantiomer become decreasingly accessible with increasing wavelength. The g-ratios measured by Kuhn and his contemporaries were often large (~ 0.1), suggesting unrealistically large separations between the groups which interacted to give the rotatory power of a chiral molecule, e.g. $d \sim 80\text{Å}$ for the chromium(III) (+)-tartrate complex. Since he had found the Drude one-particle model of optical activity deficient, Kuhn regarded its quantum-mechanical successor, the one-electron theory, as of minor significance, although the latter subsequently accounted for the 'anomalously' large g-factors which had been measured.

2.2 The quantum theory of optical activity

The one-electron theory of optical activity followed from the general quantum-mechanical reformulation by Rosenfeld (1928) of the classical ORD equations and the electromagnetic theory upon which they were based. Rosenfeld gave the quantum-mechanical analogue of the Drude equation (eqn 2.1) for the optical rotation, ϕ, in radians per cm pathlength, at the frequency, ν, in the form,

$$\phi = [16\pi^2 N/3hc)] \sum_m \nu^2 R_{om} [\nu_{0m}^2 - \nu^2]^{-1} \tag{2.2}$$

where N is the number of molecules per cm^3, and R_{om} is the rotational strength of the transition, $\psi_0 \to \psi_m$, between two stationary states of the molecule, separated by the energy, $h\nu_{om}$.

The rotational strength represents the scalar product of the electric dipole, μ_{om}, and the magnetic dipole, m_{mo}, transition moments,

$$R_{om} = Im\{<\psi_0|\hat{\mu}|\psi_m>\cdot<\psi_m|\hat{m}|\psi_0>\} \tag{2.3}$$

where Im signifies that the imaginary part is to be taken, since the magnetic dipole operator, $\hat{m}$, is a pure imaginary. The rotational strength, being signed, is a pseudoscalar quantity, not a proper scalar with magnitude only. The positive or negative sign follows from the parallel or antiparallel mutual orientation of the polar and the axial vector components, the electric and the magnetic moments, respectively, correlating with the translation and the rotation of charge, respectively. The parallel (or antiparallel) mutual orientation of μ_{0m}^{α} and m_{m0}^{α}, where α denotes the particular x, y, or z polarization, corresponds to the classical right-handed (or left-handed) helical charge-displacement in the molecule interacting with a radiation field, together with the analogous significance of a positive (or negative) Cotton effect centred on ν_{0m} and dextro (or laevo) optical rotation at lower frequencies.

Pseudoscalar quantities are invariant or symmetric to all primary symmetry operations, the pure p-fold rotations, C_p, and are antisymmetric, changing sign, under all of the secondary symmetry operations, the p-fold rotation-reflections, S_p. An observable physical property of a molecule is necessarily invariant to the operations permitted by the molecular symmetry, so that the rotational strength vanishes for all of the interstate transitions, electronic, vibrational, and rotational, of a molecule with one or more elements of rotation-reflection symmetry, S_p. A non-zero rotational strength is restricted to the interstate transitions of molecules with pure rotation point symmetry, the rotation groups, I (icosahedral), O (octahedral), T (tetrahedral), D_p (dihedral), and C_p (axial), in which pseudoscalar quantities are totally symmetric and invariant to the molecular point group operations.

A spectroscopic basis for the restriction of optical activity to molecules possessing only pure rotational symmetry is provided by the selection rules for electric-dipole and magnetic-dipole transitions. The two sets of rules are mutually exclusive with respect to molecular rotation-reflection symmetry elements for collinear electric-dipole, μ_{0m}^{α}, and magnetic-dipole, m_{0m}^{α}, transition moments, that is, moments with a common polarization. In a centrosymmetric molecule, with an inversion centre $i \equiv S_2$, the rule of Laporte (1924) restricts an electric-dipole moment to transitions connecting a *gerade* (symmetric) with an *ungerade* (antisymmetric) state, g $\leftrightarrow$ u. A magnetic-dipole transition in a centrosymmetric molecule connects only states of the same parity, however, u $\leftrightarrow$ u or g $\leftrightarrow$ g. For S_p rotation-reflection generally, taking the p-fold rotation axis as the α-axis, the selection rule for a non-vanishing μ_{0m}^{α} is that a symmetric state is connected with an antisymmetric state, (+) $\leftrightarrow$ (−), whereas the corresponding rule for a non-zero m_{m0}^{α} entails, (+) $\leftrightarrow$ (+) or (−) $\leftrightarrow$ (−). An exception to the mutual exclusion of a simultaneous electric-dipole and magnetic-dipole transition moment with a common polarization is found in the achiral systems with a S_{4n} symmetry element, with n integral. In the point groups S_{4n} the dipole component pairs, μ_x, μ_y and m_x, m_y, are each doubly-degenerate, and both pairs transform under a common E-representation. While equal in magnitude for a given transition, the scalar products are opposed in sign, $\mu_x m_x = - \mu_y m_y$, and the overall rotational strength vanishes. In the groups S_{4n} pseudoscalar properties are not totally

Fig. 2.2. An achiral molecule with a S_4 rotation–reflection symmetry axis. Although devoid of an inversion centre and a plane of symmetry, the molecule is superposable by translation and rotation alone upon its mirror-image.

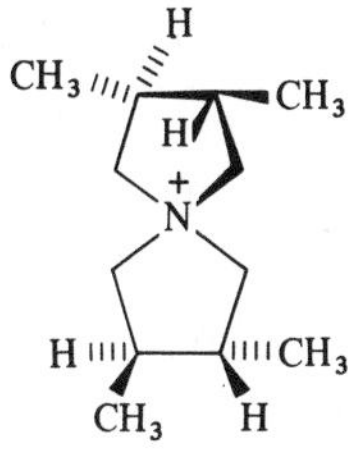

symmetric, but are spanned by the B-representation. A molecule with S_4 symmetry is superposable by rotations and translations alone upon the corresponding mirror-image structure (fig. 2.2).

2.3 The evaluation of transition strengths

The rotational strength of an interstate transition of a chiral molecule is measured either from the area of the CD band centred on an absorption frequency, ν_{om}, or from the amplitude and the peak–trough frequency-separation of the corresponding ORD curve. The absorption–dispersion relations connect the ORD amplitude $[A]$ with the CD absorption maximum through,

$$[A] = [M]_{max} - [M]_{min} = 4028 \,| \Delta\epsilon_{max} |$$ (2.4)

where the CD absorption is measured in decadic molar extinction coefficient units and the molar rotation $[M]_\lambda$ is estimated according to the prescription of Biot (eqn 1.3). The difference between the frequencies of the peak and the trough of the sigmoid ORD curve is related to the band-width at half maximum CD absorption (fig. 2.1) by,

$$|\nu_{peak} - \nu_{trough}| = 1.08 \, \Delta\nu$$ (2.5)

These relations, due to Kuhn (1958), are based on the assumption that the CD band shape is Gaussian in form.

Although the polarimeter remained the principal precision instrument available to physical chemists for over a century, mainly for analytical measurements at the sodium D-line, it proved to have limited spectroscopic utility over absorption wavelength regions. An accurate determination of the optical-rotation angle requires the polarizing and the analysing prism at a mutual orientation near to extinction, which conflicts with the requirement of an adequate transmission of radiation energy to the detector, initially the eye, subsequently the photographic plate, and, more recently, the photomultiplier. In absorption wavelength regions the precision is reduced on account of the CD absorption, resulting in elliptically-polarized transmitted radiation, and, more particularly, by the use of large monochromator slit-widths, required to attain an adequate flux of radiation. The large slit-widths limit the spectroscopic resolution, reducing the amplitude and even smoothing-out the sigmoid ORD Cotton-effect curves. The determination of the rotational strength of an interstate transition from the corresponding ORD Cotton effect requires the uncertain subtraction of the rotatory dispersion contributions from all of the transitions of the molecule, other than the particular transition considered, from the observed ORD curve (eqn 2.2). With such limitations, ORD studies employing the spectropolarimeter were progressively supplanted by CD absorption procedures during the 1960s when spectrophotometers with the capability of measuring CD Cotton effects with g-ratios down to the order $\sim 10^{-5}$ became available for the visible and quartz ultraviolet regions.

For randomly-orientated molecules in the gas phase or in solution the area of a CD absorption band and the corresponding rotational strength are related by the expression,

$$R_{om} = [3\,h\,c\,10^3\,ln10/(32\pi^3 N_A)]\int (\Delta\epsilon/\nu)\,d\nu \tag{2.6}$$

where N_A, h, and c, are Avogadro's number, Planck's constant, and the velocity of light, respectively, and the numerical factors accommodate the use of the decadic molar extinction coefficient in $dm^3\,mol^{-1}\,cm^{-1}$. Substitution of the values of the universal constants into eqn 2.6 gives the rotational strength in units of the Debye–Bohr magneton in numerical form,

$$R_{om} = 0.248 \int (\Delta\epsilon/\nu)d\nu \tag{2.7}$$

Solution measurements strictly require the multiplication of the right-hand side of eqns 2.6 and 2.7 by the factor $3/(n^2 + 2)$ to allow for the effect of the refractivity of the solvent medium upon the radiation field, although the correction is infrequently employed. The correction factor, due to Lorentz (1880), is significant, however, for chiral solvents with a large circular birefringence, such as cholesteric liquid crystals, in which achiral solutes of high symmetry display an induced CD absorption due to the difference between the effective LCP and RCP radiation field strengths (Dudley, Mason and Peacock, 1975).

The area of the corresponding isotropic absorption band measures the dipole strength, D_{om}, which represents the square modulus of the electric-dipole transition moment, $|\mu_{om}|^2$, in good approximation, as the concomitant magnetic-dipole strength, $|m_{mo}|^2$, and the higher multipole strengths, if any, make a relatively negligible contribution to the isotropic absorption intensity of polyatomic molecules. In units of the square Debye, (d^2), the dipole strength is related to the isotropic band area by the expression,

$$D_{om} = [3 \, h \, c \, 10^3 \, ln 10/(8\pi^3 N_A)] \int (\epsilon/\nu)d\nu$$

$$= 9.18 \times 10^{-3} \int (\epsilon/\nu)d\nu \tag{2.8}$$

The Lorentz field correction factor for the refractivity of the solvent is $[9n/(n^2 + 2)^2]$ for solution estimates of the dipole strength of a solute but, again, the factor is rarely employed.

The classical electromagnetic measure of isotropic absorption intensity, still commonly used, is the oscillator strength, f_{om}, which represents the effective fraction or number of charged particles in a molecule which interact resonantly with the radiation field at the frequency ν_{om},

$$f_{om} = [m_e \, c^2 \, 10^3 \, ln 10/(\pi e^2 N_A)] \int \epsilon d\nu$$

$$= 4.32 \times 10^{-9} \int \epsilon d\bar{\nu} \tag{2.9}$$

The oscillator strength is related to the corresponding dipole strength, in square Debye, by,

$$f_{om} = 4.76 \times 10^{-7} \bar{\nu}_{om} D_{om} \tag{2.10}$$

where $\bar{\nu}_{om}$ refers to the wavenumber (cm^{-1}). Kuhn, with Thomas (1925), established that the sum of the oscillator strengths of all the electronic transitions of a molecule equals the number of valency electrons. For the classical counterpart of the rotational strength, $[g_{om}f_{om}/\nu_{om}]$, Kuhn (1930) showed that the sum over all electronic transitions of an enantiomer vanishes, so that the optical rotation goes to zero not only in the limit of infinite wavelength, as Boltzmann had found, but also in the limit of infinite frequency. The two sum rules were subsequently demonstrated quantum-mechanically, notably by Condon (1937) who established that

$$\sum_m R_{om} = 0.$$

While the oscillator strength, f_{om}, and Kuhn's dissymmetry factor, g_{om}, are pure numbers, the rotational and the dipole strength are expressed in one of several different unit systems. The Debye unit for the electric-dipole moment of molecular systems is defined as 10^{-18} esu. The atomic unit of electric-dipole moment, e a_0, where e is the electronic charge and a_0 is the Bohr radius, corresponds to 2.54 x 10^{-18} esu in the cgs system, or to 8.48 x 10^{-30} C m in the SI system. The unit of the magnetic-dipole moment, the Bohr magneton, [h e/ $4\pi m_e c$], has the value of 9.27 x 10^{-21} erg gauss^{-1} in the cgs system and of 9.27 x 10^{-24} J T^{-1} in the SI system.

2.4 Chiroptical instrumentation

Despite the pioneer studies of Cotton (1895) and his successors, the measurement of optical rotatory dispersion remained the main method of investigating optical activity until the 1960s. The ORD method appeared to be the simpler procedure, applicable to all chiral substances, whereas the CD method is confined to enantiomers with absorption spectra in accessible wavelength regions. With the development of sector rules linking the stereochemistry of a chiral molecule to the sign of the Cotton effect due to a specific electronic transition of the molecular chromophore (Moffitt *et al.*, 1961) the connection between optical activity and chemical spectroscopy progressively became more significant and, with it, the CD procedure.

The visual method of Cotton (1895) for the measurement of CD, using a polarizer and a quarter-wave plate, is limited to absorption bands with a dissymmetry ratio, $g = \Delta\epsilon/\epsilon$, greater than $\sim$ 1%, which is adequate for the copper(II) and chromium(III) (+)-tartrate solutions he studied. The photographic method of Kuhn and his contemporaries (Lowry, 1935) increased the precision of CD measurements by an order of magnitude, and extended the wavelength range covered to the ultraviolet and the photographic infrared regions.

A further two orders of magnitude of precision in CD measurements were attained in the early models of current CD spectrophotometers, based upon the production of left- and right-circularly polarized radiation periodically, with phase-sensitive detection and lock-in amplification of the signal. The principle was introduced by Grosjean and Legrand (1960), who employed a Pockel's cell as the radiation modulator, constructed from a tetragonal crystal of $(NH_4)H_2PO_4$ cut perpendicular to the tetragonal crystal axis. With glycerine-based electrodes, the otherwise-isotropic crystal section becomes linearly birefringent on the application of a voltage. The sign of the birefringence follows that of the voltage applied, which is adjusted to the value required for the Fresnel quarter-wave condition (eqn 1.7).

The periodic modulation, from LCP to RCP radiation, is provided by the application of either a sine-wave or a square-wave voltage at a frequency of choice up to the kHz range. The square-wave mode is preferable, since the emergent radiation is then substantially either left- or right-circularly polarized over a modulation cycle, whereas the sine-wave mode affords all phases of polarization over a cycle, from linear through elliptical to circular at the peaks and the throughs of the sine wave. The disadvantages of the electro-optic modulator (EOM) are the transmission range of the $(NH_4)H_2PO_4$ crystal and its deuterated analogue, $0.18 - 1.0 \mu m$, and the different refractive indices of the crystal in directions along and perpendicular to the tetragonal optic axis, which limits the angular aperture, requiring good collimation of the transmitted radiation.

A modulator based upon a different principle, stress-induced birefringence, was introduced by Billardon and Badoz (1966), who constructed CD spectrophotometers for the visible and ultraviolet regions (Billardon *et al.*, 1969) and for the $1 - 6 \mu m$ infrared region (Russell *et al.*, 1972) incorporating the device. The photoelastic modulator (PEM) is made up of a transparent isotropic optical element, either vitreous, polycrystalline, or a single cubic crystal, which is periodically stressed by means of piezoelectric ceramic discs of lead zirconate titanate (Billardon and Badoz, 1966) or a single-crystal quartz transducer (Jasperson and Schnatterly, 1969, Kemp, 1969).

With an isotropic element the PEM has the advantage over the EOM of a large angular aperture. The PEM is confined to a single modulation frequency, the resonance frequency of the transducer, and to sine-wave modulation, so that the EOM is retained for square-wave modulation at frequencies of choice over the visible and ultraviolet regions. The use of a single crystal of calcium fluoride as the optical element in a PEM gives left- to right-circularly polarized modulation of radiation over the calcium fluoride transmission range, $0.13 - 8.5 \mu m$, and the zinc selenide optical element extends the infrared limit to beyond $12 \mu m$ (Cheng *et al.*, 1976). Calcium fluoride was employed as a PEM optical element initially for vacuum ultraviolet CD spectrophotometers (Schnepp *et al.*, 1970; Johnson, 1971) and subsequently for the corresponding infrared instruments (Stephens and Clark, 1979).

A standard CD spectrophotometer employing a PEM and a photomultiplier detector has the general design illustrated in fig. 2.3. The photomultiplier detector covers the vacuum and quartz ultraviolet range, and the visible and near infrared region down to $\sim 1.0 \mu m$. Measurements at longer wavelengths require a photoconductive or photovoltaic semiconductor detector which, in turn, entails a modified signal-processing system. The radiation source generally employed is a xenon-arc lamp which, equipped with a sapphire window, gives a high radiation intensity over the 0.2 – 6.5 µm range. A hydrogen-discharge source is normally used for shorter ultraviolet wavelengths, and a thermal black-body radiation source for longer infrared wavelengths.

The radiation from the monochromator exit slit is plane polarized by a Rochon prism of quartz or, better, of magnesium fluoride, which transmits over the 0.12 – 6 µm wavelength range. A wire grid polarizer is employed for longer infrared wavelengths. The PEM, following the polarizer in the optical train, transforms the radiation into left- and right-circularly polarized form sinusoidally at $\sim$ 50 kHz. After passing through the sample compartment containing a non-absorbing reference substance, the modulated radiation produces at the photo-multiplier detector only a direct current (d.c.) signal, which provides a base-line

Fig. 2.3. The general design of a circular dichroism spectrophotometer. White radiation from the source (S) enters the wavelength-scanning monochromator (MON). Radiation from the monochromator exit slit passes through a polarizing prism (P) and a photoelastic quarter-wave modulator (PEM) to a flow-cell (F), for kinetic or chromatographic studies, or a thermostatted sample-cell for spectroscopic or equilibrium studies. The transmitted radiation is detected by the photomultiplier (PM). After the preamplifier (PA), the PM signal is fed to the error-signal servo-amplifier (ESA) which governs the extra high-tension voltage supply (EHT) to the PM, in order to maintain the V_{dc} constant at a pre-set level. A signal proportional to the change in the EHT supply, regis-tered by the digital voltmeter (DVM), is fed to one channel of a multi-pen recorder to measure the absorbance, A. The V_{ac} signal from the PM, after the PA, is fed to the phase-lock amplifier (PLA) which receives a reference signal from the modulator power supply (MPS). The output from the PLA is fed to a second channel of the recorder to measure the circular dichroism absorbance differential, $\Delta A = (A_L - A_R)$, and to one input of a ratiometer, the divider (DVD). The second input to the DVD is the absorbance signal, and the DVD output measures the g-ratio, $(\Delta A / A)$, which is registered by a third channel of the recorder.

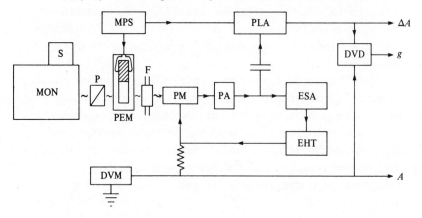

for the measurement of both the subsequent absorption (A), and differential absorption of LCP and RCP radiation $(\Delta A = A_L - A_R)$, of a chiral absorbing sample. The latter sample reduces the direct current signal from its original level, in approximate proportion to the optical density or absorbance (A), and gives rise to a small 50 kHz alternating current signal, registered as a voltage, V_{ac}. The differential optical density for LCP and RCP radiation, or the circular differential absorbance (ΔA), is measured by the ratio of the two signal voltages (V_{ac}/V_{dc}).

The voltage ratio giving the circular dichroism (ΔA) is conveniently measured, together with the overall mean absorbance (A), by maintaining the denominator (V_{dc}) at a constant level, that of the original zero-absorption d.c. signal, using an error-signal servo-amplifier (ESA in fig. 2.3) (Mondine, 1979). The fall in the d.c. signal from the photomultiplier (PM), due to the absorption of radiation by the sample, is registered by the error-signal servo-amplifier, which governs the extra high-tension voltage supply (EHT) delivered to the PM, increasing the voltage to restore the V_{dc} signal from the PM to its original zero-absorption base-line level. The response of the PM to the applied EHT voltage is nearly logarithmic, so that the increase in that voltage measures in good approximation the absorbance (A) of the sample. A signal proportional to the EHT voltage-change is fed to one channel of a multi-pen electronic recorder in order to register the absorbance.

Since the denominator of the ratio (V_{ac}/V_{dc}) is now constant, the small 50 kHz a.c. signal measures directly the circular differential absorbance (ΔA). The a.c. component from the preamplifier (PA) following the photomultiplier is fed to a phase-sensitive detector and amplifier, the phase-lock amplifier (PLA), which receives a reference signal from the photoelastic modulator (fig. 2.3). The PLA discriminates the 50 kHz signal from artefact and noise signals of different frequency, and distinguishes between positive and negative circular differential absorbance, (ΔA), which is registered by a second channel of the multi-pen electronic recorder.

A refinement allows the simultaneous measurement of the dissymmetry ratio, $g = \Delta A/A$. The individual absorbance (A) and circular differential absorbance (ΔA) signals are fed to a ratiometer, the divider (DVD), which gives an output, fed to a third channel of the recorder, measuring the g-ratio. The three quantities, A, ΔA, and g, are recorded either as a function of wavelength, in the measurement of an absorption and CD spectrum, or as a function of time, in following the kinetic change of a chiral species, or as a function of elution volume, in monitoring the progress of a chromatographic optical resolution. In the third of these applications, the recording of the g-ratio allows the measurement of the enantiomeric purity of the solute passing through the flow-cell from the chromatographic column (see § 10.1.3).

The differential interaction of a chiral molecule with the left- and the right-circular components of a radiation field is expressed, not only in absorption as circular dichroism, but also in emission as a circular intensity differential (CID) in the luminescence. If the chiral molecule is photoexcited with modulated LCP and RCP radiation, the variation in the fluorescence intensity follows the differential absorption of the two circular radiation components, giving fluorescence-detected circular dichroism (FDCD). In the simplest case, where the chiral fluorescent molecule freely rotates in solution and effectively has random orientation during the lifetime of the photoexcited state, the CID measures the dissymmetry ratio of the chiral molecule in its electronic ground state, i.e. the absorption g-ratio,

$$g_{abs} = \Delta\epsilon/\epsilon = 2(I_L - I_R)/(I_L + I_R) \tag{2.11}$$

where I_L and I_R are the respective intensities of the left- and the right-circularly polarized luminescence.

If the chiral molecule is photoexcited with unpolarized or plane-polarized radiation, the transition probabilities for the emission of left- and right-circularly polarized photons are inequivalent if the rotational strength of the transition is non-vanishing. The observed CID now registers the dissymmetry ratio of the electronically-excited state from which the luminescence originates,

$$g_{lum} = 2(I_L - I_R)/(I_L + I_R) \tag{2.12}$$

$$= 4R_{no}/D_{no}$$

The rotational strength, R_{no}, and dipole strength, D_{no}, for emission $n \rightarrow 0$, are not necessarily equivalent to the corresponding strengths, R_{on} and D_{on}, of the converse absorption process, $0 \rightarrow n$. If the chiral molecule changes in shape or size on photoexcitation, the rotational strengths, R_{no} and R_{on}, in particular, have different magnitudes, and may have opposed signs, as occurs in the 300 nm $n \rightarrow \pi^*$ absorption of some chiral ketones (Dekkers and Closs, 1976).

In general, large chiral molecules do not rotate to statistically-random orientations during the lifetime of the photoexcited state, particularly if viscous solvents are employed. Luminescence takes place preferentially from molecules photoselected by their particular orientation for the absorption of radiation, and the emission itself is not isotropic. Luminescence CID measurements provide the ground state (eqn 2.11) and the excited state (eqn 2.12) dissymmetry ratio, however, even for a chiral molecule rotationally-constrained in a glass or a crystal, under particular experimental conditions (fig. 2.4).

In the FDCD technique with modulated LCP and RCP excitation, the fluorescence is collected at an angle θ to the propagation direction of the excitation radiation, and a linear polarizer is placed in front of the photomultiplier detector with the polarization direction at an angle ϕ to the excitation-emission plane-normal (fig. 2.4a). Photoselection effects are minimised, and eqn 2.11 holds in good approximation generally, for the choice of polarization angle $\phi = 35.25°$, or if the particular luminescence collection angle $\theta = 54.75°$ or its supplement, is adopted. Similarly, in the circular-polarized luminescence technique with plane-polarized excitation (fig. 2.4b), eqn 2.12 holds in good approximation, even for solid-state samples, if the polarization direction of the excitation-polarizer is orientated at the angle $\phi = 35.25°$ to the excitation-emission plane-normal, or if the luminescence collection direction lies at the angle $\theta = 54.75°$ or its supplement to the excitation propagation direction (Richardson and Riehl, 1977; Tinoco, Ehrenberg and Steinberg, 1977).

The photoselection of appropriately-orientated molecules dissolved in a viscous solvent, or in a low-temperature glass, provides the relative polarization directions of successive electronic transitions, to higher energy, in the spectrum of the solute, whether chiral or achiral (fig. 2.4c). The incidence of radiation plane-polarized normal to the optical plane preferentially excites the molecules orientated with the transition dipole moment at a small angle to the excitation polarization direction. If the excited molecules do not rotate, the fluorescence due to that transition dipole is preferentially polarized parallel to the excitation polarization direction. If a higher-energy state is excited, with a transition dipole orientated perpendicular to that of the lowest-energy transition, the fluorescence is preferentially polarized perpendicular to the excitation polarization direction.

The fluorescence intensity is analysed into one component polarized parallel $(I_{\parallel})$ and the other polarized perpendicular $(I_{\perp})$ to the excitation polarization vector by setting the luminescence modulator, an EOM or PEM, at periodic half-wave retardation. The analysis of the fluorescence polarization gives the linear polarization anisotropy ratio, p,

$$p = (I_{\parallel} - I_{\perp})/(I_{\parallel} + I_{\perp}) \tag{2.13}$$

For a molecule with point-group rotational symmetry no higher than twofold, the polarization factor, p, has the value $+1/2$ if the electric-dipole moment of the absorption and the emission transition are parallel, or of $-1/3$ if the two transition moments have a mutually perpendicular orientation in the molecule, provided the molecule does not rotate during the lifetime of the excited state (Feofilov, 1961; Weber, 1966).

The electronic signal-processing system for the differential circular and linear polarization luminescence techniques (fig. 2.4) is analogous to that developed for the corresponding CD absorption instruments (fig. 2.3). An important difference is that the ratio of the alternating current signal to the direct current signal registered by the photomultiplier detector, (V_{ac}/V_{dc}), while measuring the circular differential absorbance ($\Delta A = A_L - A_R$) in the CD spectrophotometer (fig. 2.3), measures the dissymmetry ratio, g_{abs} (eqn 2.11), or g_{lum} (eqn 2.12), or the linear polarization ratio, p (eqn 2.13), in the luminescence techniques (fig. 2.4).

Fig. 2.4. Polarized excitation and luminescence techniques. (*a*) The fluorescence-detected circular dichroism procedure. The sample luminescence is excited by monochromatic radiation, provided by the radiation source (S) and monochromator (M), periodically LC and RC polarized by the polarizer—modulator combination (PM). The luminescence is transmitted through the linear polarizer (P) and the cut-off filter (F) to the detector (D). Linear polarization effects arising from the photoselection of solute molecules with a particular orientation are minimised by setting the transmission vector of the polarizer (P) at the angle $\phi = 35.25°$ to the excitation-emission plane normal, or by collecting the fluorescence in a direction at the angle $\theta = 54.75°$ or $125.25°$ to the excitation propagation direction.

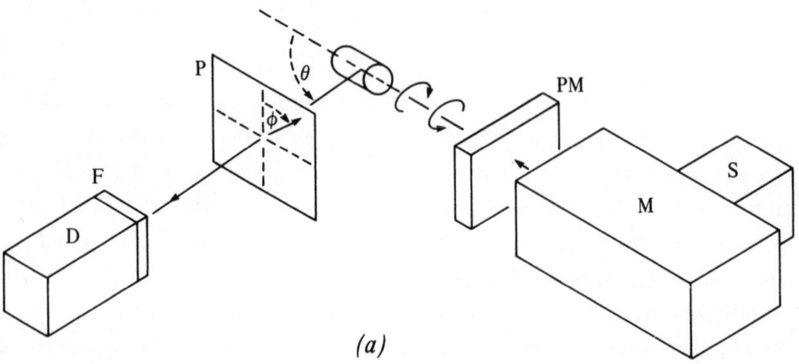

(*a*)

Fig. 2.4 (*cont.*) (*b*) The determination of the luminescence dissymmetry ratio, g_{lum} (eqn 2.12), from measurements of the circulat intensity differential in emission. The photoselection effects are minimised by excitation with the polarized radiation field vector (P_1) at the angle $\phi = 35.25°$ to the optic plane normal, and the collection of the luminescence at the angle $\theta = 54.75°$ or $125.25°$ to the excitation direction. The intensities of the LCP and RCP components of the luminescence are measured by the modulator set for quarter-wave retardation with a following polarizer (PM), and the luminescence monochromator (M_3) and detector (D). (*c*) The determination of the linear polarization anisotropy factor, *p* (eqn 2.13), for the excitation and luminescence of photoselected solute molecules in a viscous solvent or a glass, where the molecular orientation is retained during the lifetime of the photoexcited state. Radiation from the excitation monochromator (M_1) and source (S) is plane-polarized (P_1) to provide a radiation field vector ($E_\parallel$) orthogonal to the optical plane. The luminescence collected in a direction perpendicular to the excitation direction is analysed by the linear polarizer (P_2), followed by the monochromator (M_2) and detector (D), for the luminescence intensity polarized parallel ($I_\parallel$) and perpendicular ($I_\perp$) to the excitation radiation field vector. The intensities, $I_\parallel$ and $I_\perp$, are either measured individually, by separate wavelength-scans of the spectra, or the polarization anisotropy ratio, *p*, is measured with more precision directly, during a single scan, by placing a modulator set for half-wave retardation in front of the polarizer (P_2).

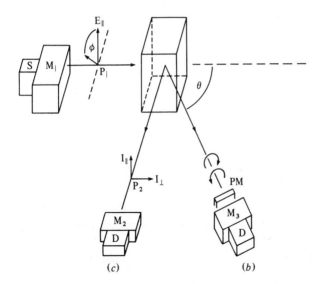

(*c*)   (*b*)

3 GENERAL MECHANISMS FOR OPTICAL ACTIVITY

3.1 The independent-systems procedure

Following the view of Boltzmann, that optical activity is an effect due to the simultaneous interaction of the radiation field with different parts of a chiral molecule, the majority of the classical electromagnetic theories were based upon a division of an enantiomer into two or more structural units, which interacted among themselves through the Coulomb potential between the dipoles induced in each group by the radiation field. Drude's one-particle model was exceptional in this respect, and the absence of additional charges interacting with the principal particle in his theory was the main basis of objections to his model. The enforced helical motion of the charged particle in a radiation field appeared to be an arbitrary assumption with no physical basis.

A similar course is adopted in the general quantum-mechanical models for optical activity. A chiral molecule is divided into a light-absorbing group, the chromophore, which generally has secondary elements of symmetry, S_p, and one or more substituent groups which form the dissymmetric molecular environment of the chromophore and reduce the overall symmetry of the molecule to that of one of the pure rotation point groups. It is assumed that the chromophore and the substituent in the molecule are quasi-independent systems, and overlap between their respective charge distributions is neglected. As in the classical theories, the chromophore and the substituent interact through the Coulomb potential between their charge distributions. If an allowance is made for electron delocalisation between the chromophore and the substituent, as in molecular orbital (MO) treatments, each enantiomer becomes an individual problem in itself. Insights into the physical basis of optical activity have been achieved by detailed MO studies of simple idealised model structures, however, such as a twisted olefin unit or chirally distorted allene, and by semiempirical MO investigations of a series of related enantiomers, notably, the chiral ketone series.

3.2 The one-electron theory

Following the quantum-mechanical reformulation of the general theory of optical activity, the one-electron model was developed by Condon, Altar and

Eyring (1937), based upon the viewpoint of the crystal-field theory of the electronic spectra of open-shell metal-ion complexes. The d → d and f → f transitions of such complexes are Laporte-forbidden, devoid of a zero-order electric-dipole moment. The multipole expansion of the radiation field in ascending powers of Boltzmann's ratio (d/λ) affords, after the principal zero-order electric-dipole component, the first-order magnetic-dipole and electric-quadrupole components dependent upon (d/λ), followed by higher multipole components proportional to higher powers of the ratio. Van Vleck (1937) drew attention to the interaction of an open-shell metal-ion complex with the magnetic-dipole and the electric-quadrupole components of the radiation field, in order to account for the observed finite probabilities of the Laporte-forbidden transitions, and to a mechanism whereby such a transition borrows a fraction of the moment of an allowed electric-dipole transition in the metal ion at a different energy. The transitions investigated by Van Vleck are of the $l \to l$ atomic type, which covered additionally the majority of the excitations for which Cotton-effect data were then available, having large and readily accessible g-ratios. These were the Cotton effects of the weak long-wavelength band systems of chiral transition metal complexes, due to d ⟷ d excitations, and of organic enantiomers containing simple hetero-unsaturated chromophores, such as a carbonyl, thiocarbonyl, nitro, nitroso, diazo, or halide group, due to n → π^* or n → σ^* transitions, with a substantial p ⟷ p component at the hetero-atom or atoms.

The one-election theory of optical activity emphasises the significance of the allowed magnetic-dipole transitions at an atomic centre, for which the selection rule is $\Delta l = 0$. The atomic centre either constitutes the chromophore of the enantiomer, as in the case of the metal ion in a chiral transition metal complex, or a principal part of the chromophore, like the oxygen atom of the carbonyl group. In addition, the one-electron model takes over the crystal-field view that the Coulombic field of the ligands or substituents is static, arising from their ground-state charge distribution, so that the substituent groups play no active role in the light-absorption process. The ligand or substituent field is taken to mix the magnetic-dipole with the electric-dipole transitions of the chromophore, giving each type of transition a non-vanishing rotational strength.

On the independent-systems basis, the electronic state functions of the enantiomer are expressed as simple products of the individual chromophore, $|M_m\rangle$, and ligand functions, $|L_l\rangle$, with the product $|M_0 L_0\rangle$ as the molecular ground-state function. For a simple three-level system, with one magnetic-dipole, $|M_0\rangle \to |M_m\rangle$, and one electric-dipole, $|M_0\rangle \to |M_n\rangle$), chromophore transition, with a common α-polarization, the molecular excited state functions are corrected to the first order of perturbation theory over the basis functions of the set,

$$|M_m L_0 > = |M_m L_0) + (E_m - E_n)^{-1} (M_n L_0 | V | M_m L_0) | M_n L_0) \tag{3.1}$$

and

$$|M_n L_0 > = |M_n L_0) + (E_n - E_m)^{-1} (M_m L_0 | V | M_n L_0) | M_m L_0) \tag{3.2}$$

where the round brackets refer to zero-order basis functions and the *kets* to corrected functions. In eqns 3.1 and 3.2 the energies of the excited states, E_m and E_n, are measured relative to that of the ground state as the zero, and V refers to the Coulomb potential between the charges, e_i, of the chromophore, M, and those, e_j, of the substituent, L, separated by the distance, r_{ij},

$$V = \sum_{i(M), j(L)} e_i e_j / r_{ij} \tag{3.3}$$

The electric moment of the chromophore transition, $|M_0) \rightarrow |M_n)$, is taken to have its zero-order value and polarization, μ_{0n}^{α}, in a chirally-substituted derivative, where the corresponding magnetic moment now becomes non-vanishing on account of the first-order mixing,

$$m_{n0}^{\alpha} = m_{m0}^{\alpha} (M_m L_0 | V | M_n L_0) (E_n - E_m)^{-1} \tag{3.4}$$

Equally, the magnetic-dipole chromophore transition, $|M_0) \rightarrow |M_m)$, while retaining its zero-order magnetic moment, m_{m0}^{α}, in the dissymmetric derivative, now acquires a first-order electric moment,

$$\mu_{0m}^{\alpha} = \mu_{0n}^{\alpha} (M_n L_0 | V | M_m L_0) (E_m - E_n)^{-1} \tag{3.5}$$

For each transition the principal and the borrowed moment are collinear, with a common α-polarization, and the two transitions in the enantiomer acquire oppositely-signed rotational strengths, equal in magnitude,

$$R_{0m} = -R_{0n} = i \mu_{0n}^{\alpha} m_{m0}^{\alpha} (M_n L_0 | V | M_m L_0) (E_n - E_m)^{-1} \tag{3.6}$$

centred upon their individual positions, ν_{0m} and ν_{0n}, on the frequency scale.

The excited state functions in the Coulombic matrix element $(M_n L_0 | V | M_m L_0)$ in eqn 3.6 refer only to the chromophore of the chiral molecule, and they may be classified under the symmetry representations of the, generally achiral, point group of the symmetric unsubstituted parent chromophore. By the initial assumption, $|M_n)$ and $|M_m)$ are excited states to which, respectively, an electric-

and a magnetic-dipole transition is allowed from the ground state, $|M_0\rangle$, with a common α-polarization. The corresponding transitional charge distributions, $(M_0|M_n)$ and $(M_0|M_m)$, transform as the collinear components of a polar and an axial vector, respectively, with a pseudoscalar product. Irrespective of the symmetry of the ground state, $|M_0\rangle$, the direct product of the symmetries of the excited states, $|M_n\rangle$ and $|M_m\rangle$, is spanned by the pseudoscalar representation of the chromophore point group. If the matrix element $(M_n L_0|V|M_m L_0)$ is non-vanishing, it is invariant to the symmetry operations of the chromophore point group, implying that the effective state-mixing component of the potential field deriving from the substituent groups in turn belongs to the same pseudoscalar representation.

The pseudoscalar property of the effective component of the substituent field affords sector rules for optical activity, based upon the one-electron mechanism, as was shown by Schellman (1966). A sector rule connects the position of a substituent group in the molecular environment of the chromophore in an enantiomer with the sign and, in favourable cases, the approximate magnitude of the Cotton effect induced by that group in a given chromophore transition. The first of such rules, the octant rule for the $n \to \pi^*$ transition of chiral ketones near 300 nm, was derived empirically and given an interpretation, based upon the chromophore orbitals and the substituent polarizability, by Moffitt and coworkers (1961). The octant rule specifies the division of the molecular environment of the carbonyl group in a chiral ketone into spatial regions defined by the nodal plane of the $2p_y$ non-bonding lone-pair orbital of the oxygen atom and the nodal planes of the carbonyl π_x^* orbital (fig. 3.1). If the product of the coordinates [XYZ] of the substituent group in the chromophore frame defined by the orbital nodal planes is negative, the Cotton effect induced in the $n \to \pi^*$ carbonyl transition is positive, provided that the mean electric-dipole polarizability of the substituent is larger than that of the hydrogen atom replaced. Thus the fluoro-substituent, with a mean polarizability smaller than that of a bonded hydrogen atom, displays antioctant or dissignate behaviour, whereas the other halogen substituents in the corresponding position are consignate and follow the octant rule, giving carbonyl $n \to \pi^*$ Cotton effects with a magnitude in approximate proportionality to the positive increment in their polarizability over the hydrogen atom replaced.

The carbonyl chromophore, in its simplest parental examples, formaldehyde, acetone, or adamantanone, belongs to the point group C_{2v}, which has the A_2 representation spanning pseudoscalar quantities. The first of the pseudoscalar functions of the coordinates of a substituent group in the C_{2v} carbonyl coordinate frame with A_2 symmetry is [XY], specifying a quadrant sector rule, but [XYZ] and all products of [XY] with higher powers of Z are similarly pseudoscalar in C_{2v}, so that the one-electron mechanism is consistent with the empirical octant rule on symmetry grounds. The observed sensitivity of the carbonyl $n \rightarrow \pi^*$ Cotton effect upon the polarizability, rather than the charge or permanent dipole, of the group in a given position of dissymmetric substitution is less compatible with the static-field one-electron mechanism. The effect of the positively charged ammonium group on the carbonyl $n \rightarrow \pi^*$ Cotton effect is small in comparison with that of the corresponding neutral amine, which has the larger polarizability, owing to the greater response of the non-bonding lone-pair electrons to an applied field.

3.3 The dynamic polarization mechanism

The static-field one-electron theory, while drawing attention to the significance of magnetic-dipole transitions for optical activity, overlooked the possible role of the even electric-multipole transition moment which accompanies the magnetic moment. The even 2^n-pole electric moment of a $l \rightarrow l$ transition at an atomic centre lies in the range, $n = 2 \cdots 2l$, that is, solely a quadrupole moment for a p $\rightarrow$ p promotion, with a hexadecapole moment additionally for a d $\rightarrow$ d excitation, and both of these, together with a 2^6-pole moment, for the f $\rightarrow$ f case. The importance of the even-multipole electric moment of a magnetic-dipole transition emerges in the dynamic-polarization theory of optical activity, due to

Fig. 3.1. The non-bonding $2p_y$ orbital of the oxygen atom and the π_x^* molecular orbital connected by the $n \rightarrow \pi^*$ transition of the carbonyl chromophore near 300 nm. The shaded and unshaded lobes represent regions of opposite phase in the orbital wave function. The XY and the YZ nodal planes of the π_x^* orbital and the XZ nodal plane of the $2p_y$ lone-pair orbital divide the molecular environment into octant sectors, such that a substituent group with the coordinates, X, Y, Z in the chromophore frame gives rise to a $n \rightarrow \pi^*$ Cotton effect with a sign dependent upon the negative of the coordinate product, [−XYZ]. The $n \rightarrow \pi^*$ transition of the carbonyl chromophore involves the rotatory displacement of charge around the bond direction, giving rise to a magnetic-dipole transition moment with z-polarization.

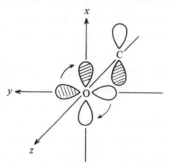

Weigang and Höhn (1966), which gave a new form to the earlier two-group treatments of optical rotation, with an emphasis on the electric-dipole polarizability of the substituent groups.

The one-electron model for optical activity had taken over the view of crystal-field theory that, in a composite molecule, the primary effect to be considered for spectroscopic purposes is the perturbation of the metal cation, or chromophore, by the ligands, or substituents, which may be anionic but, even if neutral, are the more polarizable species. The basis of crystal-field theory ran counter to the rules of Fajans (1923), according to which the polarization of a cation by an anion is negligible compared with the contrary polarization. Fajans (1923) proposed that the colour of transition metal complexes is due, in no small measure, to the polarization of the ligands by the metal ion. The blue colour of hydrated copper(II) derives largely from the polarization of the water ligands by the metal ion, since anhydrous copper(II) sulphate is colourless, and replacement of the water of coordination by the more polarizable ammonia ligand produces a substantial intensity increase and wavelength-shift in the light absorption. Kuhn and Bein (1934) analysed the absorption and CD spectrum of the tris-oxalato complex of cobalt(III) in terms of the dynamic coupling of transient metal-ion and ligand moments but, otherwise, the Fajans viewpoint remained largely undeveloped and was by-passed by the static crystal-field concept.

According to the dynamic-polarization mechanism for optical activity, the ligands or substituents in the molecular environment of the chromophore play an active role in the light-absorption process. For an allowed magnetic-dipole transition of the chromophore, the accompanying even electric-multipole transition moment induces through its Coulombic potential an electric-dipole moment in each substituent group, proportional to the dipole polarizability of the group at the radiation frequency. The resultant of the electric dipoles induced in the substituent groups is generally finite in non-centrosymmetric molecules and, being phase-locked to the multipole transition moment of the chromophore, interacts with and takes energy from the radiation field, enhancing the isotropic absorption. In chiral molecules a component of the resultant first-order electric-dipole transition moment located in the substituent groups is collinear with the zero-order magnetic-dipole transition moment of the chromophore, giving a non-vanishing rotational strength, represented by the differential absorption of RCP and LCP radiation.

On the independent-systems basis, the molecular state functions are represented by simple-product configurations containing either the ground state $|M_0\rangle$ or the magnetic-dipole excited state $|M_m\rangle$ of the chromophore and one of the complete set $|L_l\rangle$ of the substituent-group functions. Corrected to the first-order of perturbation theory, the molecular ground-state function is given by,

$$|M_0 L_0 > = |M_0 L_0) + \sum_{1 \neq 0} (-E_m - E_1)^{-1} (M_m L_1 |V| M_0 L_0) |M_m L_1) \tag{3.7}$$

and the corresponding excited state, similarly corrected, becomes,

$$|M_m L_0 > = |M_m L_0) + \sum_{1 \neq 0} (E_m - E_1)^{-1} (M_0 L_1 |V| M_m L_0) |M_0 L_1) \tag{3.8}$$

where the basis-state energies, E_m and E_1, are measured relative to that of the ground state, E_0, as the zero.

The magnetic moment of the chromophore transition, $|M_0) \rightarrow |M_m)$, is taken again, as in the static-field one-electron mechanism, to have its zero-order value and particular polarization, m_{m0}^{α}, with α denoting x, y, or z. The collinear first-order electric-dipole transition moment is now located wholly in the substituent group or groups and, from eqns 3.7 and 3.8, it is given by the expression,

$$\mu_{om}^{\alpha} = \sum_{1 \neq 0} [(E_m - E_1)^{-1} (M_0 L_1 |V| M_m L_0) \mu_{0l}^{\alpha}$$

$$- (E_m + E_1)^{-1} (M_0 L_0 |V| M_m L_1) \mu_{l0}^{\alpha}] \tag{3.9}$$

The perturbation matrix elements of eqn 3.9, with the rearranged general form $(M_0 M_m |V| L_0 L_1)$, represent the Coulomb potential (eqn 3.3) between two transitional charge distributions, $(M_0 M_m)$ centred on the chromophore origin, and $(L_0 L_1)$ located at the spatially-separated substituent group origin. An approximation to the potential is afforded by a multipole expansion of each of these transitional charge distributions, centred upon their respective origins, and the truncation of each series after the leading term. The term retained for the transitional charge distribution of the substituent group is a component of an electric dipole, μ_{0l}^{β}. The term surviving for the charge distribution of the chromophore transition is one of the $2n+1$ components of the leading even 2^n-pole electric moment, the particular component being dependent upon the form of the orbitals connected by the transition.

In the case of the carbonyl $n \rightarrow \pi^*$ transition of chiral ketones near 300 nm, the $2p_y \rightarrow 2p_x$ component of the transition at the oxygen atom gives rise to two leading moments, the z-component of a magnetic dipole, m_{m0}^{z}, and the xy-component of an electric quadrupole, θ_{om}^{xy}, where the z-axis is directed along the carbonyl bond and the oxygen $2p_y$ orbital contains the non-bonding electrons (fig. 3.2). The Coulombic matrix element of eqn 3.9 becomes, for the carbonyl $n \rightarrow \pi^*$ transition,

$$(M_0 M_m |V| L_0 L_1) = \sum_{\beta} \mu_{0l}^{\beta} \theta_{om}^{xy} G_{\beta,xy} \tag{3.10}$$

where $G_{\beta,xy}$ are geometric tensors of radial and angular factors governing the potential between the xy-component of an electric quadrupole and the β-component of an electric dipole, at a distance which is large relative to the dimensions of the dipole or the quadrupole. The substitution of eqn 3.10 into eqn 3.9 allows, after a transformation from the chromophore to the substituent coordinate frame, the extraction of the principal components of the polarizability tensor of the substituent group at the frequency of the chromophore transition,

$$\alpha_{\gamma\gamma}(\nu_{0m}) = \sum_l 2E_l \, [E_l^2 - E_m^2]^{-1} \, |\mu_{0l}^{\gamma}|^2 \tag{3.11}$$

If the substituent group is isotropic, or if only the mean polarizability of the group, $\bar{\alpha}(L)$, is taken into account, the coordinate transformation is unnecessary, as only the components of the electric dipole induced in the substituent with the same polarization in eqns 3.9 and 3.10 contribute to the sum over states in eqn 3.11. In the mean-polarizability approximation, the first-order electric-dipole moment of the carbonyl $n \to \pi^*$ transition becomes, for a single substituent group,

$$\mu_{0m}^z = -\theta_{0m}^{xy} \, \bar{\alpha}(L) \, G_{xyz}^L \tag{3.12}$$

and the geometric tensor has the form,

$$G_{xyz}^L = -15[XYZ/R^7] \tag{3.13}$$

where X,Y,Z refer to the coordinates of the substituent, L, in the chromophore frame, the group lying at a distance, R, from the chromophore origin. The scalar product of the zero-order magnetic moment of the carbonyl $n \to \pi^*$ transition with the first-order electric moment gives the dynamic polarization rotational strength,

$$R_{om} = -15 \, i \, R^{-7} \, XYZ \, \bar{\alpha}(L) \, \theta_{0m}^{xy} \, m_{mo}^z \tag{3.14}$$

Eqn 3.14 reproduces the octant sector rule for the Cotton effect of the carbonyl $n \to \pi^*$ transition of chiral ketones with correct absolute signs. The dissignate effect of a fluoro-substituent is ascribed by eqn 3.14 to the smaller mean polarizability of the group, relative to the hydrogen replaced, and other dissignate cases are explained by taking into account the anisotropy of the polarizability tensor of the substituent group.

Fig. 3.2. The electric quadrupolar charge distribution, $(2p_y \mid 2p_x)$, at the oxygen atom in the n → π* transition of the carbonyl chromophore, and the electric dipolar charge displacement in a substituent group, L, induced by the potential field of the electric-quadrupole transition moment, θ_{xy}. The electric dipole induced in the substituent group is z-polarized, or has a component with z-polarization, collinear with the magnetic-dipole moment of the n → π* transition of the carbonyl group. For a substituent located in the upper-right rear octant, viewed from the +Z direction, the induced electric-dipole in the substituent group and the magnetic-dipole transition moment in the carbonyl chromophore are antiparallel, corresponding to a charge-displacement through a left-handed helical path, which gives rise to a negative Cotton effect.

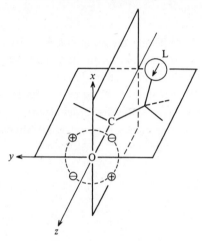

3.4 The two-group electric-dipole mechanisms

The majority of the classical electromagnetic theories of optical activity were based upon the view that, on interacting with a radiation field, linear dipolar displacements in two or more spatially-separated groups of an enantiomer formed an overall helical charge displacement in the molecule. These classical theories were placed upon a quantum-mechanical basis by Kirkwood (1937), who, like Kuhn, regarded the allowed magnetic-dipole transitions of chiral molecules as of minor significance for their optical rotation. The magnetic moment required for optical activity arises in the two-group electric-dipole mechanisms from the quasi-rotation of charge, due to the distant dipolar displacement in the second group, around the dipolar direction in the first group. At a distant point, the linear charge displacement of the electric-dipole moment of a chromophore transition, μ_{ot}, generates a magnetic-dipole moment, m_{ot}, proportional to the position vector, **R**, of the charge from the point, and to the current density, the product of the charge and its velocity. Classically and quantum-mechanically, the charge velocity of an electric-dipole transition is related to the transition frequency, ν_{ot}, and the magnetic moment produced at a distant point is given by,

$$\mathbf{m}_{ot} = -i\,\pi(\nu_{ot}/c)(\mathbf{R} \times \boldsymbol{\mu}_{ot}) \tag{3.15}$$

The electric transition moment of an isolated chromophore gives no net magnetic moment, since the contributions at two external points related by reflection through any plane containing the electric moment sum to zero. In a two-group system, the second group provides a unique external origin for the magnetic moment produced by the electric moment of the first group and *vice versa*. If the two groups or, more particularly, the two electric-dipole vectors, are not related by a rotation–reflection symmetry element, S_p, which would preclude optical activity, the dipole vectors and the position vector relating their centres form a non-vanishing triple scalar product, $[\mathbf{R}_{21} \cdot \boldsymbol{\mu}_{02} \times \boldsymbol{\mu}_{01}]$, proportional to the rotational strength.

The two-group electric-dipole mechanisms cover a range of particular applications, from the dissymmetric dimer, where the 'chromophore' and the 'substituent' are equivalent groups, to the case of an enantiomer containing a chromophore with a low energy transition and a relatively inert substituent group, which becomes chromophoric only at high energy, in the vacuum ultraviolet region. In the limiting case of the dissymmetric dimer, each excited state of one monomer group, $|M_1\rangle$, is paired with an isoenergetic excited state of the second group, $|N_2\rangle$. The individuality of a given monomer transition is lost in the dissymmetric dimer, where the excited states are formed by symmetric and antisymmetric combinations of the corresponding monomer excited configurations,

$$\psi_\pm = (1/\sqrt{2})\,\{|M_1 N_0\rangle \pm |M_0 N_2\rangle\} \tag{3.16}$$

with the dimer ground state, ψ_0, represented by the simple product, $|M_0 N_0\rangle$, on the independent-systems basis. Unlike the basic monomer configurations, the two dimer excited states (eqn 3.16) are not degenerate, the energy-interval between them being governed by the Coulomb potential between the transition dipoles,

$$V_{12} = [(R_{21})^2\,(\boldsymbol{\mu}_{01} \cdot \boldsymbol{\mu}_{02}) - 3(\boldsymbol{\mu}_{01} \cdot \mathbf{R}_{21})(\boldsymbol{\mu}_{02} \cdot \mathbf{R}_{21})]\,R_{21}^{-5} \tag{3.17}$$

The dimer transitions, $\psi_0 \rightarrow \psi_\pm$, have rotational strengths of equal magnitude and opposite sign,

$$R_{0+} = -R_{0-} = -(\pi/2)\,\bar{\nu}_{01}\,[\mathbf{R}_{21}\cdot\mu_{02}\times\mu_{01}] \qquad (3.18)$$

where $\bar{\nu}_{01}$ is the wavenumber of the monomer transition.

The CD absorption spectrum of a dissymmetric dimer in the region of a given monomer transition presents a characteristic bisignate couplet, a pair of CD bands with comparable areas and opposed signs, corresponding to the rotational strengths of eqn 3.18, centred upon the particular monomer transition frequency (fig. 3.3). The energy separation between the two dimer excited states (eqn 3.16) is $2V_{12}$ (eqn 3.17), and when this interval is comparable to, or larger than, the energy corresponding to the monomer absorption band-width at half-maximum height, (hc$\Delta\bar{\nu}$), the dimer spectrum shows two absorption bands for each monomer band, and the bisignate CD couplet has the band-areas specified by the rotational strengths of eqn 3.18. The wavenumber interval observed between the two components of a bisignate CD couplet remains large, comparable to the monomer absorption band-width, $\Delta\bar{\nu}$, when V_{12} is relatively small, but the CD band areas are reduced by the approximate ratio $(2V_{12}/hc\Delta\bar{\nu})$ below the rotational-strength values indicated by equation 3.18. The smaller band-areas of the bisignate CD couplet and the larger separation arise from the overlap on the frequency ordinate of oppositely-signed Gaussian or Lorentzian CD band forms, the mutual cancellation being complete when V_{12} goes to zero, as in the case of achiral dimers with an S_4 rotation–reflection symmetry element, such as biphenyl with a dihedral angle of $\pi/2$ between the planes of the two benzene rings.

The band-areas of the two components of a bisignate CD couplet in the spectrum of a dissymmetric dimer containing two equivalent monomer chromophores are not always equal in magnitude, on account of the coupling of the electric-dipole transition moment in one monomer group to a corresponding moment at a different energy in the second group, as well as to the isoenergetic moment of the latter chromophore. For a chiral dimer made up of two inequivalent monomer chromophores the transition dipoles in the two groups, coupled by the Coulomb potential between them (eqn 3.17), generally lie at different energies, apart from accidental degeneracies. The coupling between the two transition dipoles now gives rise to two oppositely-signed rotational strengths of equal magnitude which are located close to the corresponding monomer transition frequencies,

$$R_{01} = -(\pi V_{12}/2hc)\,[(E_{01}-E_{02})^2 + 4V_{12}^2]^{-1/2}\,(E_{01}+E_{02})\,[\mathbf{R}_{21}\cdot\mu_{02}\times\mu_{01}]$$

$$= -R_{02} \qquad (3.19)$$

Relative to the isolated monomer absorptions, the higher and the lower frequency rotational strengths are displaced respectively to higher and lower energy by the increment $|2V_{12}^2/(E_{01} - E_{02})|$.

In the second limiting case of a chiral molecule composed of a chromophore with a low-energy electric-dipole transition and an optically-inert substituent, such as an alkyl group, the two-group electric-dipole mechanism for optical activity takes the form of a Coulombic coupling between the chromophore transition moment and the electric dipole induced in the substituent by that moment. The dipole induced in the substituent group is proportional to the components of the dipole polarizability tensor of the group at the chromophore transition frequency. The two electric dipoles, the transition moment and the

Fig. 3.3. The symmetric, higher-energy, and the antisymmetric, lower-energy, coupling modes of the configurationally-isoenergetic electric-dipole excitation moments in the two individual chromophores of a chiral dimer with C_2 symmetry. In the higher-energy symmetric mode, the overall electric-dipole and magnetic-dipole transition moments are both directed along the C_2 rotation axis, but are antiparallel, giving a negative rotational strength. For the lower-energy antisymmetric mode the two overall transition moments are parallel and perpendicular to the C_2 axis, producing a positive rotational strength.

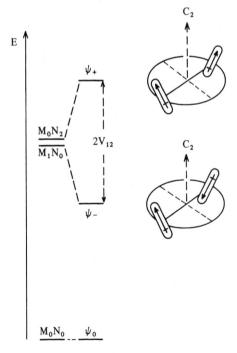

induced moment, cannot be collinear or related by a rotation–reflection symmetry element in the molecule, that is, the triple scalar product, $[\mathbf{R}_{21} \cdot \boldsymbol{\mu}_{02} \times \boldsymbol{\mu}_{01}]$, must be non-vanishing for optical activity, a condition which requires the polarizability tensor of the substituent group to be anisotropic. A number of common substituents, such as the methyl group, are effectively cylindrically symmetric, with two principal components of the group polarizability tensor, referring to the directions parallel and perpendicular to the substituent bond axis, $\alpha_{\parallel}$ and $\alpha_{\perp}$, respectively. The polarizability anisotropy represents the difference between the two components, $\beta = (\alpha_{\parallel} - \alpha_{\perp})$.

The development of the second limiting case of the two-group electric-dipole mechanism follows that of the dynamic polarization treatment of an allowed magnetic-dipole chromophore transition (Weigang, 1979). The first-order electric-dipole moment located in the substituent group (eqn 3.9) is now evaluated through the use of the dipole–dipole potential (eqn 3.17) and the principal components of the polarizability tensor of the substituent group are

Fig. 3.4. The electric-dipole transition moment of a chromophore, M, defining the Z-axis of the chromophore frame, and the electric-dipole induced by the potential field of that transition moment in the substituent group, L, which has cylindrical symmetry. The X-axis of the chromophore frame is chosen so that the induced moment in the substituent, L, lies parallel to the XZ-plane, with a projection upon the XZ-plane of the chromophore lying at an angle, θ, to the Z-axis. A positive angle, θ, corresponds to a clockwise rotation from the Z-axis to the projection, viewed from the +Y-direction, with X, Y, and Z, forming a right-handed coordinate system.

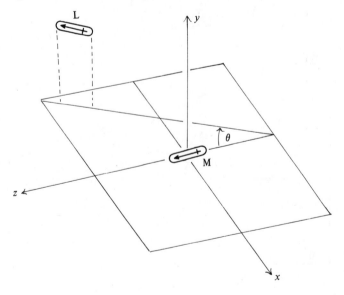

extracted by 3.11. The polarization direction of the electric-dipole moment of the low-energy chromophore transition is taken as the Z-axis of the chromophore coordinate frame, and the Y-axis is so chosen that the principal axis of the substituent group, the cylinder axis to which the polarizability component $\alpha_\parallel$ refers, lies parallel to the XZ-plane of the chromophore coordinate system, which is right handed (fig. 3.4). The particular choice of the chromophore axial system reduces the number of terms in the expression for the rotational strength, R_{01}, of the chromophore electric-dipole transition at the wavenumber, $\bar{\nu}_{01}$, with the moment, μ_{01}^Z, to the following,

$$R_{01} = (\pi/2)\bar{\nu}_{01} |\mu_{01}^Z|^2 \beta(L)[Y(3Z^2 - R^2)\sin 2\theta - 6XYZ\sin^2\theta] R^{-5} \qquad (3.20)$$

In eqn 3.20, X, Y, Z refer to the coordinates of the substituent group, at a distance R from the origin, in the chromophore frame, and θ to the angle between the Z-axis and the projection of the cylinder axis of the substituent on the XZ-plane (fig. 3.4). The angle θ is taken as positive for a clockwise rotation, and negative for an anticlockwise rotation, from the Z-axis to the projection of the principal axis of the substituent on the XZ-plane, viewed from the +Y axial aspect. The polarizability anisotropy of the substituent group, $\beta(L)$, refers to the value at the transition frequency, ν_{01}.

Eqn 3.20 expresses a sector rule relating the sign of the rotational strength of the chromophore transition to the position and the orientation of the substituent group in the chromophore axial system. The sector rule is composed of two components, an octant contribution governed by the [XYZ] coordinate function, and a conical contribution dependent upon the [Y(3Z^2 - R^2)] coordinate function, for the substituent group. To a degree, the two contributions are complementary in that the octant contribution is maximal for a substituent group placed on a body diagonal of the octants, whereas the conical contribution vanishes for a group in such a location (Weigang, 1979).

3.5   Relations between the general models for optical activity

The mechanisms for optical activity based upon the independent-systems procedure are not mutually exclusive for a given transition type, magnetic-dipole or electric-dipole allowed, and a more complete treatment is obtained by combining the different mechanisms for the particular transition type. In the two-group electric-dipole case it is generally necessary to take into account the non-degenerate coupling mechanism (eqn 3.19) as well as the corresponding isoenergetic mechanism (eqn 3.18) in an analysis of the CD absorption spectrum of a chiral molecule composed of two or more equivalent chromophores. For an allowed magnetic-dipole chromophore transition, the sector rule obtained from pseudoscalar property of the perturbing field due to the substituent groups in the one-electron model (eqn 3.6), is generally isomorphous with the corresponding sector rule given by the geometric tensor governing the multipole–dipole potential (eqn 3.13) of the dynamic polarization model in the approximation where only the mean polarizability of the substituent group is taken into account.

The two sets of mechanisms, one referring to the allowed magnetic-dipole and the other to the allowed electric-dipole transitions of the chromophore, provide the basis for a characterisation of the former transition type through determinations of the dissymmetry ratio, expressed as, $g = (\Delta\epsilon/\epsilon)$ at a given wavelength. Taken as the ratio of the band areas of a CD absorption and the corresponding isotropic absorption, the overall dissymmetry factor of the transition becomes,

$$g_{om} = 4R_{om}/D_{om} \tag{3.21}$$

which is an approximate measure of the ratio of the magnetic-dipole to the electric-dipole transition moment, $(\mathbf{m}_{mo}/\boldsymbol{\mu}_{om})$. The ratio is substantially larger for the allowed magnetic-dipole than the allowed electric-dipole type or, indeed, of the type where both transition dipoles are absent in the zero-order. A dissymmetry ratio with the magnitude, $g > 10^{-2}$, is indicative of the magnetic-dipole transition type, the assignment being the more probable the larger is the g-factor (table 3.1).

As Kuhn demonstrated classically, the dissymmetry factor for the two-group electric-dipole mechanism contains Boltzmann's ratio, (d/λ). The same holds for the overall dissymmetry ratio (eqn 3.21) based upon the quantum-mechanical reformulations of the two-group electric-dipole model, as is particularly evident for the case of the isoenergetic electric-dipole monomer transitions of a chiral dimer (eqn 3.18). The Boltzmann ratio of the molecular or chromophore dimension, d, to the wavelength of the radiation, λ, indicates a g-factor of the order of 10^{-3} for the quartz ultraviolet and visible region or of 10^{-4} to 10^{-5} for the rocksalt infrared region ($2 - 15 \ \mu m$). The g-values observed for electric-dipole transitions, electronic in the visible region or vibrational in the infrared, are typically of the order suggested by Boltzmann's ratio (table 3.1). The g-factors for magnetic-dipole transitions are not dependent upon the (d/λ) ratio, and they remain relatively large throughout the infrared region. Eqns 3.6 and 3.14 for the rotational strengths of magnetic-dipole transitions contain no reference to the corresponding transition frequency, unlike the analogous equations (eqns 3.18, 3.19 and 3.20) for the rotational strengths deriving from the two-group electric-dipole mechanisms.

A further contrast between the mechanisms refers to the relationship between the rotational strength and the electric-dipole transition moment. The proportionality is linear for the allowed magnetic-dipole chromophore transition mechanisms, indicating that, in a series of related enantiomers, the magnitude of the Cotton effect is proportional to the square-root of the isotropic absorption intensity, as is observed for the carbonyl n → π^* transition near 300 nm (Mason, 1963). In the limiting case of a low-energy chromophore transition and an optically-inert substituent, the two-group electric-dipole mechanism gives a quadratic proportionality (eqn 3.20), suggesting that the g-ratio for a particular transition is approximately constant in a series of chiral analogues with a common chromophore. While the two-group electric-dipole mechanism for the other limiting case, that of the dissymmetric dimer, also requires a quadratic relationship between the rotational strengths and the monomer electric-dipole transition moment (eqn 3.18), the frequency interval between the oppositely-signed

Table 3.1. *The frequency ($\bar{\nu}$), intensity (ϵ), circular dichroism ($\Delta\epsilon$), and the dissymmetry ratio (g), of vibrational and electronic transitions in chiral molecules*

Enantiomer	$\bar{\nu}/cm^{-1}$	ϵ	$\Delta\epsilon$	$|g|/10^{-3}$	transition	Ref.
(+)-3-methyl-	1320	38	+0.01	0.2	CH_2 bending	a
cyclohexanone	2935	130	-0.004	0.03	CH_2 stetch	b
	33600	16	+0.48	30	C=O n → π^*	c
	54000	1200	+1.0	0.8	C=O n → σ^*(3s)	c
(−)-β-pinene	1640	32	+0.003	0.09	C=C stretch	d
	2920	238	+0.009	0.04	C—H stretch	d
	50100	10800	+17.1	2 }	$\pi_x \to \pi_x^*$	
	55400	9000	-17.0	2 }	$\pi_x \to \pi_y^*$	e
(+)-2,2′diamino-	3389	136	+0.011	0.08	ν_s } NH_2	f
1,1′-binaphthyl	3482	91	+0.002	0.02	ν_a } stretch	
	40500	70000	-245	3 }	$\pi \to \pi^*$	
	43200	60000	+135	2 }	CD couplet	g
[Co(l-α-isp)Cl$_2$]	4100	24	+1.46	61 }	3d → 3d	
[(−)-α-isospartein	6450	30	-1.20	40 }	$A_2 \to T_2$	
(dichloro) cobalt (II)]	7650	75	+0.31	4 }		
	10000	32	+0.24	8 }	$A_2 \to T_1$ (F)	h
	16100	335	+0.16	0.5 }		
	18700	275	-0.29	1 }	$A_2 \to T_1$ (P)	

(*a*)   Su, Heintz and Keiderling (1980).
(*b*)   Polavarapu and Nafie (1980); Singh and Keiderling (1981).
(*c*)   Measurements from the author's laboratory.
(*d*)   Stephens and Clark (1979).
(*e*)   Drake and Mason (1977).
(*f*)   Barnett, Drake and Mason (1980).
(*g*)   Mason, Seal and Roberts (1974).
(*h*)   Drake, Hirst, Kuroda and Mason (1982).

rotational strengths additionally depends on the square of the monomer
transition moment (eqn 3.17). Overall the Cotton effects of a given transition
in a series of related chiral dimers is expected to be proportional to the square
of the isotropic absorption intensity, as is found for a series of *para*-substituted
benzoate diesters with a common chiral diol (Heyn, 1975).

4 THE SYMMETRIC CHROMOPHORE IN A CHIRAL MOLECULAR ENVIRONMENT

4.1 The symmetric and the dissymmetric chromophore

Enantiomers were divided at an early stage into two main classes, one covering the molecules with an inherently-dissymmetric chromophore which in itself is primarily responsible for the optical activity, and the other referring to the optical isomers with a symmetrical chromophore in a chiral molecular environment due to substituent groups. The general mechanisms for optical activity, based on the independent-systems procedure, have been applied mainly to the latter class, although they are equally applicable to the treatment of substituent effects in the former cases. The two classes divide a range of intermediate cases. The polycyclic aromatic helicene series is representative of the limiting inherently-dissymmetric class, but the optical activity of a steroid containing a 1,3-diene group often owes as much to the particular substituent groups as to the helicoidal diene chromophore. The dissymmetric torsion of the ethylene chromophore in a chiral olefin appears to be generally even less important than the effect of the substituent groups in regard to the optical activity, and chiral olefins are frequently considered from the symmetric-chromophore viewpoint. The limiting case of a symmetric chromophore in a chiral molecular environment is represented by the carbonyl group of chiral ketones, although an inherent electronic dissymmetry in the π-delocalised lone-pair non-bonding orbital emerges from molecular orbital treatments, on abandoning the independent-systems procedure, where the lone-pair is taken as localised in the $2p_y$ orbital of the oxygen atom or, at most, delocalised over the carbonyl group.

4.2 The carbonyl chromophore

The most extensively investigated Cotton effects are those of the chiral ketones due to the carbonyl $n \rightarrow \pi^*$ transition near 300 nm. The CD and isotropic absorption arising from the $n \rightarrow \pi^*$ transition is well-separated on the frequency scale from the corresponding absorptions due to other transitions of the carbonyl chromophore (fig. 4.1), so that other band systems do not generally overlap the $n \rightarrow \pi^*$ bands, and small changes in the frequency, CD and isotropic absorption of the latter, due to the change in the position or the nature of the substituent, are readily detectable. The frequency-isolation of the carbonyl n → π* Cotton effect was of particular importance in the 1950s when ORD methods were primarily employed, allowing a precise determination of ORD amplitudes for the characterisation of substituent effects. The accessibility of the Cotton effect is additionally favoured by the relatively large dissymmetry ratio of the carbonyl n → π* transition, of the order of 10^{-1} for chiral ketones with C_2 symmetry.

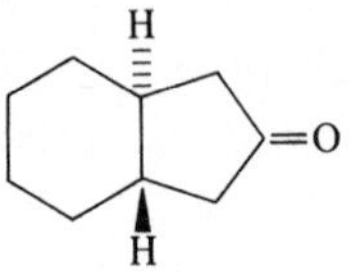

1

The carbonyl compounds with C_2 symmetry, such as, (+)-(1S:6S)-bicyclo-[4.3.0]nonan-8-one, **1**, have the significance that the additional electric-dipole moment of the n → π* transition, over and above that of the parent achiral ketone, cyclopentanone, is directed along the symmetry axis of the carbonyl bond, collinear with the magnetic-dipole n → π* transition moment. The collinearity accounts for the large g-ratio, and it allows the evaluation of the magnetic-dipole transition moment. The subtraction of the n → π* dipole strength of cyclopentanone from the corresponding strength of **1** gives the square modulus of the additional z-polarized electric moment, from which, through the rotational strength of **1**, the n → π* magnetic-dipole transition moment is determined as approximately one Bohr magneton, in agreement with the value calculated from the SCF-MO wavefunctions of formaldehyde (Mason, 1962). A similar value is obtained by the direct measurement of the n → π* magnetic-dipole strength of formaldehyde, separated from the predominant electric-dipole component, and is estimated additionally from the microwave Zeeman effect upon the rotational levels of formaldehyde (Callomon and Innes, 1963).

4.3   The octant rule for the carbonyl n → π* Cotton effect

In ketones of low symmetry the magnitude of the additional electric-dipole n → π* transition moment, and the frequency of the transition, are sensitive to the position and the orientation of the substituents, the wavelength and intensity changes being employed diagnostically in the steroid and terpenoid α-halo-ketone series (Cookson, 1954). The n → π* ORD Cotton effect of such ketones proved to be even more discriminatory, distinguishing not only between equatorial and axial α-halo-substituents in cyclohexanones, but also between the alternative axial positions (Djerrassi and Klyne, 1957). The ORD and subsequently the CD absorption study of chiral ketones containing substituents more remote from the carbonyl chromophore, and groups alternative to the halogens, culminated in the general octant rule (Moffitt *et al.*, 1961). The octant rule relates the location of a substituent group in the coordinate frame of the carbonyl chromophore to the sign of the n → π* Cotton effect through the sign of the position-function, [-XYZ] (fig. 3.1).

Initially the majority of the substituent-locations investigated referred to the rear octants in the -Z hemisphere of the carbonyl coordinate space, and the simpler [XY] quadrant rule appeared to be equally acceptable. The [XY] position- function corresponds to the simplest pseudoscalar basis functions of the point group C_{2v}, to which the carbonyl chromophore belongs. In addition the quadrant [XY] sector rule is required by the particular one-electron mechanism in which the carbonyl n → π* transition borrows an electric-dipole moment from the π → π* transition of the chromophore (Schellman and Oriel, 1962). Subsequent studies demonstrated that the carbonyl n → π* sector rule is of an octant type, the CD absorption near 300 nm changing sign when a given substituent

Fig. 4.1. The circular dichroism (upper curves) and absorption spectrum (lower curves) of (R)-(+)-3-methylcyclopentanone, **4**, in the vapour phase (full line), in pentane solution (dashed line) and in trifluoroethanol solution (dash-dot line) (Drake and Mason, 1978; and A. F. Drake, unpublished measurements).

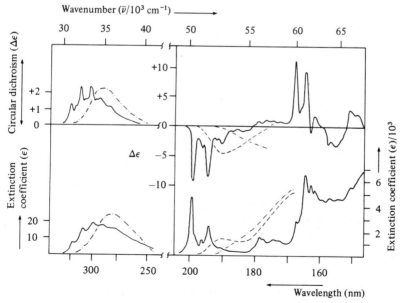

group is moved from a rear to the corresponding front octant in which the [XY] position product retains its sign (Kirk, Klyne and Mose, 1972). Both the front-octant ketone, **2**, and the related rear-octant analogue, **3**, for example, give a negative carbonyl n → π* Cotton effect (Lightner and Chang, 1974).

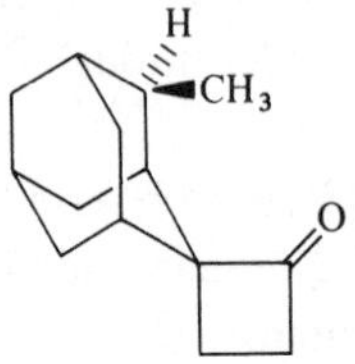

2 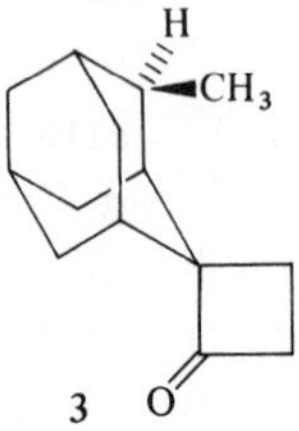   3

4.4   The octant rule and the one-electron mechanism

The octant rule for the carbonyl n → π* Cotton effect emerges from the one-electron model on the basis of the particular mechanism that the n → π* transition mixes principally with an allowed electric-dipole Rydberg transition of the oxygen atom, $n(2p_y) \rightarrow 3d_{yz}$, under the Coulombic field due to the incomplete screening of the nuclear charges of the substituent atoms by the valency electrons (Watanabe and Eyring, 1964; Bouman and Moscowitz, 1968). Extensions of CD absorption measurements to the vacuum ultraviolet region allow the particular mechanism to be investigated spectroscopically through the expectation (eqn 3.6) that the carbonyl n → π* and n → 3d transitions have rotational strengths of comparable magnitude and opposed sign for a given chiral ketone.

Rydberg bands are characterised, in the vapour phase, by a sharp line-shape which collapses in the corresponding solution spectrum, where any broad absorption remaining is shifted substantially to higher frequency (blue-shift). Bands due to valence-state electronic transitions undergo a small red-shift between the vapour phase and solution in an inert solvent, where any sharp vibrational structure is generally retained. The structure is lost, however, and the valence-state band undergoes a blue-shift, if hydrogen bonding between the solute prevails. The distinction between the Rydberg and the valence-state bands arises from the larger size of the Rydberg orbitals, the 3s, 3p, and 3d orbitals of the atoms of the carbonyl group, relative to the valence 2s and 2p orbitals. An electron in a Rydberg orbital, with a charge-distribution remote on the average from the atomic nucleus, has an energy sensitive to environmental charge distributions, the Rydberg-orbital energy being raised and broadened by neighbouring solvent molecules.

On the basis of these characteristics, the bands of (R)-(+)-3-methyl-cyclopentanone, **4**, at 195, 177, and 165 nm (fig. 4.1) are attributed to the Rydberg

4

transitions, n → 3s, n → 3p, and n → 3d, respectively, with a contribution from
the valence shell transition, n → σ_{CO}*, to the 195 nm band system (Robin, 1975).
As the CD absorption due to the n → π* transition at 300 nm and the n → 3d
Rydberg transition at 165 nm have the same, positive, sign (fig. 4.1) the mixing
of the two electronic promotions by the one-electron (eqn 3.6), or any other
mechanism, cannot make a significant contribution to the observed rotational
strengths of **4**.

The CD and isotropic absorption spectrum of **4** does not rule out other
particular mechanisms afforded by the general one-electron model. The mixing
of the carbonyl n → π* transition at 300 nm with the corresponding n → 3s, σ_{CO}*
transition is precluded, since the electric-dipole moment of the latter excitation
is known, from polarized single-crystal and electrochromism studies, to be
directed perpendicular to the carbonyl bond along the *y*-axis of the non-bonding
orbital. Moreover, the 300 and 195 nm Cotton effects, while of opposite sign in
the case of **4** (fig. 4.1), have the same sign in other cases (table 3.1). The specific
one-electron mechanism based upon the mixing of the n → π* with the π → π*
transition of the carbonyl chromophore gives the octant rule for the Cotton
effects of the former excitation if the non-bonding electrons in the $2p_y$ orbital
of the oxygen atom are delocalised over the carbonyl group. As yet the carbonyl
π → π* transition of ketones remains incompletely characterised, but it is likely
to lie near 155 nm where, in the CD absorption spectrum of **4**, a negative CD
band with a magnitude comparable to that of the positive n → π* CD band is
observed (fig. 4.1), in conformity with eqn 3.6 for the carbonyl n → π* and
π → π* mixing mechanism (Wagnière, 1966; Drake and Mason, 1978).

4.5   The octant rule and the dynamic polarization mechanism

The dynamic coupling mechanism for allowed magnetic-dipole
transitions was initially developed in connection with the carbonyl n → π*
Cotton effect and it yielded, in the mean-polarizability approximation for the
substituent groups, the octant sector rule for chiral ketones with the correct
absolute sign (eqn 3.14). The known antioctant or dissignate exception, the
fluoro-substituent group, was accommodated by eqn 3.14, on the grounds of
the smaller mean polarizability of the C—F bond relative to that of the C—H
bond, and some subsequent exceptions, notably, the heavy-isotope substituents,

C—D, C—CD$_3$, C—^{13}CH$_3$, which are dissignate relative to the corresponding light-isotope counterpart, are accounted for on the same basis (table 4.1).

A different type of dissignate behaviour occurs in some β-axially substituted ketones, notably, the series of 2-axially substituted adamantan-4-ones, **5**, where the corresponding 2-equatorially substituted analogues, **6**, show normal consignate substituent effects, following the octant rule (table 4.1). The dissignate effect of β-axial groups is not general, and in cases where the parent ketone is chiral and effectively polysubstituted, as in the steroid series, the differential

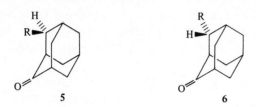

5   **6**

Table 4.1 *The n → π* Cotton effects of the 2-axial, 5, and the 2-equatorial, 6, substituted adamantan-4-one derivatives (fig. 4.2). The mean polarizability of the substituent $\Delta\bar{\alpha}(R)$ is taken relative to that of hydrogen as zero ($\bar{\alpha}(H) = 0.41$ Å^3). The mean polarizability and the polarizability anisotropy of the R—C bond $\beta(R-C) = (\alpha_\parallel - \alpha_\perp)$ in Å^3 are quoted from Le Fèvre (1965) and Bridge and Buckingham (1966). The electric-dipole polarizability unit, 1 Å$^3 = 1.112 \times 10^{-40}$ C^2 m^2 J^{-1} (SI)*

| | | | β-axial **5** | | β-equatorial **6** | | |
R—C[a]	$\Delta\bar{\alpha}(R)$	$\beta(R-C)$	λ(nm)	$\Delta\epsilon$	λ(nm)	$\Delta\epsilon$	Ref.
H—^{13}C			295	-0.006			b
D—C	-0.01		294	-0.017	295	-0.09	c
F—C	-0.09		278	+0.033	294	-0.19	d
CH$_3$—C	+1.83	+0.72	303	-0.093	300	+0.78	d
Cl—C	+1.91	+1.0	303	-0.41	306	+4.15	d
Br—C	+3.05	+1.5	303	-0.45	305	+7.88	d
I—C	+5.07	+2.1	304	-0.76	306	+14.62	d

(a) The perturbation of the n → π* Cotton effect of carbonyl compounds by H/D or by ^{12}C/^{13}C isotopic substitution is reviewed, together with the electronic circular dichroism of other isotopically chiral molecules, by Barth and Djerassi (1981).

(b) Sing, Numan, Wynberg and Djerassi (1979).

(c) Numan and Wynberg (1978).

(d) Snatzke and Eckhardt (1968); Snatzke, Ehrig and Klein (1969).

effect of a β-axial methyl group is often consignate. For a cylindrically-symmetric substituent group, the 2-axial and the 2-equatorial position in an adamantan-4-one are distinguished by the orientation of the principal axis of the group in the carbonyl chromophore coordinate frame. The difference between the components of the polarizability tensor of the substituent group along and perpendicular to the principal axis, $(\alpha_\| - \alpha_\perp) = \beta$, makes a contribution to the dynamic-coupling rotational strength of the carbonyl $n \to \pi^*$ transition additional to that afforded by the mean polarizability of the group, expressed in the octant rule equation (eqn 3.14). The additional contributions due to the polarizability anisotropy of the substituent group, β, differ for the 2-axial and the 2-equatorial position in an adamantan-4-one. The signed sector-rule diagrams for the polarizability-anisotropy contributions are conical in form (fig. 4.2), and they indicate that for a positive anisotropy, β, the contribution of a 2-axial group is dissignate, subtracting from the mean-polarizability octant contribution, whereas the anisotropy contribution of a 2-equatorial group is consignate, enhancing the mean-polarizability octant contribution (fig. 4.2) (Weigang and Höhn, 1968).

Fig. 4.2. The sector rules for the additional contributions, due to the polarizability anisotropy of the substituent, to the $n \to \pi^*$ Cotton effect of the carbonyl chromophore, due to, (*a*) a 2-axial group, **5**, and, (*b*) a 2-equatorial group, **6**, in adamantan-4-one, taking the anisotropy, $(\alpha_\| - \alpha_\perp)$, to be positive (Weigang and Höhn, 1968).

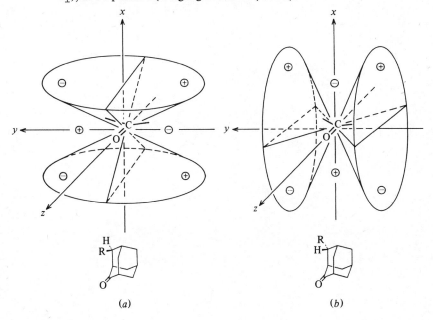

(*a*)   (*b*)

Fig. 4.3. The coordinate frame of the olefin chromophore, defining the octant sector rule, according to which a substituent group, L, located at X, Y, Z in the frame produces a Cotton effect with a sign corresponding to that of the product, [+XYZ], in the frequency-region of the lower-energy $\pi \to \pi^*$ transition. The main transitions of the olefin chromophore below $\sim 60\ 000$ cm^{-1} connect the occupied π_x orbital with the unoccupied π_x^*, π_y^*, and the Rydberg 3s orbital. The shaded and the unshaded lobes of the orbitals illustrated represent regions of opposite phase in the orbital wavefunction.

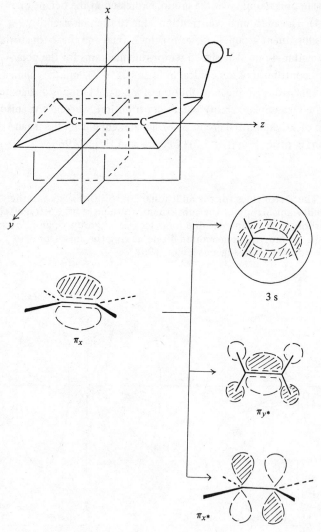

$3s$

π_x

π_{y^*}

π_{x^*}

4.6   The olefin chromophore

The ethylene chromophore of chiral olefins has a manifold of electronic transitions which is less favourable for Cotton-effect studies than that of the carbonyl group. The electric-dipole allowed transition, $\pi_x \to \pi_x^*$, lies near to the shorter-wavelength limit of the quartz ultraviolet region, and it is overlapped by an electric-quadrupole transition with a similar energy, characterised by electron-impact spectroscopy, probably $\pi_x \to \pi_y^*$, and the Rydberg transition, $\pi_x \to 3s$, at the longer wavelength band-edge (fig. 4.3). The Rydberg transition is charac-terised by sharp vibrational structure in the vapour-phase spectrum, with a spacing of $\sim 1400\,cm^{-1}$ corresponding to the $C{=}C$ stretching mode, and a shift to higher frequencies by $\sim 1000\,cm^{-1}$ in the presence of 10^2 atmosphere of an inert gas. The vibrational structure of the Rydberg band is lost in the corresponding solution spectrum, and the band shifts to the blue by some 4000 cm^{-1} in hydrocarbon solvents, where the cohesive internal pressure amounts to $\sim 10^3$ atmosphere, and by more than $7000\,cm^{-1}$ in hydrogen-bonding solvents where the effective internal pressure is larger ($\sim 10^4$ atmosphere) (fig. 4.4). The blue-shifts arise from the destabilisation of the Rydberg excited state by the environmental inert-gas or solvent molecules, due to the large size of the 3s orbital, which has a radial electron-density maximum at 2.1 Å from the centre of the olefin chromophore (Robin, 1975).

In contrast the more intense $\pi \to \pi^*$ bands of chiral olefins and other ethylene derivatives undergo a small red-shift on passing from the vapour-phase to the corresponding solution spectrum (fig. 4.4). In general two bands are observed in the $\pi \to \pi^*$ region, associated in the case of chiral olefins with CD absorption bands of opposite sign and comparable magnitude (table 3.1, fig. 4.4). The single-crystal spectrum of bicyclohexylidene, 7, measured with plane-polarized

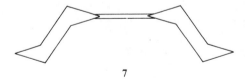

7

radiation, shows that both of the bands in the $\pi \to \pi^*$ region are z-polarized along the direction of the $C{=}C$ bond (fig. 4.3). The common polarization of the two bands indicates that the z-polarized electric-dipole moment of the $\pi_x \to \pi_x^*$ transition is shared by the virtually isoenergetic electric-quadrupole transition, while the CD spectra of chiral olefins suggest that the latter transition has additionally a z-polarized magnetic-dipole moment, as expected for a $\pi_x \to \pi_y^*$ excitation or a $\pi_y \to \pi_x^*$ promotion.

With a small energy-separation, the electric-dipole $\pi_x \rightarrow \pi_x{}^*$ transition and the neighbouring magnetic-dipole and electric-quadrupole transition became extensively mixed by small perturbations due to substituent groups in the molecular environment of the olefin chromophore. The mixing of an electric-dipole and a magnetic-dipole transition with a common polarization gives two rotational

Fig. 4.4. The absorption spectrum (upper curves) and the circular dichroism (lower curves) of (−)-α-pinene, 8, in the vapour phase (full curves), in pentane solution (dash-dot curve), and in 2,2,2-trifluoroethanol solution (dashed curves) (Drake and Mason, 1977).

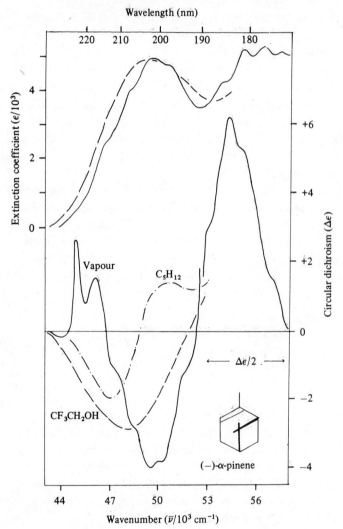

strengths with opposed signs and equal magnitudes, as is indicated by the one-electron mechanism (eqn 3.6), although the effective perturbation may be the dynamic polarization of the substituent group by the fields of the $\pi_x \to \pi_x{}^*$ electric-dipole moment and of the $\pi_x \to \pi_y{}^*$ electric-quadrupole moment, or a torsion of the olefin chromophore produced by the substitution.

An empirical sector rule connects the position of a substituent group in the coordinate frame of the olefin chromophore (fig. 4.3) with the sign of the lower-energy $\pi \to \pi^*$ Cotton effect induced by the substitution. The higher-energy $\pi \to \pi^*$ Cotton effect is not normally accessible with CD instruments restricted to the quartz ultraviolet region, with a shorter wavelength limit of 185 nm, although the onset of the second component of the oppositely-signed CD couplet may be observable. The sector rule for chiral olefins has an octant form (fig. 4.3), and it associates a positive lower-frequency Cotton effect in the $\pi \to \pi^*$ region of the spectrum with a positive value of the coordinate function [XYZ] of the substituent group in the olefin-chromophore frame (Scott and Wrixon, 1970).

The function [XYZ] provides the simplest basis for the pseudoscalar representation, A_u, of the point group D_{2h}, to which the ethylene chromophore belongs, so that the octant rule for chiral olefins is consistent with the one-electron mechanism (eqn 3.6). If the static-field perturbation of the one-electron mechanism is taken to be a net positive charge in the substituent group, such as that due to the incomplete screening of the nuclear charges of the atoms in the group by the valency electrons, the correct absolute signs of the octant rule for chiral olefins are given by the mechanism. Quantitatively, however, a positively-charged substituent group, such as quaternary nitrogen, is found to have a minor effect upon the CD absorption of a chiral olefin, relative to that of a neutral alkyl group in the corresponding position.

The dynamic coupling mechanism applied to the bisignate CD absorption couplet observed in the spectra of chiral olefins takes the $\pi_x \to \pi_x{}^*$ and the $\pi_x \to \pi_y{}^*$ transition to be coupled together through the Coulombic potential between the dipole induced in the substituent group and the leading electric moment of each of the two transitions, a dipole for the former and a quadrupole for the latter. Taking into account only the mean polarizability of the substituent group, the dynamic-polarization mechanism gives the octant rule for chiral olefins with the correct absolute signs, the lower-frequency component of the bisignate CD couplet being positive, and the higher-frequency component negative, for a positive value of the coordinate function [XYZ] in the chromophore axial system (fig. 4.3).

On the independent-systems basis, mechanisms of the one-electron or the dynamic-polarization type are required to account for the optical activity of

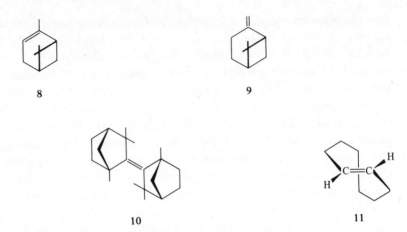

8   9

10   11

chiral olefins in which the torsion of the C=C bond on a molecule- and a time-average is small, such as (-)-α-pinene, **8**, or absent, as in the case of (-)-β-pinene, **9**. Crystal structure determinations by X-ray diffraction show that the C=C bond is twisted through 11.8° in (-)-syn-2,2'-bifenchylidene E, **10**, and through 43.5° in (R)-(-)-*trans*-cyclooctene, **11**, while electron diffraction of **11** in the vapour phase gives a twist angle of 23° about the double bond. Both **10** and **11** give the characteristic olefin bisignate CD absorption couplet in the $\pi \rightarrow \pi^*$ region of the spectrum, the magnitudes of the two components of the couplet being substantially larger than their counterparts observed in the corresponding spectra of **8** and **9**, although the associated isotropic absorption bands are of comparable intensity in the two sets.

The torsion of the double bond in an alkene through a non-zero twist angle θ produces an inherently dissymmetric olefin chromophore and virtually all transitions of the chromophore acquire an intrinsic rotational strength. In particular, the $\pi_x \rightarrow \pi_x^*$ transition attains a magnetic-dipole moment and the $\pi_x \rightarrow \pi_y^*$ excitation an electric-dipole moment, each of the torsion-induced moments

Fig. 4.5. A negative π-bond torsion angle, θ, of the olefin chromophore, giving the lower- and the higher-energy component of the olefin bisignate circular-dichroism couplet a positive and a negative rotational strength, respectively.

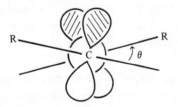

being proportional to sinθ. More significantly the two transitions are mixed, on account of the Coulomb potential between the transitional charge distributions in the twisted chromophore, independently of any other substituent effect than that producing the torsion of the double bond. A negative π-bond torsion angle (fig. 4.5) requires the connection of a positive [XYZ] substituent position-function of the groups producing the twist of the C=C bond with a positive lower-frequency and a negative higher-frequency component in the olefin bi-signate CD couplet. In the particular cases of 10 and 11 the π-bond torsion angle is positive, and the lower- and the higher-frequency components of the olefin CD couplet are negative and positive, respectively.

In chiral olefins with, or without, a twisted double bond the Cotton effects accessible in the quartz ultraviolet region are sensitive to axial-allylic substituents, that is, groups with an optimum magnitude of the position-function [XYZ] for a given distance of the group from the origin of the olefin coordinate frame (fig. 4.3). A number of exceptions to the olefin octant rule have been found, particularly in exocyclic methylene and alkylidene cases, and oxy-substituted olefins. The proportion of exceptions to the octant rule for chiral olefins is rather larger than the corresponding proportion of octant-rule anomalies observed in the case of the n $\rightarrow$ π^* transition of chiral ketones. The difference between the degree of correspondence of the two sets of Cotton effects to an octant rule is due, at least in part, to the isolation of the n $\rightarrow$ π^* carbonyl transition on the frequency scale, compared to the crowded set of olefin transitions near to the limit of the quartz ultraviolet region. A Cotton effect near 300 nm in the spectrum of a chiral ketone may be ascribed with confidence to the n $\rightarrow$ π^* transition, but the corresponding CD absorption of a chiral olefin near 200 nm may arise from the Rydberg π $\rightarrow$ 3s transition rather than the lower-frequency component of the bisignate olefin π $\rightarrow$ π^* couplet (Hudec and Kirk, 1976; Drake and Mason, 1977).

5 INHERENT DISSYMMETRY

5.1 Helical four-centre systems

The simplest chiral molecular system consists of four atoms linked by three bonds forming a segment of a left- or a right-handed helix. Hydrogen peroxide is a racemate in the fluid phase and it forms enantiomorphous crystals, all of the molecules in a given crystal having a common chirality. The 1,2-dithianes are chiral for all non-coplanar conformations of the C—S—S—C group, and they give a weak ($\epsilon \sim 10^2$) lowest-energy absorption band at a wavelength sensitive to the particular conformation. The band undergoes a red-shift from 250 to 370 nm as the dihedral angle, θ, between the two C—S—S planes is reduced from $\pi/2$ to zero (fig. 5.1). The circular dichroism associated with the lowest-energy absorption band is relatively strong ($\Delta\epsilon/\epsilon \sim 10^{-2}$), and it has a sign dependent upon $\sin 2\theta$, giving a quadrant sector rule (fig. 5.1) (Neubert and Carmack, 1974).

Fig. 5.1. The P-configuration of the inherently-chiral disulphide chromophore, producing a positive rotational strength in the lower-energy $n \rightarrow \sigma^*$ transition for a dihedral angle, $0 < \theta < (\pi/2)$.

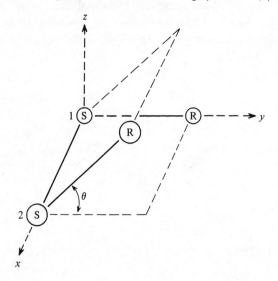

The lowest-energy absorption is ascribed to a n → σ* transition from an antibonding combination of the 3p lone-pair orbitals of the two sulphur atoms to the σ* orbital composed of an antisymmetric combination of a hybrid 3s, 3p orbital of each sulphur atom (Woody, 1973). The n→ σ* transition involves, at each sulphur atom, the generation of a mutually orthogonal electric- and mag-netic-dipole moment, e.g. at the S_1 centre the electric moment, μ_z, arises from the p_z→ s component of the excitation and the magnetic moment, m_y, from the $p_z → p_x$ component (fig. 5.1). For general non-coplanar conformations of the C—S—S—C chromophore, the electric moment at the S_1 centre is collinear with a component of the magnetic moment at the S_2 centre, and *vice versa*, giving overall a non-vanishing rotational strength. The corresponding bonding combi-nation of the 3p lone-pair orbitals of the two sulphur atoms gives a second n → σ* transition at a shorter wavelength ($\sim$ 240 nm) with a rotational strength of opposite sign and of comparable magnitude for a given 1,2-dithiane conform-ation. At the particular conformation where the dihedral angle θ (fig. 5.1) has the value of $\pi/2$ the bonding and antibonding combinations of the 3p lone-pair orbitals become degenerate, and the overall rotational strength of the two n → σ* transitions vanishes.

A number of inherently-dissymmetric chromophores are composed of a con-jugated, four-centre, skewed unit in which a π → π* electronic transition involves an overall helical charge displacement over the chiral σ-bond molecular framework. The significant charge displacement over each individual bond in the π → π* transition is linear, that is, electric dipole in character. Although the central 2,3-bond of such chromophores as the skewed 1,3-diene, and its hetero-atom ana-logues, is twisted, the torsion of the bond makes only a small contribution to the rotational strength of the π → π* transition relative to that due to the two-group coupled electric-dipole mechanism, where the groups are now the individual bonds of the conjugated system.

The lowest-energy π → π* transition of a 1,3-diene, or of its hetero-atom ana-logues, and of conjugated polyenes generally, gives rise to component-bond transition moments with the electric dipoles orientated head-to-tail along the successive bonds of the conjugated chain (fig. 5.2). The overall charge displace-ment in the transition is thus helicoidal, with the same chirality as that of the σ-bond structural frame over which the π-electrons are delocalised. A helicity rule for both *cisoid* and *transoid* skewed 1,3-dienes for the lowest-energy π-electron transition, located near to 260 nm, connects a positive or a negative Cotton effect in that wavelength region with the P (right-handed) or the M (left-handed) helical morphology of the diene chromophore (Charney, 1979).

The helicity rule holds for simple chiral 1,3-dienes and the hetero-atom ana-logues, but allylic axial substituent groups strongly influence the 260 nm Cotton effect, and may reverse the sign of the effect required by the particular helical form of the diene chromophore. The positive or negative contribution of allylic axial substituent groups to the $\pi \to \pi^*$ Cotton effect of both 1,3-dienes and olefins is connected with the P or the M helical relationship between the direction of the electric-dipole transition moment and the electric dipole induced by the field of that moment in the bond to the allylic axial group (Gawronski and Gawronski, 1980; Lightner *et al.*, 1981).

5.2   Dipole-length and dipole-velocity rotational strengths

An estimate of the rotational strength of a $\pi \to \pi^*$ electronic transition in a helical conjugated system requires the vector summation of the individual components of the electric-dipole moment along each bond, in order to obtain the overall electric transition moment, and the summation of the contributions to the magnetic transition moment from each bond, given by the vector product of the bond dipole and the position-vector of the bond centre from a fixed, common, origin. The helical π-systems, notably the skewed 1,3-diene and the helicene series of polycyclic aromatic hydrocarbons, do not have a unique, symmetry-determined, origin to the molecular coordinates, possessing at most a twofold rotational axis of symmetry (fig. 5.2). Two methods are commonly employed for an estimate of the electric-dipole moment of an electronic transition and one of these, the dipole-length procedure, gives rotational strengths which are sensitive to the choice of an origin for the molecular coordinates in cases where

Fig. 5.2. The charge displacements along the individual carbon-carbon bonds produced by the lowest-energy $\pi \to \pi^*$ transition of a 1,3-diene. The overall charge displacement is helically left-handed for a chiral *cisoid* or *transoid* diene with the M-configuration, as illustrated, giving a negative rotational strength. For a chiral diene with the P-configuration, the lowest-energy $\pi \to \pi^*$ Cotton effect is positive.

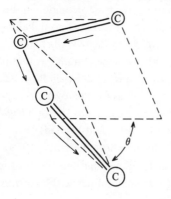

the origin is not symmetry-determined. The second method, the dipole-velocity procedure, gives origin-independent rotational strengths (Moscowitz, 1965).

The dipole-length method calculates the effective distance through which an electron is displaced during the course of an electronic transition, whereas the dipole-velocity procedure computes the effective current-flow, which is required additionally for an estimate of the associated magnetic moment at an external point (eqn 3.15). For an electron, e_i, the operator for the position vector, r_i, and the momentum, $(\hbar/i)\nabla_i$, are employed for the length and the velocity method, respectively. In the orbital approximation, an electronic transition between two stationary energy-states is reduced to a promotion from an occupied, ψ_i, to an unoccupied, ψ_j, molecular orbital, or to a combination of such orbital-promotions. The one-electron electric-dipole transition moment is expressed in dipole-length form by,

$$\mu_{ij}(DL) = -e <\psi_i|\mathbf{r}|\psi_j> \tag{5.1}$$

where the position-vector of the electron, $\mathbf{r}$, is directed from an arbitrary origin. The corresponding dipole-velocity form is given by the relation,

$$\mu_{ij}(DV) = -(\beta_m/\pi\tilde{\nu}_{ij}) <\psi_i|\nabla|\psi_j> \tag{5.2}$$

where β_m is the Bohr magneton and $\tilde{\nu}_{ij}$ is the transition wavenumber.

The $\pi \rightarrow \pi^*$ transition moment of a conjugated molecule is evaluated in both procedures by expanding the molecular orbitals connected by the transition, ψ_i and ψ_j, over the constituent atomic orbitals linearly combined to form the π–MO, i.e.,

$$\psi_i = \sum_{\nu} C_{\nu i}\phi_{\nu} \tag{5.3}$$

where $C_{\nu i}$ is the amplitude coefficient of the $2p_{\pi}$ atomic orbital, ϕ_{ν}, of the carbon atom, ν, in the π–MO, ψ_i. Neglect of differential overlap between the atomic orbitals gives the transition moment in dipole-length form as,

$$\mu_{ij}(DL) = -e \sum_{\nu}^{atoms} C_{\nu i}C_{\nu j}\mathbf{R}_{\nu} \tag{5.4}$$

where $\mathbf{R}_{\nu}$ is the position vector of the atom ν from the origin chosen, and the transitional charge density on that atom is measured by the amplitude-coefficient product, $(C_{\nu i}C_{\nu j})$. A similar development gives the corresponding dipole-velocity moment in the form,

$$\mu_{ij}(DV) = - (\beta_m/\pi\bar\nu_{ij}) \sum_{\mu\nu}^{\text{bonds}} P^{ij}_{\mu\nu} <\nabla_{\mu\nu}> \qquad (5.5)$$

where $P^{ij}_{\mu\nu}$ is a transitional bond-order change,

$$P^{ij}_{\mu\nu} = (C_{\mu i}C_{\nu j} - C_{\nu i}C_{\mu j}) \qquad (5.6)$$

which is directed from atom μ to atom ν and measures the magnitude and the direction of the charge displacement along the bond linking the two atoms in the transition connecting the π-MO ψ_i with ψ_j. The expectation value of the dipole-velocity element,

$$\nabla_{\mu\nu} = <\phi_\mu|\nabla|\phi_\nu> \qquad (5.7)$$

has a magnitude dependent upon the particular atoms bonded and the bond length. Values of $<\nabla_{\mu\nu}>$ for C—C, C—N, and C—O bonds have been tabulated for a range of bond lengths (Inskeep, Miles and Eyring, 1970).

Employing molecular wavefunctions which are inevitably inexact, the electric-dipole moments of a given transition calculated by eqns 5.5 and 5.6 are not generally congruent and differ in magnitude, direction, and the mean, centre-of-gravity, location in the molecule. The π-MOs of the 1,3-diene chromophore connected by the lowest-energy $\pi \rightarrow \pi^*$ transition have the form,

$$\psi_{i/j} = 0.60 \, (\phi_1 \mp \phi_4) \pm 0.37 \, (\phi_2 \mp \phi_3) \qquad (5.8)$$

where the upper signs refer to the highest-occupied and the lower signs to the lowest-unoccupied π-MO of a 1,3-diene. The charge displacements in the lowest-energy $\pi \rightarrow \pi^*$ transition of a 1,3-diene obtained by the dipole-velocity procedure (fig. 5.2) differ even qualitatively from the corresponding dipole-length displacements in that the component moment over the 2-3 bond is relatively reversed in direction in the latter case.

The determination of the absolute stereochemical configuration of an enantiomer by the anomalous X-ray scattering method (Bijvoet, Peerdeman and van Bommel, 1951) and by comparing the observed circular dichroism spectrum with the corresponding theoretical rotational strengths are generally congruent provided that the rotational strengths are calculated by the dipole-velocity procedure. The use of the dipole-length method gave rise to an anomaly which led to some doubts in regard to the foundations of the X-ray method (Tanaka,

1972). The doubts were dispelled by a recalculation of the relevant rotational strengths by the dipole-velocity method, and by scattering experiments employing beams of charged particles, helium and neon ions, rather than X-ray photons (see § 9.5).

5.3   The helicene series

The polycyclic aromatic hydrocarbons of the helicene series owe their optical activity principally to the inherently dissymmetric π-electron system, and the influence of substituent groups is minor. Substituents in the sterically-hindered positions affect the optical activity by increasing the pitch of the helical π-system, through enhanced steric overcrowding, and such substituents are necessary in the lower members of the helicene series to ensure an adequate optical stability. The smallest unsubstituted member of the series with an appreciable optical stability is [5]-helicene, which has a half-life of some three hours in hydrocarbon solution at ambient temperature.

The absorption and the CD spectra of [4]-, [5]-, [6]-, and [7]-helicene indicate that, in each case, the (+)-isomer has the right-handed helical P-configuration, as is found additionally for [6]-helicene by the anomalous X-ray scattering method. The spectra of (+)-[5]-helicene (fig. 5.3) are typical of the series. The lowest-energy absorption and CD band of high intensity, lying near 310 nm, are due to the transition of an electron from the highest-occupied to the lowest-unoccupied π-MO, $\pi_{11} \to \pi_{12}$ in [5]-helicene with the π-MOs numbered serially in order of increasing energy.

The amplitude coefficients of the carbon $2p_\pi$ atomic orbitals in the π-MOs of [5]-helicene, required for an estimate of the charge displacement over the individual bonds of the molecule in a particular $\pi \to \pi^*$ transition, are given in tabulations of π-electron calculations (Coulson and Streitwieser, 1965). Only the coefficients of π_{11} are required for the $\pi_{11} \to \pi_{12}$ transition, as the highest-occupied and lowest-unoccupied π-MOs form a conjugate pair, with coefficients of the same magnitude and sign on the starred atoms and of the same magnitude but opposite sign on the unstarred atoms, as in other even neutral alternant hydrocarbons, such as 1,3-butadiene (eqn 5.8). For the $\pi_{11} \to \pi_{12}$ transition in [5]-helicene, and analogous transitions between conjugate bonding and antibonding π-MOs in alternant hydrocarbons, eqn 5.6 for the transitional bond-order charge reduces to,

$$P_{\mu\nu}^{ij} = -2C_{\mu i}^* C_{\nu i}^\circ \qquad (5.9)$$

The transitional charge displacement is directed from the starred atom, μ, to the bonded unstarred atom, ν, if $P_{\mu\nu}^{ij}$ is positive and in the converse direction if negative. The bond charge-displacement diagram, from eqn 5.9, for the $\pi_{11} \to \pi_{12}$ transition of [5]-helicene (fig. 5.4) indicates that the overall transitional charge displacement has a right-handed helical form for the P-configuration of the molecule, correlating with the positive CD band near 310 nm in the spectrum of the (+)-isomer (fig. 5.3).

Quantitatively the rotational strength of each $\pi \to \pi^*$ transition in the manifold is estimated for a given helicene enantiomer from the dipole-velocity expression,

$$R_{ij} = (2\beta_m{}^2/\pi\bar{\nu}_{ij}) \langle\psi_i|\nabla|\psi_j\rangle \cdot \langle\psi_i|\mathbf{r} \times \nabla|\psi_j\rangle \tag{5.10}$$

Fig. 5.3. The observed absorption (upper curve) and circular dichroism spectrum (lower full curve) of (+)-[5]-helicene and the theoretical circular dichroism spectrum (broken line), calculated for the P-configuration (Brickell *et al.*, 1971).

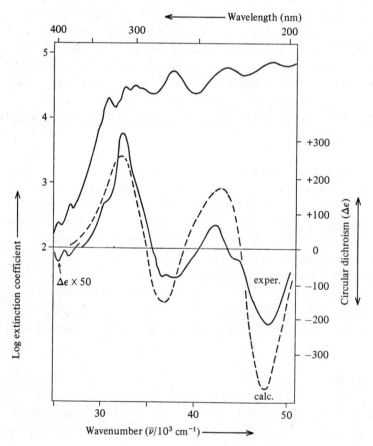

where the factor of 2 allows for the promotion of either of the two electrons occupying the π-MO ψ_i in the electronic ground state. The fitting of the computed rotational strengths to Gaussian band-shapes gives, for a particular helical configuration, a theoretical CD spectrum which may be directly compared with the corresponding observed spectrum of the helicene enantiomer over the frequency range investigated (fig. 5.3), (Brickell *et al.*, 1971; Wagnière, 1971).

Fig. 5.4. The direction and the magnitude ($\times 100$) of the individual transitional bond-order changes (eqn 5.9) produced by the lowest-energy allowed π-electron transition, $\pi_{11} \rightarrow \pi_{12}$, of [5]-helicene (*a*). Overall the charge-displacement is helically right-handed for the P-configuration of [5]-helicene (*b*).

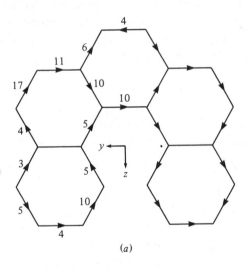

(*a*)

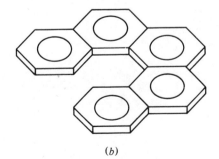

(*b*)

6   DIMERS AND POLYMERS

Chiral dimers and polymers are commonly produced biosynthetically in optically-pure form, and a number of synthetic oligomers are either resolvable optically or may be generated optically-active by asymmetric synthesis or by the condensation of chiral monomers. The optical activity of an oligomer composed of chiral monomer units is rarely represented by a single sum of contributions from those units, even in conformationally-labile macromolecules. Indeed the optical activity of such systems may be diagnostic of the main conformation prevailing under given conditions, as in the case of polypeptides. The limited additive character of optical activity is evident in the chiral dimers and polymers based upon achiral monomers, where the steric relationship between the con-stituent units and the electronic properties of the monomer chromophores govern the optical rotation. Molecular dissymmetry in the biaryl series derives primarily from an axis of asymmetry along the internuclear bond between achiral aromatic ring-systems with one or more planes of symmetry.

6.1   The biaryl series

The biaryls and other chiral oligomers may be considered as inherently-dissymmetric chromophores with restricted electron delocalisation between the monomer units. Electron delocalisation, strictly speaking, is not negligible be-tween the two naphthalene chromophores of the dihydro-[5]-helicene 1, which has a dihedral angle of 49° between the planes of the two aromatic nuclei. How-ever, the CD spectrum of the (S)-(+)-isomer of 1 more closely resembles that of

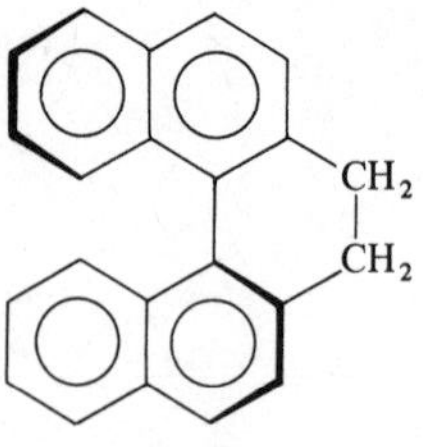

1

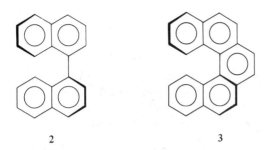

2   3

(S)-(+)-1,1′-binaphthyl, **2**, than that of (P)-(+)-[5]-helicene, **3**, itself. The dihedral angle between the planes of the naphthalene nuclei in 1,1′-binaphthyl is 68° in the racemic crystal and 103° in the active crystal, supporting the theoretical view that little energy is required to change the dihedral angle over the 60–120° range, and that the mean dihedral angle in solution approximates to $\pi/2$ where π-delocalisation between the two naphthalene nuclei is minimal. Thus the absorption spectrum of 1,1′-binaphthyl in solution is approximately additive over the two constituent naphthalene units, whereas the corresponding CD spectrum is a particular feature of the dimer assembly, reflecting the mutual steric relationship of the two naphthalene chromophores. While the optical activity of 1,1′-binaphthyl and its derivatives, and of other dimer and higher oligomer systems, has been investigated from the viewpoint of the inherently-dissymmetric chromophore, through a comparison of the calculated (eqn 5.10) and the observed CD spectra, the physical basis of the optical activity of such systems becomes the more apparent from the use of the two-group dipole-coupling mechanisms.

The application of the two-group electric-dipole mechanism of § 3.4 to the optical activity of chiral dimers and polymers generally requires a prior knowledge of the polarization direction in the monomer chromophore of each electronic transition considered, although the CD spectrum of the dimer may indicate those polarization directions in some cases, and in others the outcome of the application of the mechanism is independent of a possible choice between two alternative polarization directions in the monomer chromophore. The $\pi \rightarrow \pi^*$ transitions of aromatic substances are polarized in the molecular plane, and the particular in-plane polarization directions are given by spectroscopic studies of single crystals (Craig and Walmsley, 1968), stretched polymer films containing the aromatic substrate (Thulstrup, Michl and Eggers, 1970; Thulstrup and Michl, 1980), and other techniques for probing orientated molecules with polarized radiation (Meier, Sackmann and Grabmaier, 1975). The results indicate that, over the quartz ultraviolet region, the lowest-energy absorption band with a high intensity in the electronic spectra of anthracene, naphthalene, and benzene derivatives with strongly-conjugating substituents, such as phenol and aniline, arises from a $\pi \to \pi^*$ transition with a moment directed along the shorter in-plane axis of the molecule, while the following intense band at higher frequency is polarized in the direction of the longer in-plane axis (fig. 6.1). The spectrum of naphthalene shows, in addition, a weak long-wavelength absorption of mixed overall polarization, and the short-axis polarized band of toluene, and other benzene derivatives with a weakly-conjugating substituent, is of low intensity.

In the derived dimer, the electric-dipole transition moment of a low-intensity monomer absorption band is generally too small to give an observable optical activity through the isoenergetic dipole–dipole coupling mechanism, due in part to the minor rotational strengths produced (eqn 3.18), and in part to the small frequency-separation between those strengths (eqn 3.17). The large dipole moments of the high-intensity monomer transitions give, if chirally disposed in the derived dimer, the bisignate couplet of CD absorption bands characteristic of the isoenergetic dipole-coupling mechanism. The higher-frequency long-axis polarized $\pi \to \pi^*$ transitions of anthracene and naphthalene conform to this condition in $1,1'$-bianthryl and $1,1'$-binaphthyl, respectively, giving a bisignate CD absorption couplet of large amplitude in the 260 nm, and in the 220 nm wavelength region, respectively, (fig. 6.2).

For the (S)-configuration of $1,1'$-binaphthyl, **2**, the transition from the ground state to the symmetric combination of the individual naphthalene ex-cited configurations (ψ_+, eqn 3.16), with A-symmetry in the C_2 point group of the dimer, gives a negative rotational strength, from the overall left-handed helical charge-displacement (fig. 6.3). The corresponding transition to the anti-symmetric combination of individual naphthalene excited configurations (ψ_-, eqn 3.16), with B-symmetry in the C_2 group, involves an overall right-handed helical charge-displacement, giving a positive rotational strength.

The two dimer transitions, $\psi_0 \to \psi_\pm$, have different energies on account of the different Coulomb potential between the individual monomer excitation dipoles (eqn 3.17), positive in the one case and negative in the other. The wave-number interval between the rotational strengths of the two dimer transitions, ($\bar{\nu}_A - \bar{\nu}_B = 2V_{12}/hc$), is positive for a dihedral angle between the molecular

Fig. 6.1. The in-plane polarization-direction of the lower-energy, (short axis, s), and the higher-energy, (long-axis, l), high-intensity $\pi \to \pi^*$ tran-sition of anthracene, naphthalene, and conjugated monosubstituted benzene derivatives.

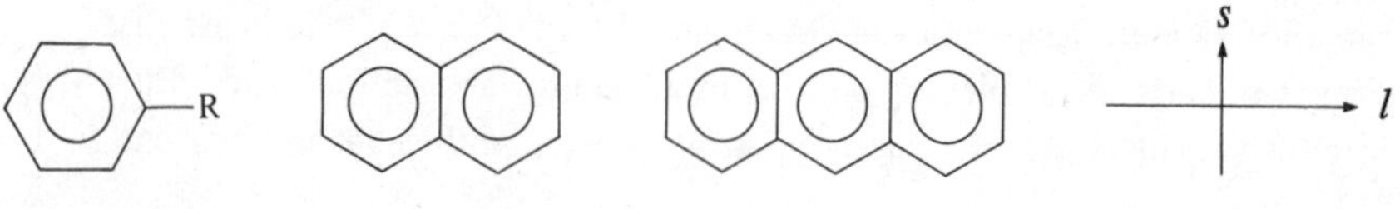

Fig. 6.2. The absorption (upper curves) and circular dichroism spectrum (lower curves) of the dication of (R)-(+)-2,2'-diamino-1,1'-binaphthyl in aqueous acid (full line, and the circular dichroism of (S)-(+)-1, 1'-binaphthyl in ethanol solution (broken line) (Mason, Seal and Roberts, 1974).

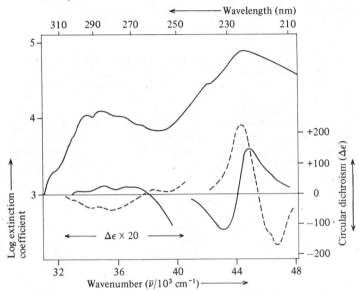

Fig. 6.3. The symmetric, higher-energy, (*a*), and the antisymmetric, lower-energy (*b*), coupling mode of the $\pi \to \pi^*$ electric-dipole transition moments polarized in the direction of the longer in-plane axis of each naphthalene nucleus in a chiral 1,1'-binaphthyl with the (S)-configuration. The modes give a bisignate circular dichroism couplet, positive at the lower frequency and negative at a higher frequency, for a dihedral angle, $\theta = (\pi - 2\alpha)$, in the range, $0 < \theta < 110°$.

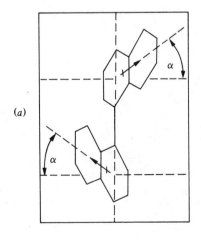

(*a*)

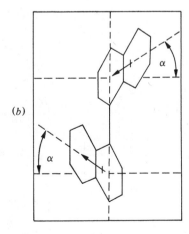

(*b*)

planes of the two naphthalene nuclei in the range, $0 < \theta < 110°$, covering $1,1'$-binaphthyl and its $2,2'$-disubstituted derivatives (fig. 6.4). Over this range of the dihedral angle, the sequence of bands in the bisignate CD absorption couplet, positive followed by negative at a higher frequency in the $250 - 210\,\text{nm}$ region, is indicative of the (S)-configuration for a $1,1'$-binaphthyl, and for a $1,1'$-

Fig. 6.4. The relationships connecting the dihedral angle, θ, between the individual aromatic nuclei of a biaryl with the (S)-configuration, with the rotational strengths, R_A and R_B, of the symmetric and antisymmetric coupling-modes, respectively, of the $\pi \to \pi^*$ excitations long-axis polarized in each of the aromatic nuclei (top curve), with the frequency interval between the two coupling modes $(\nu_A - \nu_B)$ (middle curves); and with the observable lower-frequency circular dichroism at maximum height, $\Delta\epsilon_{\text{max}}$, (bottom curves). The relations refer to (a) the $1,1'$-binaphthyls; (b) the $1,1'$-bianthryls; and (c) the $9,9'$-bianthryl derivatives.

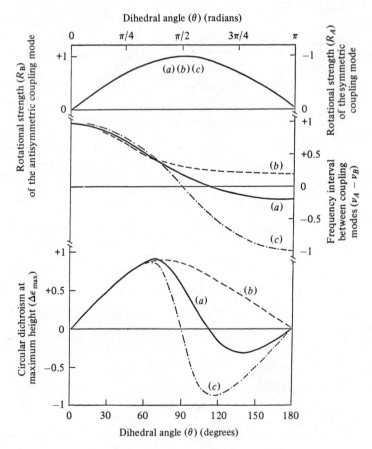

bianthryl if a corresponding sequence is observed in the 280 – 230 nm region, irrespective of the value of the dihedral angle in the latter case (fig. 6.4).

Although the lower-energy absorption band of naphthalene, in the 290 nm region, and of anthracene, near 400 nm, has a moderate to high intensity, the CD in the corresponding regions of the 1,1'-binaphthyl and the 1,1'-bianthryl spectra is weak and predominantly of a single sign (fig. 6.2). The monosignate character of the longer wavelength CD absorption suggests that the optical activity is due here primarily to the non-degenerate two-group electric-dipole coupling mechanism (eqn 3.19). The two electric dipoles of a short-axis polarized $\pi \to \pi^*$ transition in the individual naphthalene nuclei are not dissymmetrically arrayed in 1,1'-binaphthyl for any value of the dihedral angle between the naphthalene molecular planes. The two transition dipoles with short-axis polarization in each of the naphthalene nuclei are antiparallel for the A-coupling mode in 1,1'-binaphthyl (ψ_+, eqn 3.16), and parallel for the B-coupling mode (ψ_-, eqn 3.16). The vector sum of the two constituent excitation dipoles vanishes for the A-coupling mode, so that the dimer transition, $\psi_0 \to \psi_-$, carries all of the intensity of the 290 nm absorption in 1,1'-binaphthyl.

In addition, the B-coupling mode, $\psi_0 \to \psi_-$, near 290 nm, carries the rotational strength arising from the non-degenerate coupling of a short-axis polarized excitation dipole in one naphthalene nucleus in 1,1'-binaphthyl with a long-axis polarized excitation in the other nucleus (fig. 6.5). For the (S)-configuration of 1,1'-binaphthyl, the non-degenerate coupling mechanism (eqn 3.19) gives a

Fig. 6.5. The coupling, in a 1,1'-binaphthyl with the (S)-configuration, of the lower-energy, short-axis-polarized excitation in one naphthalene chromophore with the higher energy, long-axis-polarized excitation in the other, generating a negative rotational strength in the lower-frequency absorption region.

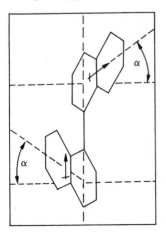

negative rotational strength at the lower energy (fig. 6.5), associated with the 290 nm absorption due to the $\psi_0 \rightarrow \psi_-$ dimer transition, and a positive rotational strength at the higher energy, near 230 nm (fig. 6.2), where it is superimposed on the positive rotational strength, larger by a factor of ~ 100, arising from the iso-energetic B-mode coupling of the two long-axis polarized excitation dipoles in 1,1'-binaphthyl (fig. 6.3).

Only the dimer transitions with the same symmetry are mixed by the non-degenerate coupling, and the mixing requires a non-zero Coulomb potential be-tween the dimer transition dipoles (eqn 3.17). The dimer A-mode combinations of the two long-axis and the two short-axis polarized $\pi \rightarrow \pi^*$ excitations do not effectively mix, as the resultant transition dipole of the short-axis A-mode combination vanishes, and the non-degenerate Coulomb potential between the resultant A-mode transition dipole goes to zero. The non-degenerate mixing of the dimer B-mode combinations of the two long-axis and the two short-axis polarized excitations is primarily responsible for the longer-wavelength CD ab-sorption of 1,1'-binaphthyl near 290 nm, and of 1,1'-bianthryl near 400 nm.

The longer-wavelength CD absorption necessarily has a sign opposite to that of the particular component of the bisignate CD absorption couplet at higher frequency due to the B-mode combination of the two long-axis polarized mono-mer excitations (eqn 3.19). Thus the two components of a bisignate CD ab-sorption couplet given by a 1,1'-biaryl are assigned to the respective A or B isoenergetic-coupling mode, not only by their sequence on the frequency ordinate, $(\bar{\nu}_A - \bar{\nu}_B) = 2V_{12}/hc)$, from the Coulomb energy (eqn 3.17), but also from the opposed-sign relationship between the CD absorption arising from the non-degenerate coupling of the B-mode combinations of the two short-axis and the two long-axis polarized excitations (eqn 3.19). The CD spectra of 1,1'-bianthryl and 1,1'-binaphthyl over the quartz ultraviolet region support, in themselves, through the isoenergetic and the non-degenerate two-group electric-dipole coupling mechanisms, the well-established assignment of the lower- and the higher-frequency absorption bands of the monomer aromatic nucleus to short-axis and long-axis polarized $\pi \rightarrow \pi^*$ transitions, respectively.

The CD absorption of chiral 9,9'-bianthryl derivatives, notably the 2,2'-dicarboxylic acid and the corresponding esters, is very small relative to that of the 1,1'-bianthryl analogues. The dihedral angle between the molecular planes of the two anthracene nuclei in the 9,9'-bianthryl derivatives is probably close to a mean value of $\pi/2$ in solution and, at this dihedral angle, 9,9'-bianthryl itself is achiral, with D_{2d} symmetry. The isoenergetic coupling of the long-axis polarized excitations in the two anthracene nuclei of 9,9'-bianthryl give the dimer transitions, $\psi_0 \rightarrow \psi_\pm$ (eqn 3.16), rotational strengths with an optimum magnitude at the dihedral angle of $\pi/2$ (fig. 6.4). However, the Coulomb potential between the long-axis excitation dipoles in the individual anthracene nuclei goes to zero when the nuclei are mutually perpendicular (eqn 3.17). The dimer transitions, $\psi_0 \rightarrow \psi_\pm$, are then degenerate, and the sum of the two rotational strengths vanishes (eqn 3.18).

The non-degenerate coupling of a long-axis polarized excitation in one anthracene nucleus with a short-axis polarized excitation in the other similarly gives a vanishing rotational strength in 9,9-bianthryl. A minor change of the dihedral angle of 9,9′-bianthryl from the mean value of $\pi/2$, such as that produced by association with the molecules of a chiral solvent, reduces the symmetry of the biaryl to that of the chiral point group, D_2, and gives rise to a bisignate CD absorption couplet in the 280 – 240 nm region of the long-axis polarized $\pi \rightarrow \pi^*$ anthracene transition (Mason, Seal and Roberts, 1974).

The isoenergetic and the non-degenerate two-group electric-dipole coupling mechanism has been extended to the analysis of the CD spectra of chiral dimers containing two non-conjugated aromatic chromophores, such as the naphthalene dimer, **4**, and the anthracene dimer analogue, **5** (Harada, Takuma and Uda,

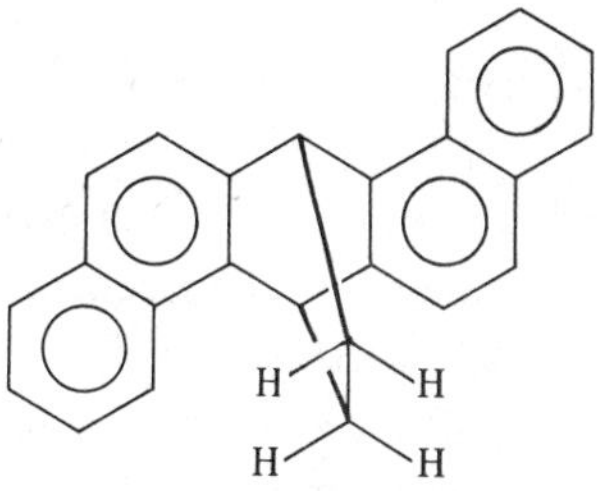

4

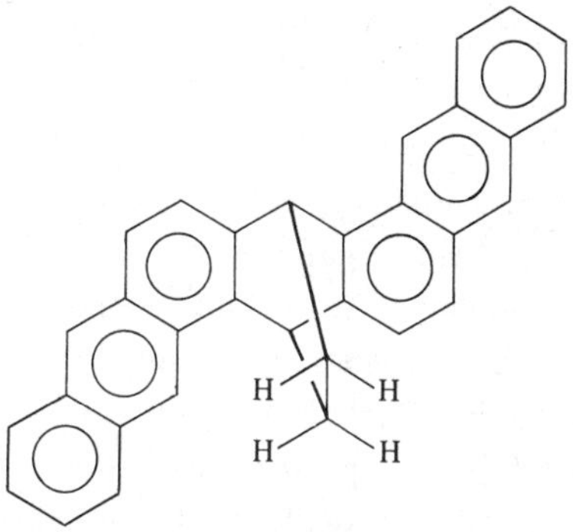

5

1977, 1978). In these cases the longer wavelength short-axis polarized $\pi \rightarrow \pi^*$ transition moments of the two aromatic chromophores are no longer anti-parallel and parallel in the respective A and B coupling modes, and a bisignate CD couplet appears in the lower-frequency region of the spectra of **4** and **5**, as well as the higher frequency region of the long-axis polarized $\pi \rightarrow \pi^*$ transitions of the aromatic chromophores. Indeed, each vibronic component of the longer wavelength band system of the anthracene dimer, **5**, covering the 320 – 390 nm range, appears as a bisignate CD couplet. The analysis is extended further to the CD spectra of a range of annelated triptycene derivatives (Harada, Tamai, Takuma and Uda, 1980; Harada and Nakanishi, 1982).

6.2   Dimeric alkaloids

,   A number of alkaloids have a dimeric structure, and some of these, such as calycanthine, **6**, have twofold rotational point symmetry. In calycanthine and analogous alkaloids, two aniline chromophores are related by the twofold rotation axis. The lower- and the higher-energy $\pi \rightarrow \pi^*$ transition of aniline,

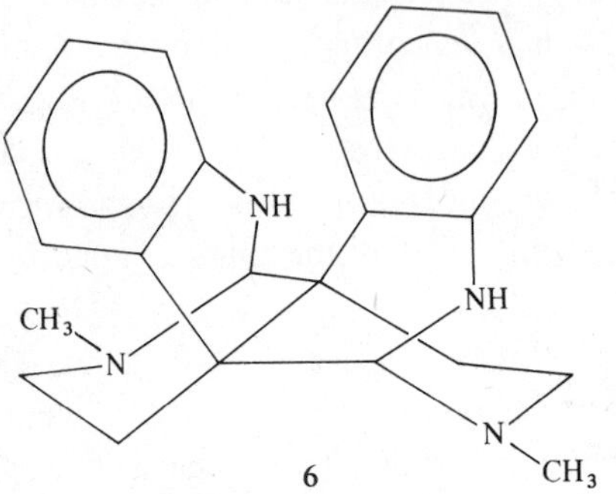

located near 310 and 250 nm, respectively (fig. 6.6), have an electric-dipole moment directed along the shorter and the longer in-plane axes of the chromophore, respectively (fig. 6.1). In the dimer **6** the two lower-excited aniline configurations are Coulombically coupled to give dimer excited states of A and B symmetry in the C_2 group, represented by ψ_+ and ψ_-, respectively (eqn 3.16), and a similar isoenergetic coupling holds additionally for the two higher-excited aniline configurations.

In both sets, the dimer A and B state with a common monomer origin have an energy separation dependent upon the potential between the isoenergetic excitation dipoles of the two constituent chromophores (fig. 3.3). The isoenergetic coupling gives rotational strengths of equal magnitude and opposite sign to the dimer transitions to the A and the B excited states in both the lower- and the higher-energy set (eqn 3.18), expressed in the bisignate CD absorption couplet associated with each of the two absorption band systems of **6** (fig. 6.6). Although oppositely-signed, the two components of a given CD couplet do not have equal band areas (fig. 6.6), owing to the superposition of the non-degenerate two-group electric-dipole coupling mechanism (eqn 3.19), which intermixes the excited states of B symmetry in **6** and, equally, mixes the states in the A symmetry class.

For the absolute stereochemical configuration **6** of calycanthine the sign-order of the rotational strengths, expected from the dipole–dipole potential (eqn 3.17) and the spatial relationship between the monomer excitation dipoles in the dimer, is positive followed by negative at a higher frequency for each of the two sets of dimer transitions. That is, the sign-order expectation is independent of the particular long-axis or short-axis polarization of the monomer chromophore excitation, and the determination of absolute configuration is not dependent upon the prior determination of those polarization directions.

With the polarization directions established for the first two $\pi \rightarrow \pi^*$ excitations of the aniline chromophore, the dimer excited states in **6** have the symmetry sequence B, A, A, B, to higher energy. The transitions to the excited

Fig. 6.6. The isotropic and the polarized electronic spectra of calycanthine, **6**, referring to the emission (full curves) and the absorption (broken curves). The topmost curves represent the isotropic spectra, and the middle curves give the circular differentials, $\Delta \epsilon = (\epsilon_L - \epsilon_R)$, and $\Delta I = (I_L - I_R)$, both sets referring to ethanol solution. The bottom curves refer to glycerol solution and record the linear polarization factor, $p = (I_\parallel - I_\perp)/(I_\parallel + I_\perp)$, for the fluorescence excited at 260 and 320 nm, and for the excitation spectrum monitored at the fluorescence maximum, 350 nm (Barnett, Drake and Mason, 1979) (see § 2.4).

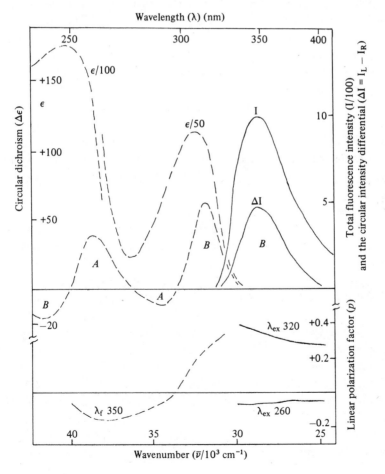

states of A symmetry are polarized along the C_2 rotation axis of the dimer **6**, and those to the B excited states have an orthoaxial polarization (fig. 3.3). The symmetry-sequence of dimer excited states, required by the two-group dipole-coupling mechanism, is supported by the linearly-polarized excitation spectrum of the fluorescence of **6** in glycerol solution (fig. 6.6).

The linear polarization anisotropy ratio, p (eqn 2.13), of the fluorescence monitored at 350 nm, the wavelength of the maximum luminescence intensity, is expected to be positive for the excitation of a transition with a dipole moment parallel to the emission dipole, and to have a negative value if the excitation dipole is orientated in the molecule perpendicular to the emission dipole (Dörr, 1971). For calycanthine, **6**, the polarization ratio, p, measured at 350 nm as a function of the excitation wavelength, is positive over the absorption region corresponding to the transition to the lowest-energy dimer excited state, with B symmetry, and negative over the shorter-wavelength region of the transitions to the next two dimer excited states at higher energy, so that the latter two states have A symmetry (fig. 6.6). The luminescence of calycanthine is relatively homogeneous, arising from a single electronic transition, since, with a fixed excitation wavelength, the polarization ratio, p, remains single-signed across the frequency region of the fluorescence band, positive for excitation at 320 nm into the lowest-energy dimer excited state with B symmetry, or negative for excitation into the A-excited dimer state at 260 nm (fig. 6.6).

The circularly-polarized components of the fluorescence of calycanthine, **6**, excited with unpolarized radiation have unequal intensities, and the dissymmetry ratio of the luminescence, g_{lum} (eqn 2.12), is comparable in peak height to the corresponding absorption dissymmetry ratio, g_{abs} (eqn 2.11), of the lowest-energy dimer transition, measured either by fluorescence-detected CD or directly (fig. 6.6). The width of the luminescence circular intensity differential (CID) band is, however, appreciably larger than that of the lowest-energy CD absorption band (fig. 6.6).

If a chiral molecule does not change in size or shape on photoexcitation to the lowest-energy excited state, the luminescence CID band and the corresponding absorption CD band are expected to have the same sign, magnitude, and band-shape, provided that the lowest-energy rotational strength is not overlapped by a higher-energy rotational strength in the absorption process. Aniline itself does not change appreciably in shape or size on photoexcitation (Cvitas, Hollas and Kirby, 1970), and in calycanthine the excitation energy is shared between the two aniline chromophores, so that the net distortion of each is very small. However, the transitions to the A- and the B-excited dimer states arising from the lowest-energy aniline chromophore excitation have overlapping rotational strengths with opposed signs, so that the observed band-width of the corresponding CD absorptions is smaller than that of the luminescence CID band, which is due to a single transition, that from the *B*-excited dimer state (fig. 6.6).

Relatively small CD absorption band-widths are observed generally in bisignate CD couplets due to the isoenergetic electric-dipole coupling mechanism when the energy interval (eqn 3.17) between the two oppositely-signed rotational strengths produced (eqn 3.18) is smaller than the corresponding absorption bandwidth. The two components of the bisignate CD absorption couplet then have a larger observed separation than the calculated energy interval, and smaller band areas than indicated by the theoretical rotational strengths. The CD bands associated with the calycanthine absorption at 308 nm, derived from the lowest-energy $\pi \to \pi^*$ excitation of the two aniline chromophores, have the rotational strengths, $R(B) = +0.71$ and $R(A) = -0.09$ Debye–Bohr magneton, and they are separated by 2700 cm^{-1}, whereas the theoretical rotational strengths are, $R(B) = +1.37$ and $R(A) = -1.14$ Debye–Bohr magneton, separated by the calculated interval, $(\nu_A - \nu_B)$ of +700 cm^{-1} (Barnett, Drake and Mason, 1979).

The CD spectra of a number of dimeric alkaloids have been analysed with the object of determining the stereochemical configuration of the substance (Mason, 1967), and the procedure has been extended to natural products, such as steroid diols, which themselves do not absorb radiation in the visible or the quartz ultraviolet region. Chiral diols form benzoate diesters which give a characteristic bisignate CD couplet in the 220 – 280 nm region, due to the two chiral coupling modes of the benzoate $\pi \to \pi^*$ transitions, long-axis polarized in each chromophore. The analysis of the frequency and sign order of the two oppositely-signed CD bands, based upon the isoenergetic two-group electric-dipole mechanism, gives the absolute configuration of the original chiral diol (Harada and Nakanishi, 1972, 1982; Harada, Chen and Nakanishi, 1975).

6.3   Dihedral coordination compounds

The most readily-accessible chiral molecules with high point symmetry are the tris-chelate coordination compounds formed by bidentate ligands with twofold rotational symmetry (C_2) and non-metal or metal ions. The C_2 axes of the ligands are interrelated by the principal threefold rotation axis (C_3) of the coordination compound, giving the dihedral symmetry, D_3, overall. If the ligand is a conjugated system, the $\pi \to \pi^*$ electronic transitions of the free ligand are polarized in the molecular plane, in a direction either parallel or perpendicular to the C_2 axis of the ligand (fig. 6.7).

The triple degeneracy of a given ligand $\pi \to \pi^*$ excitation in the trischelate complex is lifted by the Coulomb potential between the three individual ligand excitation dipoles. The potential couples the three ligand excited configurations, χ', to give the trimer excited states, ψ, of the coordination compound. The trimer excited states corresponding to a given ligand excitation are either symmetric with respect to the C_3 operation of the complex, and have A_1 or A_2 symmetry in D_3, or are doubly-degenerate and antisymmetric under the C_3 operation, with E symmetry in D_3.

The trimer ground state, ψ_0, is totally symmetric (A_1), and it is represented by the simple product of the ground configurational functions, χ, of the individual ligands,

$$\psi_0(A_1) = [\chi_1 \chi_2 \chi_3] \tag{6.1}$$

The totally-symmetric trimer excited state, $\psi_1(A_1)$, derives only from a single-ligand excitation which is polarized parallel to the C_2 axis of the ligand. A perpendicularly-polarized single-ligand excitation gives a symmetric trimer excited state with A_2 symmetry, $\psi_1(A_2)$ (fig. 6.8). In each case the trimer excited state is represented by a common linear combination of single-ligand excited configurational functions, χ',

$$\psi_1(A_1)(A_2) = [\chi_1' \chi_2 \chi_3 + \chi_1 \chi_2' \chi_3 + \chi_1 \chi_2 \chi_3'] / \sqrt{3} \tag{6.2}$$

Similarly the antisymmetric doubly-degenerate excited trimer state has a functional form common to the alternative polarization directions of the single-ligand excitation,

$$\psi_1(E) = [\chi_1' \chi_2 \chi_3 - \chi_1 \chi_2' \chi_3] / \sqrt{2} \tag{6.3}$$

and,

$$\psi_2(E) = [2\chi_1 \chi_2 \chi_3' - \chi_1 \chi_2' \chi_3 - \chi_1' \chi_2 \chi_3] / \sqrt{6} \tag{6.4}$$

With an octahedral coordination polyhedron, the molecular planes of the three chelate rings are mutually orthogonal to one another, and the single-ligand exci-

Fig. 6.7. The $\pi \to \pi^*$ transitions of unsaturated bidentate ligands with twofold symmetry have in-plane polarizations directed either parallel or perpendicular to the C_2 rotation axis.

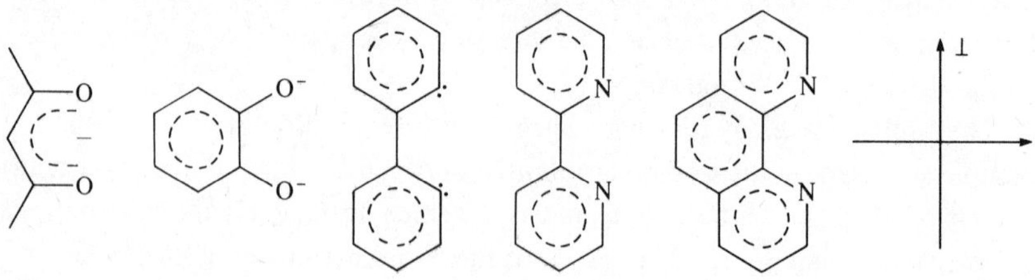

tation dipoles polarized parallel to the ligand C_2 axis ($\mu_{0l}^{\parallel}$) lie in a common plane perpendicular to the C_3 axis of the coordination complex. If the ligand excitation is polarized perpendicular to the ligand C_2 axis ($\mu_{0l}^{\perp}$), each of the three ligand moments lies at the angle $\cos^{-1}(2/3)^{1/2}$ to the C_3 axis of the chelated complex (fig. 6.8). These geometric relationships give simple expressions for the first-order dipole and rotational strengths of the transitions from the ground state (eqn 6.1) to the excited states (eqn 6.2, 6.3 and 6.4) of the coordination compound.

For the parallel-type single-ligand excitation, the trimer transition, $\psi_0(A_1) \rightarrow \psi_1(A_1)$, is forbidden, since the coplanar symmetric coupling of the three ligand excitation dipoles gives a vanishing resultant (fig. 6.8a). All of the dipole strength is centred upon the corresponding doubly-degenerate transition, $\psi_0(A_1) \rightarrow \psi(E)$,

Fig. 6.8. The coupling modes of the ligand $\pi \rightarrow \pi^*$ excitation moments in a tris-chelated coordination compound. The ligand excitations polarized parallel to a C_2 axis give rise to a symmetric coupling mode (a) and a doubly-degenerate antisymmetric mode (b). Only the latter have a net resultant trimer electric-dipole moment, and no magnetic moment results from either (a) or (b). The ligand excitations polarized perpendicular to the local C_2 axis give rise to the symmetric coupling mode (c) and the doubly-degenerate mode (d). The resultant electric and magnetic moment of the trimer transition are non-vanishing for both the (c) and the (d) modes (Mason, 1968).

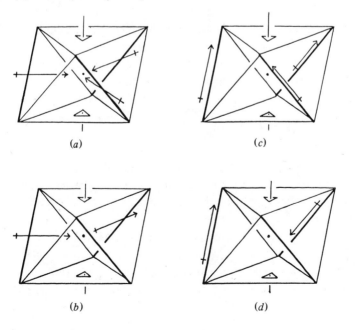

(a)   (c)

(b)   (d)

$$D^{\parallel}(E) = 3|\mu_{0l}^{\parallel}|^2 \tag{6.5}$$

but the first-order rotational strength goes to zero, since the degenerate E-coupling mode of the three $\mu_{0l}^{\parallel}$ gives a vanishing magnetic-dipole transition moment, that is, no overall helical charge displacement is involved (fig. 6.8b).

In contrast, the trimer transition to the symmetric excited state resulting from the perpendicular-type of single-ligand excitation, $\psi_0(A_1) \to \psi_1(A_2)$ (fig. 6.8c) carries the major part of the total dipole strength,

$$D^{\perp}(A_2) = 2D^{\perp}(E) = 2|\mu_{0l}^{\perp}|^2 \tag{6.6}$$

Both of these trimer transitions have a magnetic moment and a non-vanishing rotational strength. In the case of a D_3 coordination compound with the Λ-configuration, the overall helical charge displacement is left handed for the A_2 coupling mode of the three ligand excitation moments, $\mu_{0l}^{\perp}$ (fig. 6.8c) and right handed for the corresponding E coupling mode (fig. 6.8d). The rotational strengths of the two trimer transitions are given by the relation,

$$R^{\perp}(E) = -R^{\perp}(A_2) = (2/3)^{1/2} \pi \bar{\nu}_{0l} r_{12} |\mu_{0l}^{\perp}|^2 \tag{6.7}$$

where $\bar{\nu}_{0l}$ and $\mu_{0l}^{\perp}$ are, respectively, the wavenumber and the electric-dipole moment of the single-ligand excitation, and r_{12} is the distance between the point excitation dipoles of two of the ligands.

The Coulomb potential between the single-ligand excitation dipoles, $\mu_{0l}^{\perp}$, gives the trimer transition, $\psi_0(A_1) \to \psi_1(A_2)$, a higher energy than the corresponding $\psi_0(A_1) \to \psi(E)$ transition. For an octahedral coordination geometry in a D_3 tris-chelate complex, the wavenumber interval between the two transitions due to the dipole-dipole potential (eqn 3.17) follows the relation,

$$hc[\bar{\nu}(A_2) - \bar{\nu}(E)] = + 3|\mu_{0l}^{\perp}|^2/(4r_{12}^3) \tag{6.8}$$

A comparison of the observed CD spectrum of a tris-chelate complex with the expectations of eqns 6.7 and 6.8 provides an indication of the absolute stereochemical configuration of the coordination compound (Mason, 1968; Bosnich, 1969). A bisignate CD couplet associated with the ligand $\pi \to \pi^*$ excitations in a D_3 complex following the sign order, positive to negative, from lower to higher frequency, indicates that the compound has the Λ-configuration (table 6.1), e.g. **7** for the tris(1,10-phenanthroline) complexes, **8** for the (-)-[P(biph)$_3$]$^-$ and analogous tris(2,2'-bipyridyl) coordination compounds, and **9** for (+)-[As(cat)$_3$]$^-$ and other tris-catechyl complexes.

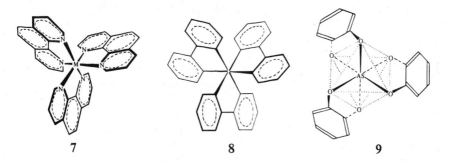

7   8   9

The CD spectrum of a tris-chelate coordinate compound provides in addition the polarization direction of the single-ligand excitation moments. In the spectrum of Λ-(+)-$[As(cat)_3]^-$ (fig. 6.9), the CD associated with each of the two longer-wavelength band systems, at 275 and 225 nm, is monosignate, and only the third system at 196 nm shows the bisignate CD couplet form. Thus the 275 and 225 nm bands arise, in each case, from a parallel-type single-ligand excitation, with the dipole strength confined to the trimer state transition, $\psi_0(A_1) \rightarrow \psi(E)$ (eqn 6.5) (fig. 6.8b). The third band at 196 nm is due to a perpendicular-type single-ligand

Fig. 6.9. The absorption spectrum (upper curve) and circular dichroism (lower curve) of Λ-(+)-$K[As(cat)_3]$ in ethanol solution (Peart, 1970).

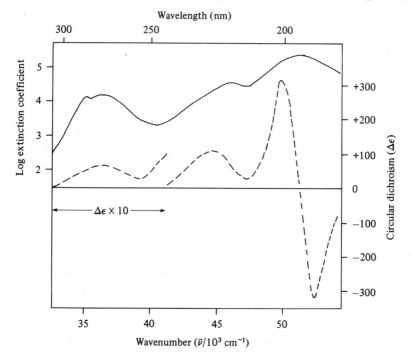

Table 6.1. *The absorption and circular dichroism spectra of tris-chelate coordination compounds with the Λ-stereochemical configuration in the region of the lowest-energy ligand $\pi \to \pi^*$ excitation polarized in-plane perpendicular to the two-fold rotation of the ligand*

Compound[a]	E-mode CD		A_2-mode CD		$(A_2 + E)$Abs		Ref.
	$\bar{\nu}/10^3\,cm^{-1}$	$\Delta\epsilon$	$\bar{\nu}/10^3\,cm^{-1}$	$\Delta\epsilon$	$\bar{\nu}/10^3\,cm^{-1}$	$\epsilon/10^3$	
$(+)$-$[Si(pd)_3]^+$	33.2	+30	36.3	-20	34.7	35	b
$(+)$-$[As(cat)_3]^-$	49.8	+320	51.9	-320	51.0	250	c
$(-)$-$[P(biph)_3]^-$	45.8	+325	52.7	-290	49.0	150	d
$(+)$-$[Cr(pd)_3]$	31.0	+56	34.3	-37	33	10	e
$(+)$-$[Co(bipy)_3]^{3+}$	31.2	+90	33.2	-50	32.3	35	f
$(+)$-$[Ni(bipy)_3]^{2+}$	32.5	+250	35.1	-50	33	48	f
$(+)$-$[Ru(bipy)_3]^{2+}$	34.4	+240	36.1	-105	34.9	56	f
$(+)$-$[Cr(phen)_3]^{3+}$	36.5	+145	38.6	-105	37.3	44	f
$(-)$-$[Fe(phen)_3]^{2+}$	36.8	+625	38.5	-360	37.5	89	f
$(+)$-$[Co(phen)_3]^{3+}$	35.3	+240	36.9	-150	36.1	67	f
$(+)$-$[Ni(phen)_3]^{2+}$	36.5	+550	38.3	-260	37.2	87	f
$(+)$-$[Ru(phen)_3]^{2+}$	37.4	+540	38.9	-410	38.2	89	f
$(+)$-$[Os(phen)_3]^{2+}$	37.3	+470	39.0	-300	37.8	93	f

(a) The ligand abbreviations are:- pentane-2,4-dionate (pd); catechol (1,2-benzenediol)dianion (cat); biphenylene-2,2'-dianion (biph); 2,2'-bipyridine (bipy); 1,10-phenanthroline (phen).
(b) Larsen, Mason and Searle (1966).
(c) Ito, Kobayashi, Marumo and Saito (1971); Peart, (1970).
(d) Hellwinkel and Mason (1970).
(e) Mason, Peacock and Prosperi (1977).
(f) Mason and Peart (1973).

excitation, and the isoenergetic coupling of the three excitations gives the rotational strengths of equal magnitude and opposite sign (eqn 6.7) (fig. 6.8c,d). The trimer excited states of E symmetry, derived from the parallel-type and the perpendicular-type single-ligand excitations, mix Coulombically, in an inverse proportionality to their energy-separation, to give the monosignate positive CD observed for the longer wavelength band systems, minor for the 225 nm band and very small for the 275 nm band more distant in energy from the 196 nm band (fig. 6.9).

6.4   Biopolymers

Polypeptides, polysaccharides, and polynucleotides, together with chiral synthetic macromolecules, have an optical activity which is generally sensitive to environmental conditions of temperature and the solvent medium, notably, ionic strength and pH in aqueous solution, due to changes in the chain conformation. The optical activity of a random-coil conformation, where there is no regular stereochemical relationship between successive monomer units, approximates to the sum of monomer contributions, but even in the random-coil case the additivity is rarely quantitative and there are often new features. The CD spectrum of a polypeptide derived from the L-amino acids in the random-coil conformation shows a negative CD absorption at 198 nm with a band area, on an individual amide chromophore basis, almost twice as large as that of a monomeric analogue, and a new positive CD feature near 220 nm which has no counterpart in the monomer spectrum (fig. 6.10) (Johnson and Tinoco, 1972).

The optical activity of a chiral macromolecule undergoes substantial and distinctive changes on assuming a stereoregular conformation. In the β-pleated sheet polypeptide conformation, where the chain backbone is extended with adjacent carbonyl groups either parallel or antiparallel to one another and hydrogen bonded to the N—H groups of neighbouring chains, the CD spectrum becomes virtually enantiomorphous to that of the corresponding random-coil conformation (fig. 6.10). The optical-activity changes in the stereoregular conformations originate primarily from the pairwise coupling of excitation dipoles in the individual monomer chromophores, summed over the n ordered units in the macromolecule. Such optical-activity changes become large in the helical conformations, where the collective excitation of the n chromophore units by the radiation field gives rise to an overall helical charge-displacement in the polymeric system. In the polypeptide α-helix conformation, where the carbonyl groups of each turn of the helix hydrogen bond to the N—H groups of a succeeding turn to give the macromolecule a rigid rod-like structure, the observed CD band areas become twice as large as those of the corresponding β-sheet structure (fig. 6.10) (Brahms and Brahms, 1980).

The analysis of the optical activity of helical macromolecules was initiated by Moffitt (1956a,b), employing the coupled excitation-dipole mechanism to interpret the unusual optical rotatory dispersion of the polypeptides composed of L-amino acid residues in the α-helix conformation. Moffitt considered a helix composed of n amide chromophores in the limit that n goes to infinity and showed, that of the n possible polymer transitions corresponding to a particular monomer excitation, only three have an appreciable dipole and rotational strength. The three allowed polymer transitions with a common monomer origin have mutually perpendicular polarization-directions. The collective in-phase excitation of the n monomer chromophores gives a polymer transition with an electric- and a magnetic-dipole moment directed along the helix axis (fig. 6.11). Each of the two other allowed polymer transitions involve a phase relationship between the individual monomer excitation dipoles giving a non-zero resultant electric and magnetic moment, collinear with one another, in the plane perpendicular to the helix axis (fig. 6.11). The latter two polymer transitions have a mutually-orthogonal polarization in the plane perpendicular to the helix axis and, for the infinite polymer, the two transitions are isoenergetic. In the case of a finite polymer the latter two polymer transitions are no longer degenerate, and both parallel and perpendicularly-polarized transitions are allowed to a degree over a wider range of monomer-coupling modes.

For a particular monomer chromophore excitation with the dipole strength, D_{01}, at the wavenumber, $\bar{\nu}_{01}$, the resultant rotational strengths, per monomer residue, of the polymer transitions polarized parallel and perpendicular to the helix axis are given by Moffitt's expression,

$$R_{\parallel} = -R_{\perp} = \pi \bar{\nu}_{01} D_{01} \rho \cos t \cos \nu \qquad (6.9)$$

where $\cos t$ and $\cos \nu$ are, respectively, the tangential and the vertical cosines of the excitation-dipole vector in each individual monomer chromophore, lying at a radial distance, ρ, from the helix axis. The terms vertical and tangential refer to a cylindrical coordinate system relating the monomer residues of a helical poly-

Fig. 6.10. Representative absorption spectra (upper curves), circular dichroism (lower curves), and conformations of polypeptides composed of L-amino acids: (*a*) the α-helix conformation (full line); (*b*) the β-sheet conformation (dash-dot line); and (*c*) the random-coil conformation (broken line). Adapted from Brahms and Brahms (1980).

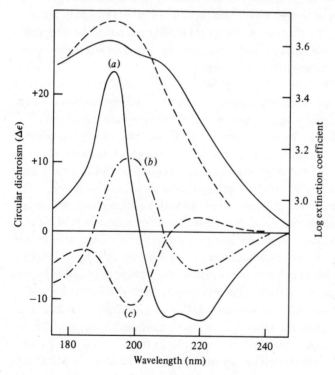

Fig. 6.10. Continued.

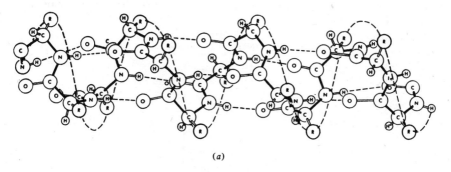

(a)

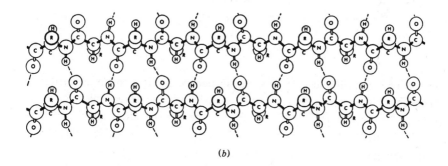

(b)

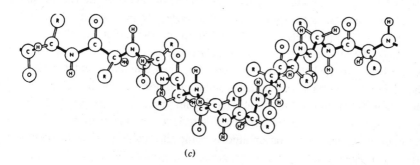

(c)

mer in which the outward radial vector, the tangential vector, and the vertical director of the advancing screw form a right-handed axial set (fig. 6.12).

The polarized single-crystal spectrum of myristamide, **10** indicates that the lowest-energy absorption of the amide chromophore with a high intensity, due

$$CH_3(CH_2)_{12}CONH_2$$

10

Fig. 6.11. The coupling modes of the $\pi \to \pi^*$ excitation moments in the individual amide groups of a polypeptide in the α-helix conformation, giving a resultant polymer transition polarized (*a*) parallel, and (*b*) perpendicular to the helix axis (Moffitt, 1956*a*).

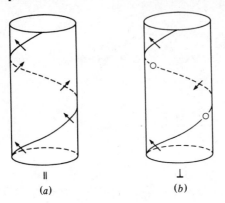

$\parallel$

(*a*)

$\perp$

(*b*)

to a $\pi \to \pi^*$ transition near 185 nm, has an electric-dipole moment directed in the plane of the amide group at a small angle (9°) to the oxygen–nitrogen direction, in the sense that a parallel line through the nitrogen atom cuts the carbonyl bond (fig. 6.13). The polypeptide α-helix contains some 3.6 amide residues per turn, through which the helix, with a radius of 1.6 Å, rises by 5.4 Å. The particular disposition of the amide chromophore in the cylindrical coordinate frame of a right-handed α-helix (fig. 6.12) gives the excitation dipole of the 185 nm amide absorption the radial, vertical and tangential angles the approximate values of 90°, 50°, and 140°, respectively (Peterson and Simpson, 1955; Moffitt, 1956*a*).

The relative energies of the allowed polymer transitions polarized parallel and perpendicular to the helix axis are governed by the Coulomb potential between the excitation dipoles of neighbouring amide chromophores in the polypeptide chain (eqn 3.17), and the potential is dependent upon the stereochemical relationship in the helix between the amide residues and the phase relationship between the monomer excitation dipoles in the allowed polymer coupling modes (fig. 6.11). The location of the amide group in the polypeptide α-helix and the orientation of the excitation dipole of the 185 nm amide absorption in the chromophore place the polymer transition polarized parallel to the helix axis at the lower energy, Moffitt estimating a wavenumber separation, $(\bar{\nu}_{\parallel} - \bar{\nu}_{\perp})$, of $-2800\ \text{cm}^{-1}$. The polarized spectra of orientated films of poly-L-alanine and poly-γ-methyl-L-glutamate locate the parallel- and the perpendicularly-polarized polymer transitions of the α-helix at 206 and 191 nm, respectively, corresponding to a wavenumber separation of $-2700\ \text{cm}^{-1}$ (Holzwarth and Doty, 1965; Mandel and Holzwarth, 1972).

Fig. 6.12. The orientation of the amide $\pi \to \pi^*$ transition moment in the polypeptide α-helix at the origin, 0, in a cylindrical coordinate system, defined by the radial, r, tangential, t, and vertical angle, v. The amide group at O is related to its neighbour at M by the rise along the helix axis, Z, the helix radius, ρ, and the angle, $(2\pi/P)$, about the helix axis, where P is the number of amide residues per unit turn of the helix (Moffitt, 1956a).

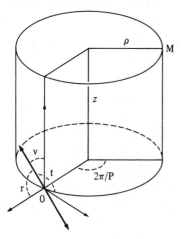

For a right-handed polypeptide α-helix, the values of the terms in eqn 6.9 require the parallel- and the perpendicularly-polarized polymer transitions to have a negative and a positive rotational strength, respectively, as was supported by the particular form of the optical rotatory dispersion given by the α-helix conformation of polypeptides derived from the L-amino acids and, subsequently, by CD measurements (fig. 6.10). The CD spectra of polypeptides in the α-helix conformation show, in addition to the positive and negative CD bands due to the perpendicularly- and the parallel-polarized polymer $\pi \to \pi^*$ transitions, respect-

Fig. 6.13. The orientation of the electric-dipole moment of the lowest-energy $\pi \to \pi^*$ transition in the amide chromophore (Peterson and Simpson, 1955, 1957).

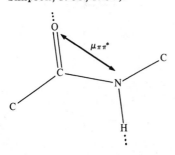

ively, a negative CD band at longer wavelength, near 222 nm, where the isotropic absorption intensity is relatively low. The latter CD band is ascribed to the carbonyl n → π* transition of the amide chromophore. The interchromophore coupling of the amide n → π* transition is weak in the polypeptides, since the electric-dipole moment of the transition is small. The principal negative CD absorption of the L-amino acid polypeptides in the random-coil conformation, and of monomeric amide analogues, is attributed to the amide n → π* transition (Brahms and Brahms, 1980).

Fig. 6.14. The schematic structure of double-stranded B-DNA and the absorption spectrum (upper curves) and the circular dichroism (lower curves) of DNA in aqueous solution when stationary (full line) and under flow along the direction of propagation of the radiation (broken line). In the schematic structure, the hydrogen-bonded base pairs are adenine–thymine (A–T), and guanine–cytosine (G–C). Each base is covalently bonded to a sugar unit (S) and the latter are linked by phosphate (P). Adapted from McCaffery and Mason, (1964); and Chung and Holzwarth, (1975).

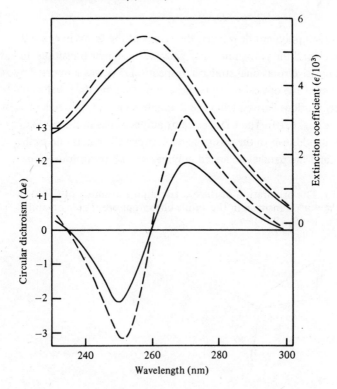

Fig. 6.14.
Continued.

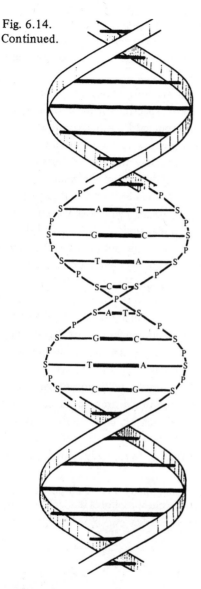

The initial study by Moffitt of the optical activity of helical macromolecules gave polymer rotational strengths dependent solely upon the radius of the helix and the vertical and the tangential components of the monomer excitation dipole (eqn 6.9). These radius-dependent rotational strengths are absent in helical polymers made up of monomer chromophores in which the excitation dipole is directed in the plane perpendicular to the helix axis, with radial and tangential components only, and no vertical component (fig. 6.12). The helix axis of double-stranded DNA and of a number of synthetic polynucleotides is directed perpendicular, or nearly so, to the molecular planes of the constituent purine and pyrimidine bases, and the $\pi \rightarrow \pi^*$ transitions of each base are polarized in the molecular plane (Tinoco, 1973, 1979).

The spectra of the nucleic acids in the helical conformation show a bisignate CD absorption couplet which is absent from the spectrum of the corresponding random-coil conformation. The long double-stranded DNA helices are statistically orientated in solution under flow-conditions along the direction of the stream-lines, and radiation propagated along the flow-direction preferentially excites the electronic transitions polarized perpendicular to the helix axis of the macromolecule. The area of each component of the bisignate CD absorption couplet given by helical DNA over the 290 - 230 nm region is enhanced when the macromolecule is statistically orientated under flow conditions, and the isotropic absorption also increases (fig. 6.14). Thus the optical activity of helical DNA arises primarily from coupling modes of monomer excitation dipoles polarized perpendicular to the helix axis (Chung and Holzwarth, 1975).

These coupling modes are dependent upon the pitch of the helix and may involve solely monomer excitation dipoles directed radially from the helix axis (fig. 6.15) or both the radial and the tangential components of excitation dipoles directed in the plane perpendicular to the helix axis. As in the radius-independent case (eqn 6.9), the pitch-dependent helical coupling modes are conservative, giving rise to rotational strengths of equal magnitude and opposite sign. For each pitch-dependent mode giving a positive rotational strength, there is a complementary coupling with a converse phase relationship between the constituent monomer excitation dipoles producing a compensatory negative rotational strength at a different energy (fig. 6.15), (Tinoco, 1973, 1979; Woody, 1977).

Fig. 6.15. The coupling modes of the radially-directed dipole transition moments of the individual monomer chromophores in a helical polymer giving rise to, (*a*) a negative rotational strength, and (*b*) a positive rotational strength.

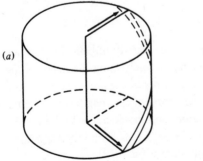
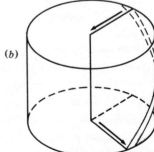

(*a*)   (*b*)

6.5 Microscopic and macroscopic helices

A stereoregular chiral macromolecule may have a chain length of the same order as the wavelength of visible radiation, or larger, giving a Boltzmann-Kuhn ratio, d/λ (§ 2.1) apparently favourable for a large optical activity. The observed optical activity generally remains relatively small, at the molecular magnitude level, and approximately constant per monomer beyond a minimum length, determined by the repeat-unit, which is the effective radiation-absorbing group. A molecule of microbial DNA has a length of $\sim$ 1 mm when fully extended, but the dissymmetry ratio, $\Delta\epsilon/\epsilon$, in the quartz ultraviolet region has a value similar to that ($\sim 4 \times 10^{-4}$) of a single turn of the B-DNA helix, containing ten nucleotides, with a pitch-length of 33.6 Å. A similar dissymmetry ratio is found in the compact form, chromatin, where the right-handed double-stranded DNA forms a left-handed superhelix around a histone protein core, although the CD of the DNA partly overlaps that of the histone polypeptide α-helix (Cowman and Fasman, 1978).

Similarly, the effective radiation-absorbing group of a macroscopic molecular or ionic crystal, the unit cell, has molecular dimensions, and the ratio, d/λ, generally is small. Achiral molecules in an enantiomorphous crystal lattice exhibit a unit-cell optical activity with a dissymmetry ratio which is large for a magnetic-dipole transition, but small for an electric-dipole excitation. In the latter case, the optical activity arises from the Coulombic interactions between the transitional charge distributions of the inequivalent molecules in a unit cell, and in the former, from the static-field or dynamic-polarization mechanism (§ 3.2 and 3.3).

The optical activity of a unit cell in a macroscopic single crystal is directly accessible spectroscopically in an enantiomorphous cubic crystal, or a uniaxial crystal with radiation propagated along the optic axis. The α-[Ni(H$_2$O)$_6$]SO$_4$ crystal and the isomorphous α-[Zn(H$_2$O)$_6$]SeO$_4$ crystal belong to the tetragonal enantiomorphous space groups, P4$_1$2$_1$2 or P4$_3$2$_1$2. In the unit cell, containing four formula units, a set of 24 hydrogen-bonded water molecules form a right-handed 4$_1$ or a left-handed 4$_3$ helical array around the tetragonal axis with a pitch-length of 18.3 Å (Beevers and Lipson, 1932; O'Connor and Dale, 1966). The infrared CD spectrum of the α-[Ni(H$_2$O)$_6$]SO$_4$ crystal, with radiation propagated along the tetragonal optic axis, gives a dissymmetry ratio, $g = \Delta\epsilon/\epsilon$, in the range 2×10^{-4} to 6×10^{-4} for the water overtone and combination vibrational bands at 2.0, 2.5 and 4.3 μm, which are electric dipole in origin (Hsu and Holzwarth, 1973). In contrast, $g \sim 10^{-1}$ for the magnetic-dipole d $\rightarrow$ d transition at 1.2 μm of the nickel(II) ion in the chiral hydrate helix of the crystal (Grinter, Harding and Mason, 1970). Thus the ratio of the g-factors for the two types of transition in the crystal, electric and magnetic dipole, ($\sim 10^{-3}$), approximates to the Boltzmann-Kuhn ratio, (d/λ), of the unit-cell pitch-length to the wavelength of near-infrared radiation.

Model systems with a dimension ratio (d/λ) of the order of unity have a large and readily-detected optical activity. Lindman (1925) found that copper wire models of chiral organic molecules, with the form of asymmetric tetrahedra, rotate the plane of polarized electromagnetic radiation in the 10 cm wavelength region. In similar experiments, Tinoco and Freeman (1957) employed copper wire helices, with a pitch-length of 3.3 mm and a radius of 2.5 mm, and tuneable polarized microwave radiation over the 2 - 4 cm wavelength range, observing large ORD Cotton effects near 2.35 and 3.10 cm. The Cotton effects measured support the theory of the optical activity of chiral macromolecules with a chain length comparable to the wavelength of visible and ultraviolet radiation (Tinoco, 1979).

The dimension ratio of the helix pitch-length to the radiation wavelength is adjustable to unity over a range from the quartz ultraviolet to the rocksalt infrared region in cholesteric liquid crystals. Some five per cent of all organic compounds transform at the melting point into an opaque anisotropic liquid phase which clears at a higher temperature to an isotropic transparent liquid (Steinstrasser and Pohl, 1973). Friedel (1922) distinguished three types of liquid crystal, or mesophase, as he termed the state. The lower-temperature layered smectic phase, if formed, transforms at a higher temperature to either the thread-like nematic phase or the twisted cholesteric phase, which subsequently clear at a more elevated temperature. A given compound does not form both the linearly-birefringent nematic phase and the circularly-birefringent cholesteric phase, given only by chiral molecules, typically derivatives of cholesterol. A nematic phase is transformed into a cholesteric phase, however, by the addition of a small mole percent of a chiral solute (fig. 6.16).

Small pathlengths (10 - 50 μm) of a cholesteric mesophase between polished transparent plates give the plane texture in which the helix axis of the statistically-orientated molecules lies perpendicular to the faces of the cover plates. The cholesteric plane texture is analogous to a uniaxial enantiomorphous crystal which has a large circular birefringence but is isotropic for radiation propagated along the optic axis, namely, the axes of the cholesteric helices (fig. 6.16). Friedel (1922) and others found that the cholesteric plane texture gives a large optical rotation and is often coloured, owing to the selective reflection of a band of visible radiation.

Employing CD spectroscopy, Mathieu (1939) discovered that unpolarized radiation is transmitted by a cholesteric plane texture, outside absorption regions, except for a band centred on the wavelength equal to the effective pitch-length,

Fig. 6.16. The arrangement of cylindrically anisotropic molecules in liquid crystals; (*a*) the smectic phase, (*b*) the nematic phase, and (*c*) the cholesteric phase.

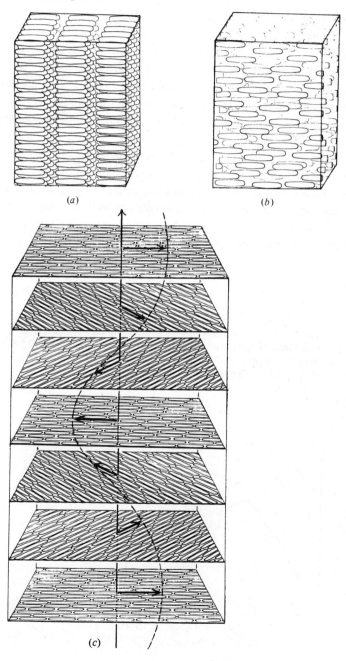

P, of the cholesteric helix. When $\lambda \sim P$, a right-handed cholesteric preparation transmits LCP radiation but reflects RCP radiation, giving an apparent CD absorption, the pitch-band, and an ORD Cotton effect of the same form as that exhibited by an isotropic solution of chiral molecules in an absorption wavelength region. The reflection of RCP radiation by a right-handed cholesteric plane texture at $\lambda \sim P$ is incomplete unless the pathlength includes some 20 – 40 complete turns of the cholesteric helix, when the apparent dissymmetry ratio, $g = \Delta\epsilon/\epsilon$, attains its maximum value of two (Mathieu, 1939). The condition is not realised for long cholesteric pitch-lengths, corresponding to infrared wavelengths, where the pitch-band appears with less than 50% absorption (Baessler and Labes, 1970) (fig. 6.17).

For wavelengths close to, but outside, the region of the cholesteric reflection pitch-band, the dimension ratio, P/λ, remains of the order of unity, and large Cotton effects are observed in the absorption bands of the molecules composing the mesophase, due to the intermolecular coupling of the chirally arrayed electric-dipole transition moments of different molecules. Vibrational Cotton effects in cholesteric liquid crystals were discovered by Schrader and Korte

Fig. 6.17. The infrared spectrum in the 3 μm region of p-n-butyl-N-(p-methoxybenzylidene)aniline (MBBA), **11**, in the nematic phase (broken line) and in the cholesteric phase, induced by 6 mole % of cholesteryl chloride (full line). The reflection pitch-band of the cholesteric plane texture at 3600 cm^{-1}, where the transmission is 60%, corresponds to a pathlength of $\sim$ 10 complete turns of the cholesteric helix (Dudley, Mason and Peacock, 1975).

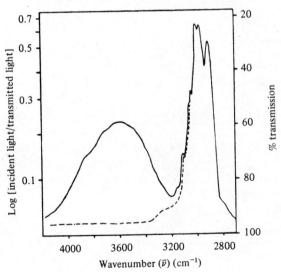

(1972) from ORD measurements, employing an infrared spectropolarimeter, and subsequently these effects were investigated by the corresponding CD technique (Chabay, 1972; Dudley *et al.*, 1972, 1975; Holzwarth, Chabay and Holzwarth, 1973).

$$CH_3O-\text{⟨O⟩}-CH=N-\text{⟨O⟩}-C_4H_9$$

11

The cholesteric liquid crystals employed are a room-temperature nematic mesophase, such as, *p-n*-butyl-N-(*p*-methoxybenzylidene)aniline (MBBA), **11**, containing a few mole percent of a chiral solute. The vibrational CD spectrum of such a cholesteric plane texture is related to the corresponding linear dichroism

Fig. 6.18. The infrared spectrum of *p-n*-butyl-N-(*p*-methoxybenzylidence) aniline (MBBA), **11**, over the rocksalt region. The upper curves refer to the isotropic spectrum (broken line) and the linear dichroism (full line) of orientated nematic MBBA, and the lower curve to the circular dichroism of MBBA in the cholesteric phase, induced with 2 mole % of (-)-menthol, giving a cholesteric pitch-length of ∿ 15 μm (Dudley *et al.*, 1975).

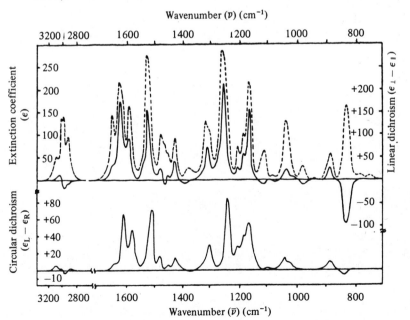

(LD) spectrum of the parent nematic mesophase, a CD band of the former having the same sign as the corresponding LD band of the latter (fig. 6.18). The correlation supports analyses of the vibrational Cotton effects of cholesteric liquid crystals, based upon an extension to absorption frequencies of the classical treatment of de Vries (1951) of the optical rotation of a cholesteric mesophase over transparent wavelength regions (Holzwarth, Chabay and Holzwarth, 1973; Dudley *et al.*, 1975). The application of cholesteric vibrational ORD to the determination of the absolute stereochemical configuration of a chiral solute employed to transform a nematic mesophase into a cholesteric plane texture is reviewed by Korte and Schrader (1981).

7 CHIRAL TRANSITION-METAL COMPLEXES

7.1 The CD spectroscopy of tris-chelate complexes

Circular dichroism in the fluid phase was discovered by Cotton (1895) who employed aqueous solutions of chiral coordination compounds. The extensive series of chelate transition-metal complexes resolved into enantiomers by Werner (1911) provided the materials for the pioneer photographic CD spectroscopy of the 1930s, performed notably by Lowry (1935), Mathieu (1957), and Kuhn (1958). The principal chiral complexes investigated were the six-coordinate compounds of the open-shell transition-metal ions with the (d^3) and the spin-paired (d^6) configuration, notably, chromium(III), cobalt(III), rhodium(III), iridium(III), and platinum(IV), containing either three bidentate ligands of the diamine or dicarboxylate series, or two such chelating ligands and two unidentate ligands in *cis* coordinate positions.

These complexes have in common an orbitally non-degenerate metal-ion ground state, arising from the (t_2^3) or the (t_2^6) ground configuration, with $^1A_1$ octahedral symmetry in the latter case. Two principal d-electron excited states arise from the octahedral orbital promotion, $t_2 \rightarrow e$, the excited configuration ($t_2^5 e$) giving the triply degenerate $^1T_1$ and $^1T_2$ states (fig. 7.1). The lower-energy $^1T_1$ state derives from orbital promotions of the $d_{xy} \rightarrow d_{x^2-y^2}$ type and the higher energy $^1T_2$ state from excitations of the $d_{xy} \rightarrow d_{z^2}$ type. The former, but not the latter, type of orbital promotion is magnetic-dipole allowed, and a relatively strong CD is generally associated with the longer wavelength $d \rightarrow d$ absorption band of chiral chelate complexes, observed in the visible region for the chromium(III) and cobalt(III) cases.

From an investigation of the chiral Werner complexes, Mathieu (1936) proposed that the dihedral tris-chelate coordination compounds giving a predominantly positive CD in the region of the lowest-energy absorption band have the same stereochemical propeller configuration as the ethylenediamine complex, (+)-[Co(en)$_3$]$^{3+}$. The first determination of the absolute configuration of a metal complex by the anomalous X-ray scattering method, due to Saito and coworkers (1954), showed that (+)-[Co(en)$_3$]$^{3+}$ has the Λ-configuration, 1, and subsequently the absolute configuration of the majority of the chiral Werner

Fig. 7.1. The relative d-electron energies in a cubic ligand field. (*a*) The lifting of the five fold degeneracy of a d-orbital set in a tetrahedral or an octahedral field to give a triply-degenerate t_2 -set, d_{xy}, d_{yz}, and d_{xy}, and a doubly degenerate e-set, d_{x2} and d_{x2-y2}. (*b*) The $^4A_2$ d-electron ground state of octahedral chromium (III) from the ($t_2{}^3$) electron-con-figuration, or of tetrahedral cobalt(II) from the ($t_2{}^3$) hole-configur-ation, and the manifold of quartet d-electron excited states. (*c*) The $^1A_1$ ground state of the spin-paired ($t_2{}^6$) configuration of octahedral cobalt (III) and its analogues, together with the lower excited singlet d-electron states. In (*b*) and (*c*) representative single-orbital promotions are indi-cated for the lower and the higher energy d-electron transitions, which are triply-degenerate. The leading moments of the particular lower-energy transition, in both cases, are the z-component of a magnetic dipole and the $[xy(x^2 - y^2)]$-component of an electric hexadecapole, whereas the leading moment of the higher energy transitions is the xy-component of an electric quadrupole.

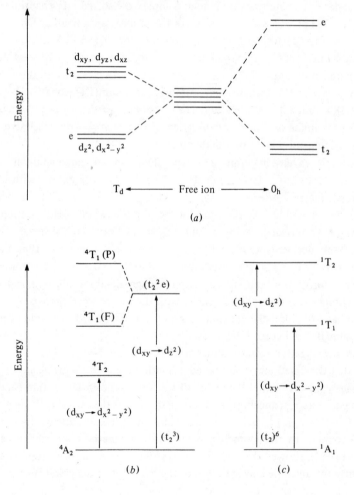

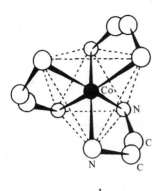

1

complexes and a number of their analogues were determined by the same procedure (Saito, 1979). These determinations indicate that, while generally valid, Mathieu's rule is not invariably correct, as the long-wavelength d → d CD of tris-chelate complexes often consists of two oppositely-signed components, and the one or the other component may dominate the overall sign.

The two components arise from the splitting of the triply-degenerate lowest-energy octahedral d-electron excited state, $^1T_1$ in the (d^6) complexes or $^4T_2$ in the (d^3) cases, by the lower dihedral symmetry, D_3, of the ligand field in tris-chelate complexes. The d-electron transition $^1A_1 \rightarrow {}^1T_1$ of octahedral cobalt(III) breaks down in the corresponding tris-chelate complex into the D_3 components, $^1A_1 \rightarrow {}^1A_2$, polarized parallel to the principal C_3 axis of the complex, and the doubly-degenerate transition, $^1A_1 \rightarrow {}^1E$, with the two mutually orthogonal components polarized in the plane perpendicular to the C_3 rotation axis. The two D_3 components of the lowest-energy octahedral d → d transition are distinguished by single-crystal circular dichroism and linear dichroism measurements. A number of the salts of the complex ion (+)- $[Co(en)_3]^{3+}$ and its analogues crystallise in a uniaxial class, trigonal, tetragonal, or hexagonal, which are optically isotropic in the plane perpendicular to the principal axis and do not change the state of polarization of radiation propagated along the axis. The crystal and molecular structure of these salts shows that the C_3 rotational axis of each complex ion lies parallel, or at a determinate angle, to the optic axis of the crystal. In the crystal, $2\{\Lambda$-(+)- $[Co(en)_3]Cl_3\}.NaCl.6H_2O$, and other cases where the C_3 axes of the individual complex ions are parallel to the crystal axis, the propagation of radiation along the optic axis excites only those transitions of the complex ion polarized in the plane perpendicular to the C_3 axis of the ion (table 7.1).

The axial single-crystal spectrum of Λ-(+)- $[Co(en)_3]^{3+}$ shows in the lowest-energy absorption region a single positive CD band with an area larger by an order of magnitude than that of the bisignate CD couplet in the same region given by the complex ion randomly orientated in solution (fig. 7.2). The axial crystal CD band at 475 nm measures the rotational strength, $R(E)$, of the doubly-degenerate D_3 component, $^1A_1 \rightarrow {}^1E$, of the lowest-energy d $\rightarrow$ d transition of octahedral cobalt(III). The rotational strength, $R(A_2)$, of the corresponding non-degenerate D_3 component, $^1A_1 \rightarrow {}^1A_2$, is found to be negative and of a comparable magnitude from the orthoaxial crystal CD spectrum, corrected for the linear dichroism and birefringence involved by the non-axial propagation of the radiation (Jensen and Galsbøl, 1977). The solution CD spectrum of Λ-(+)-[Co(en)$_3$]$^{3+}$ in the lowest-energy absorption region thus represents the bisignate residual wing CD absorptions of the two large and oppositely-signed D_3 rotational strengths,

Table 7.1. *The axial single-crystal circular dichroism spectra of tris-chelate complex ions, and the corresponding CD of the randomly orientated complex. The assignment refers to the upper state in D, symmetry of the d – d transition from the $^1A_1$ ground state of the d^6 complexes, or from the corresponding $^4A_2$ or $^3A_2$ state of the d^3 or d^8 complexes, respectively*

Crystal[a]	Single crystal		Random orientation		Assignment	Ref.
	λ(nm)	$\Delta\epsilon$	λ(nm)	$\Delta\epsilon$		
2{Λ-(+)-[Co(en)$_3$]Cl$_3$}.NaCl.6H$_2$O	475	+23	490	+1.9	^1E	b,c,d,e,f
			430	−0.3	$^1A_2$	
2{Λ-(+)-[Cr(en)$_3$]Cl$_3$}.NaCl.6H$_2$O	458	+15	456	+1.4	^4E	g
Λ-(+)-[Co(S-pn)$_3$]Br$_3$	485	+17	491	+1.9	^1E	e,h
			436	−0.5	$^1A_2$	
Λ-(+)-[Co(S,S-chxn)$_3$]Cl$_3$.5H$_2$O	488	+22	512	+2.4	^1E	e
			450	−0.6	$^1A_2$	
Δ-[Co(R,R-cptn)$_3$]Cl$_3$.4H$_2$O	510	−13	534	−0.58	^1E	i
			476	+1.89	$^1A_2$	
Δ-(+)-[Co(tn)$_3$]Cl$_3$.4H$_2$O	483	+4.6	530	−0.06	($^1A_2$)	h,j
			475	+0.12	(^1E)	
Δ-[Co(R,R-ptn)$_3$]Cl$_3$.2H$_2$O	478	+2.7	533	−0.13	^1E	e
			477	+0.79	$^1A_2$	
Λ-(+)-NaMg[Cr(ox)$_3$].9H$_2$O	561	+20	546	+2.9	^4E	c,g
Λ-(−)-NaMg[Co(ox)$_3$].9H$_2$O	620	+39	617	+3.0	^1E	c,g
Λ-(+)$_{546}$-NaMg[Rh(ox)$_3$].9H$_2$O	395	+49	388	+4.1	^1E	c,g
Λ-[Ni(en)$_3$](NO$_3$)$_2$	885	+3.1	950	+0.2	^3E	k,l
			830	−0.1	$^3A_1$	

(a) The ligand abbreviations are: 1,2-ethylenediamine (en); (S)-(+)-1,2-propylenediamine (S-pn); (1S,2S)-(+)-*trans*-1,2-cyclohexanediamine (S,S-chxn); (1R,2R)-(−)-*trans*-1,2-cyclopentanediamine (R,R-cptn); trimethylenediamine (tn); (2R,4R)-(−)-2,4-diaminopentane (R,R-ptn); oxalato (ox).
(b) McCaffery and Mason (1963).
(c) Ballard, McCaffery and Mason (1965).
(d) Denning (1967).
(e) Kuroda and Saito (1976).
(f) Jensen and Galsbøl (1977).
(g) Mason (1973b).
(h) Judkins and Royer (1974).
(i) Toftlund and Pedersen (1972).
(j) Judkins and Royer (1970). The angle between the optic axis of the crystal and the C$_3$ axis of the complex ion is uncertain. The corresponding angle for Δ-[Co(R,R-ptn)$_3$]Cl$_3$.2H$_2$O is 69.5° (e).
(k) Harding, Mason and Peart (1973).
(l) Bernal and Palmer (1981).

$R(E)$ and $R(A_2)$, with only a small mean frequency-separation. The linearly-polarized single-crystal spectrum of $[Co(en)_3]^{3+}$ at low temperature shows that the D_3 components, $^1A_1 \rightarrow {}^1A_2$ and $^1A_1 \rightarrow {}^1E$, have in fact a common band origin, although the distribution of dipole and rotation strength over the vibronic progressions built up on that origin differ for the two components, rising to a maximum at the lower frequency for the E component, and extending over the larger frequency range in the case of the A_2 component (Dingle and Ballhausen, 1967).

The axial single-crystal CD studies extended to other chelating ligands and to other metal ions show that, for the Λ-configuration of the tris-chelate complex, the rotational strength $R(E)$ of the doubly-degenerate D_3 component of the lowest-energy octahedral d $\rightarrow$ d transition is positive, provided that the L–M–L angle within a chelate ring is less than $\pi/2$, as is generally the case in complexes containing five, six, and even seven-membered chelate rings. An exception is the trimethylenediamine complex, Λ-(-)-$[Co(tn)_3]^{3+}$, **2**, where the ring angle is 91° at the metal ion, and the axial single-crystal CD spectrum gives $R(E)$ a negative

Fig. 7.2. The axial single-crystal circular dichroism spectrum ($\Delta\epsilon/10$) of $2\{\Lambda$-(+)-$[Co(en)_3]Cl_3\}.NaCl.6H_2O$ (solid curve), and the absorption and circular dichroism spectrum of Λ-(+)-$[Co(en)_3](ClO_4)_3$ in aqueous solution (upper and lower broken curves, respectively).

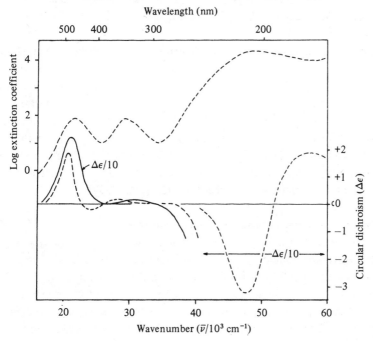

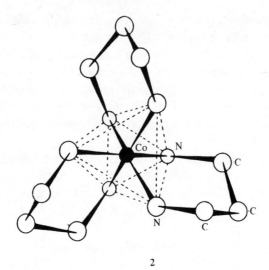

2

value. Less generally, the rotational strength, $R(E)$, is predominant in the lowest-energy region of the corresponding solution CD spectrum, accounting for the scope and the limitations of Mathieu's rule. The (S,S)-*trans*-1,2-diaminocyclo-pentane complex, Λ-(–)-$[Co(S,S\text{-cptn})_3]^{3+}$, for example, has a positive $R(E)$, from the axial single-crystal CD spectrum, but the lowest-energy CD absorption of the complex ion in aqueous solution is negative overall (Toftlund and Pedersen, 1972).

The axial single-crystal CD spectra of the tris-chelate complexes further show that the optical activity of the D_3 components of the higher-energy octahedral d → d transition, $^1A_1 \rightarrow {}^1T_2$ in the (d^6) cases, is intrinsically weak, and that the minor CD absorption observed in the higher-frequency region of the spectrum is not the result of the overlapping of substantial rotational strengths with opposed signs (fig. 7.2). Thus the single-crystal and the solution CD spectra of the chiral tris-chelate complexes, taken together, support the attribution of the lower- and the higher-energy parent octahedral absorptions to allowed and forbidden magnetic-dipole d → d transitions, respectively (Mason, 1973b).

7.2 Conformational and configurational optical activity

The chelate rings formed by 1,2-diamines were shown by Corey and Bailar (1959) to exist preferentially in one of two enantiomerically-puckered conformations. In the IUPAC convention (1970) the conformation has the δ form if the C—C bond forms a segment of a right-handed helix with respect to the N···N direction in the chelate ring, or the λ form if the helical segment is left-handed. In a bis- or tris-chelate complex of 1,2-ethylenediamine the two conformations are energetically non-equivalent on account of the steric interaction between adjacent pairs of chelate rings, and the particular interaction is dependent upon the configurational disposition of the chelate rings around the metal ion. The most stable conformation of a tris-ethylenediamine complex contains chelate rings in which the C—C ring bonds are parallel, or nearly so, to the C_3 rotation axis of the complex, forming the collective *lel*-conformation, whereas the ring C—C bonds are obliquely inclined with respect to the C_3 axis of the complex in the least stable overall *ob*-conformation.

Whether the overall conformation is made up of individual δ or λ ring forms is dependent upon the stereochemical configuration of the complex as a whole. For the Λ-configuration, the *lel*-conformation is represented by the individual ring contributions ($\delta\delta\delta$) and the *ob*-conformation by ($\lambda\lambda\lambda$), and the order of decreasing stability follows the sequence,

$$\Lambda(\delta\delta\delta) > \Lambda(\lambda\delta\delta) > \Lambda(\lambda\lambda\delta) > \Lambda(\lambda\lambda\lambda) \tag{7.1}$$

with an enthalpy increment of approximately 2.5 kJ mol^{-1} for each δ to λ ring inversion. Statistically the intermediate forms ($\lambda\delta\delta$) and ($\lambda\lambda\delta$) are favoured over the limiting *lel* and *ob* forms by $R\ln3$ or 2.9 kJ mol^{-1} at ambient temperature. In most of the Λ-(+)-$[Co(en)_3]^{3+}$ crystals as yet analysed the complex ion has the *lel*-conformation 1, but a number of the salts of $[Cr(en)_3]^{3+}$ with large hydrogen-bonding anions contain the latter complex ion in all forms of the sequence (eqn 7.1). The nuclear magnetic resonance (NMR) spectra of $[M(en)_3]^{n+}$ complex ions in solution indicate that the most abundant conformation is generally $\Lambda(\lambda\delta\delta)$ and its antipode at ambient temperature (Sudmeier, Blackmer, Bradley and Anet, 1972).

A particular ring conformation is obtained by the use of a chiral C-substituted 1,2-diamine ligand. The δ chelate ring conformation is favoured by (S)-(+)-1,2-propylenediamine, owing to the equatorial preference of the methyl group with respect to the ring, and (S)-(+)-*trans*-1,2-cyclohexanediamine gives a locked δ chelate ring form, whereas the corresponding (R)-(−)-diamines give the λ chelate ring conformation. Crystal and molecular structure X-ray diffraction analyses indicate that (S)-(+)-1,2-propylenediamine forms the *lel* complex, $\Lambda(\delta\delta\delta)$-(+)-$[Co(S-pn)_3]^{3+}$, 3, and the *ob* isomer, $\Delta(\delta\delta\delta)$-(−)-$[Co(S-pn)_3]^{3+}$, 4 (Saito, 1979). The CD spectrum of 4 and of other *ob* isomers have the same general form as that of 1, 3, and other *lel* isomers, but differ in that the lowest-energy CD absorption consists of a single band with a sign reflecting the overall stereochemical configuration of the complex, lacking the bisignate couplet form generally observed for the *lel* forms (fig. 7.2).

The *lel* and the *ob* isomers differ further in that the d → d optical activity of

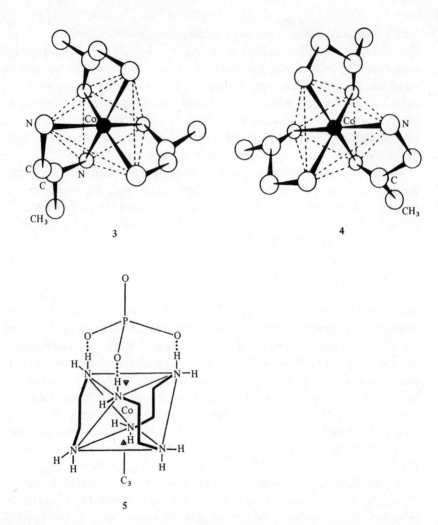

3

4

5

the former is particularly sensitive to outer-sphere coordination of the complex by polarizable oxyanions. One of the N—H bonds of each amino group in a tris-1,2-diamine complex is orientated parallel to the C_3 rotation axis of the complex in the *lel* conformation, but each N—H bond is obliquely inclined with respect to the C_3 axis in the *ob* conformation. The array of three N—H groups perpendicular to an exposed octahedral face of a *lel* isomer is favourably disposed sterically for triple hydrogen-bonding to a tetrahedral or a trigonal oxyanion, forming an outer-sphere complex in which the cation and the anion share a common C_3 rotation axis, **5**. A structure analogous to **5** is found by the X-ray diffraction analysis of the crystal $[Co(en)_3]_2[HPO_4]_3 \cdot 9H_2O$ (Duesler and Raymond, 1971).

The addition of phosphate or analogous oxyanions to an aqueous solution of **1**, **3** or other *lel* complexes changes the form of the lowest-energy bisignate CD absorption couplet, enhancing the rotational strength of one D_3 component, $R(A_2)$, with an apparent diminution of the other, $R(E)$, in the region of the octahedral d → d transition, $^1A_1 → {}^1T_1$ of cobalt(III) (fig. 7.3). Concurrently, absorption due to interionic charge transfer is observed in the ultraviolet region, near 250 nm, associated with a new CD absorption which has a sign and magnitude compensating for the CD changes found in the visible region (fig. 7.3). It is probable that the interionic charge transfer absorption has an electric moment directed along the C_3 axis of the outer-sphere complex, **5**, and mixes

Fig. 7.3. The absorption spectrum (top curves) and circular dichroism (middle curves) of Λ-(+)-[Co(en)$_3$](ClO$_4$)$_3$ in water (full line) and in 0.1M aqueous phosphate (broken line). The bottom curves refer to the circular dichroism of the corresponding 1,1,1-tris(2'-aminoethylamino-methyl)ethane complex, (+)-[Co(sen)]$^{3+}$, **6**, in water (full line) and in 0.1M aqueous phosphate (broken line).

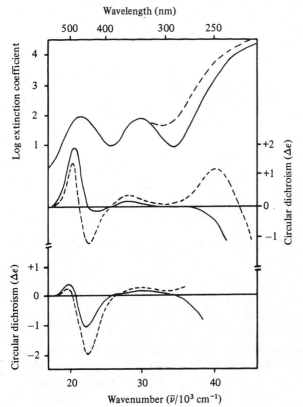

with the parallel-polarized D_3 component, $^1A_1 \rightarrow {}^1A_2$, of the lowest-energy $d \rightarrow d$ transition, which has a magnetic moment, producing the positive CD near 250 nm and the enhancement of the negative CD absorption in the 450 nm region (fig. 7.3).

The changes in the relative intensities of the two components of the bisignate CD couplet in the visible region of the Λ-(+)-$[Co(en)_3]^{3+}$ spectrum produced by ion-pairing with an oxyanion as in **5** appear also in the covalently bridged analogue **6** (+)-$[Co(sen)]^{3+}$, obtained from the ligand 1,1,1-tris(2′-aminoethylamine)

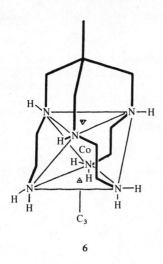

6

ethane (sen). The complex **6** has a remaining exposed octahedral face from which three N—H groups project parallel to the C_3 rotation axis of the complex, and the addition of phosphate to an aqueous solution of **6** enhances further the negative component of the long-wavelength CD couplet (fig. 7.3) (Sarneski and Urbach, 1971).

Although similar in general form, the CD spectra of corresponding *lel* and *ob* isomers are not identical and a pseudoracemate of the two forms is optically active. The CD spectrum of an equimolecular mixture of $\Lambda(\delta\delta\delta)$-(+)-$[Co(S-pn)_3]^{3+}$, **3**, and $\Delta(\delta\delta\delta)$-(-)-$[Co(S-pn)_3]^{3+}$, **4**, records the Cotton effects induced by three puckered 1,2-propylenediamine chelate rings with the δ conformation, the optical activity due to the Λ and the Δ configuration of the three chelate rings around the metal ion being internally compensated in the pseudoracemic mixture (fig. 7.4). The optical activity induced by a chiral chelate ring conformation in a series of complexes containing a common octahedral chromophore, notably the $[Co^{(III)}N_6]$ cluster, is found to be additive over the number of such chelate rings in the absence of configurational optical activity due to the chiral relationship between the mean planes of two or more chelate rings coordinated to the metal ion. Thus the long-wavelength CD absorption couplet of $[Co(NH_3)_4(S\text{-}pn)]^{3+}$, **7**, and its N-methyl derivatives **8** and **9**, has a total band area approximately equal to one-half of that of the corresponding CD absorption given by *trans*-$[Co(NH_3)_2(S\text{-}pn)_2]^{3+}$, **10**, and its N-methyl derivatives **11** and **12**, and to one-third of the CD absorption band area of the pseudoracemic mixture of **3** and **4** (fig. 7.4).

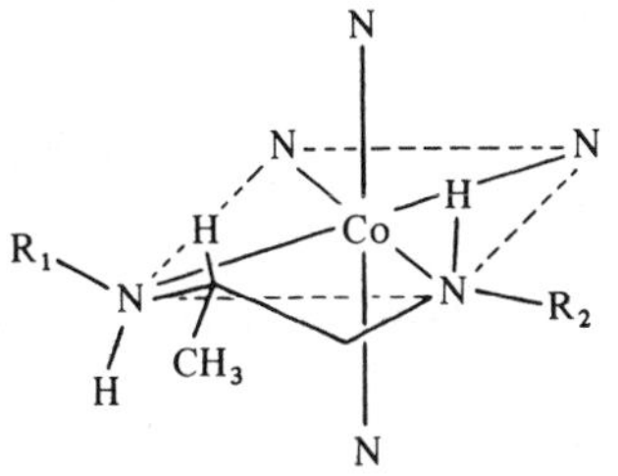

7 $R_1 = R_2 = H$
8 $R_1 = Me, R_2 = H$
9 $R_1 = H, R_2 = Me$

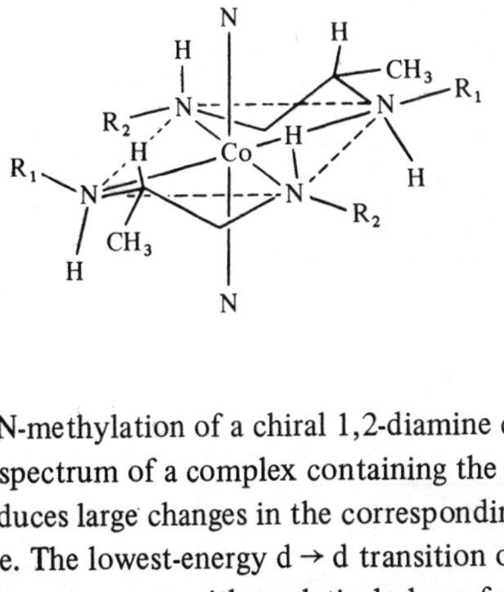

10 $R_1 = R_2 = H$
11 $R_1 = Me, R_2 = H$
12 $R_1 = H, R_2 = Me$

The N-methylation of a chiral 1,2-diamine chelate ring has little effect upon the CD spectrum of a complex containing the octahedral $[Co^{(III)}N_6]$ cluster, but produces large changes in the corresponding spectrum of a tetragonal analogue. The lowest-energy d → d transition of octahedral cobalt(III) splits into two components with a relatively large frequency separation in the tetragonal chromophore, *trans*-$[Co^{(III)}N_4Cl_2]$, the D_4 components, $^1A_1 \to {}^1E$ and $^1A_1 \to {}^1A_2$, lying at the lower and the higher frequency, respectively. The CD spectrum of *trans*-$[Co(S\text{-}pn)_2Cl_2]^+$, **13**, shows that the $^1A_1 \to {}^1E$ component has a negative rotational strength and the $^1A_1 \to {}^1A_2$ component a smaller positive rotational strength, whereas the converse holds for the corresponding N-methyl derivative, *trans*-$[Co(N_1Me\text{-}S\text{-}pn)_2Cl_2]^+$, **14**, and for the N-methylethylenediamine complex with the same δ chelate ring conformation, $(-)$-*trans, trans*-$[Co(Meen)_2 Cl_2]^+$, **15** (fig. 7.5).

Fig. 7.4. The absorption spectrum (*a*) of [Co(pn)$_3$]$^{3+}$, and the circular dichroism of (*b*) the *lel*-isomer, Λ-(+)-[Co(S-pn)$_3$]$^{3+}$, **3**, (*c*) the *ob*-isomer, Δ-(−)-[Co(S-pn)$_3$]$^{3+}$, **4**, (*d*) an equimolar mixture of (+)- and (−)-[Co(S-pn)$_3$]$^{3+}$, giving the conformational optical activity of three δ-puckered diamine chelate rings, (*e*) *trans*-[Co(S-pn)$_2$(NH$_3$)$_2$]$^{3+}$, **10**, (*f*) *trans*-[Co(N$_1$ Me-S-pn)$_2$(NH$_3$)$_2$]$^{3+}$, **11**, (*g*) [Co(S-pn)(NH$_3$)$_4$]$^{3+}$, **7**, and (*h*) [Co(N$_2$ Me-S-pn)(NH$_3$)$_4$]$^{3+}$, **9**.

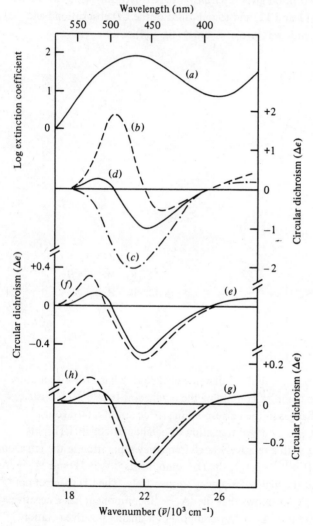

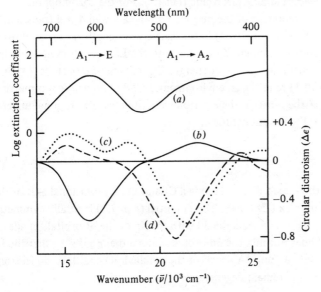

13 $R_1 = H, R_2 = Me$
14 $R_1 = R_2 = Me$
15 $R_1 = Me, R_2 = H$

In all three cases the sum of the rotational strengths over the tetragonal components of the lowest-energy octahedral cobalt(III) transition, $^1A_1 \rightarrow {}^1T_1$, is negative and comparable in magnitude as well as sign to the corresponding sum, $R(T_1)$, for the analogues, **10**, **11** and **12** containing the octahedral $[Co^{(III)}N_6]$ chromophore (fig. 7.4). The sum, $R(T_1)$, represents the optical activity induced by two (S)-(+)-propylenediamine chelate rings with the δ conformation and the additional optical activity arising from N-methyl substitution in the diamine ligand becomes apparent when the frequency-separation between the components of the octahedral $^1A_1 \rightarrow {}^1T_1$ cobalt(III) transition is large, as in the cases of **13**, **14** and **15**, but it is not evident when the energy-splitting is small, as in the cases of **10**, **11** and **12** since the N-methyl contribution

Fig. 7.5. The absorption spectrum (*a*) and the circular dichroism (*b*) of *trans*-[Co(S-pn)$_2$Cl$_2$]$^+$, **13**; and the circular dichroism of (*c*) (−)-*trans*, *trans*-[Co(N-Meen)$_2$Cl$_2$]$^+$, **15** and (*d*) *trans*-[Co(N$_1$ Me-S-pn) Cl$_2$]$^+$, **14**.

is oppositely-signed and of similar magnitude for the two components (Mason, 1973*b*).

7.3   Crystal-field theory of d-electron optical activity

Crystal-field theory originated with Bethe (1929) to account for the line-spectra of crystalline lanthanide(III) complexes. The appearance of each gas-phase Ln(III) line as a multiplet in the corresponding crystal spectrum was interpreted as a Stark effect, due to the electrostatic field of the ligands with a symmetry dependent upon the particular crystal structure. Van Vleck (1937) proposed the magnetic-dipole, the electric-quadrupole, and the forced electric-dipole mechanisms to account for the finite intensities of the Laporte-forbidden f → f transitions and, with Finkelstein (1940), extended crystal-field theory to the d → d spectra of transition metal complexes. Attention was confined, however, to a line-like spectrum, the spin-forbidden bands of chrome alum near 690 and 476 nm, but the broad absorption bands at 555 and 417 nm remained un-characterised until the 1950s, owing to the expectation that the corresponding spin-allowed d → d bands would be line-like too.

The application of crystal-field theory to the d-electron optical activity of chiral transition metal complexes was introduced by Moffitt (1956*c*), employing the one-electron mechanism (§ 3.2). The D_3 tris-chelate complexes investigated were divided into a symmetric octahedral chromophore, such as the $[Co^{(III)}N_6]$ cluster of **1**, and a dissymmetric chelate-ring environment providing an electro-static field. The field of the groups forming the chelate rings mix the magnetic-dipole with the electric-dipole electronic transitions of the chromophore. Specifically each component of the magnetic-dipole allowed d → d transition $^1A_1 \rightarrow \, ^1T_1$ of octahedral cobalt(III) in the visible-wavelength region acquires a collinear electric-dipole moment by mixing with the Laporte-allowed 3d → 4p transitions of the metal ion at higher energy. The mixing of d-orbitals ($l = 2$) with p-orbitals ($l = 1$) requires an environmental field of *ungerade* symmetry of the third-order, at the most, in the electronic coordinates. The trigonal ungerade potential, V_3, has the functional form,

$$V_3 \approx X(X^2 - 3Y^2) \qquad\qquad (7.2)$$

with the Y and the Z axes directed along a C_2 and the C_3 rotational axis of the D_3 tris-chelate complex (fig. 7.6*a*). The potential (eqn 7.2) is totally symmetric in the group D_3, and it connects the 3d with the 4p or the 4f orbitals of the metal ion, but it does not mix the 3d orbitals among themselves, so that the D_3 components, $^1A_1 \rightarrow \, ^1A_2$ and $^1A_1 \rightarrow \, ^1E$ of the octahedral cobalt(III) d-electron transition, $^1A_1 \rightarrow \, ^1T_1$, remain degenerate.

Fig. 7.6. Sector rules relating the position of substitution to the sign of the lowest-energy d-electron optical activity in chiral six-coordinate complexes of cobalt(III): (*a*) the sextant rule for the configurational optical activity of trigonal tris-chelate complexes (eqn 7.2); substitution in an unhatched region gives the $^1A_1 \rightarrow {}^1E$ component a positive rotational strength; (*b*) the octahedral rule for the conformational optical activity (eqn 7.4); substitution in a hatched region gives a net positive circular dichroism over the region of the $^1A_1 \rightarrow {}^1T_1$ octahedral cobalt (III) absorption; and (*c*) the rule for the specific tetragonal optical activity (eqn 7.5); substitution in a hatched sector gives the tetragonal components, $^1A_1 \rightarrow {}^1A_2$ and $^1A_1 \rightarrow {}^1E$ a positive and a negative rotational strength, respectively.

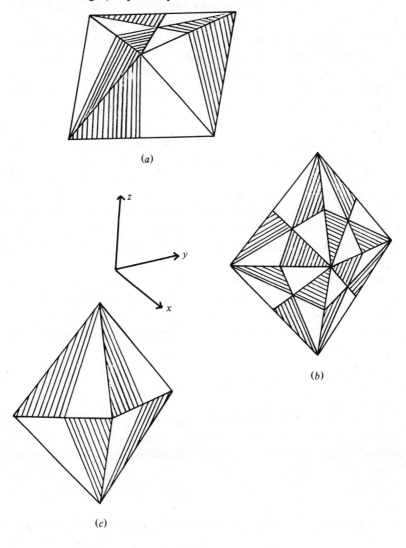

Subsequently, Sugano (1960) showed that the trigonal potential (eqn 7.2) gives first-order rotational strengths of equal magnitude and opposite sign to the D_3 components, and since the component transitions remain degenerate, the overall rotational strength, $R(T_1)$, of the randomly-orientated complex ion in solution goes to zero,

$$R(T_1) = R(A_2) + R(E) = 0 \tag{7.3}$$

The trigonal potential (eqn 7.2) has T_{2u} symmetry in the octahedral point group, O_h, as is required for the operation of the one-electron mechanism for optical activity in a chiral complex ion orientated in a single crystal, but it lacks the pseudoscalar form necessary for the effective operation of the mechanism in a randomly-oriented chiral molecule. Taken together, the axial single-crystal and solution CD spectra of Λ-(+)-$[Co(en)_3]^{3+}$, **1**, over the visible-wavelength region indicate that eqn (7.3) holds only at an order-of-magnitude level, the solution rotational strength, $R(T_1)$, being as large as 10% of the single-crystal strength, $R(E)$ (fig. 7.2).

The trigonal potential (eqn 7.2) provides a sector rule connecting the position of the ligand groups in a tris-chelate complex with the sign of the axial single-crystal rotational strength $R(E)$ and, by extension, with the sign of the corresponding component of the bisignate CD absorption couplet in the solution spectrum, in the cases where that component is identified. Ligand groups substituted into the regions of the chromophore space where the potential function (eqn 7.2) is positive induce a positive rotational strength, $R(E)$ and, correspondingly, a negative rotational strength, $R(A_2)$ (fig. 7.6a).

Sugano showed that, for a randomly-orientated chiral complex ion containing an octahedral chromophore, a pseudoscalar crystal field is required for a first-order one-electron optical activity, with A_{1u} symmetry in the O_h group. The simplest octahedral pseudoscalar potential is ninth-order with respect to the electronic coordinates (fig. 7.6b),

$$V_9 \approx [XYZ(X^2 - Y^2)(Y^2 - Z^2)(Z^2 - X^2)] \tag{7.4}$$

With a ninth-order form, the potential (eqn 7.4) mixes 3d-orbitals ($l = 2$) only with metal orbitals of large angular momentum ($l = 7, 9, 11$), to which transitions from d-orbitals are electric-dipole forbidden. Thus d $\rightarrow$ d optical activity through the one-electron crystal-field mechanism is ruled out to the first-order of perturbation theory in chiral complexes containing an octahedral chromophore.

The mechanism remains feasible, however, in chiral complexes containing a tetragonal chromophore, such as the *trans*-$[Co^{(III)}N_4Cl_2]$ cluster. The simplest tetragonal pseudoscalar potential, with A_{1u} symmetry in the D_{4h} group, has the functional form (fig. 7.6c),

$$V_s \approx [XYZ(X^2 - Y^2)] \tag{7.5}$$

With a fifth-order form, the tetragonal potential (eqn 7.5) mixes the 3d with the 4f orbitals of the metal ion, and a magnetic-dipole d → d transition acquires a collinear electric-dipole moment from the d → f component introduced by the mixing.

The potential functions (eqn 7.4) and (eqn 7.5) provide sector-rules relating the ligand group positions to the sign of the optical activity induced by a group in a given electronic transition for chiral complexes containing an octahedral and a tetragonal chromophore, respectively (fig. 7.6), although the one-electron basis of the former rule does not have a first-order validity. The changes in the CD spectrum of *trans*-$[Co(S-pn)_2Cl_2]^+$, **13**, over the visible-wavelength region produced by N-methyl substitution, **14** (fig. 7.5), require a hexadecant rule (fig.

Fig. 7.7. The projection of the chelate ring and the ligand atoms of $[Co(N_1 Me-S-pn)(NH_3)_4]^{3+}$, 8, on the YZ planar section of the octahedral pseudoscalar function (eqn 7.4, and fig. 7.5b).

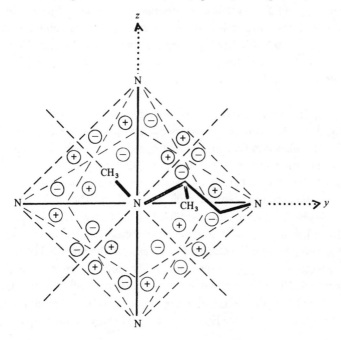

7.5c) with the functional form of the tetragonal potential (eqn 7.5), as opposed to a simpler rule of octant form, suggested by the analogy of the octant rule for the carbonyl n → π* transition. The absence of a corresponding effect of N-methyl substitution upon the CD spectra in the visible-wavelength region of the complexes $[Co(S-pn)(NH_3)_4]^{3+}$, **7**, and *trans*-$[Co(S-pn)_2(NH_3)_2]^{3+}$, **10**, which contain the octahedral $[Co^{(III)}N_6]$ chromophore (fig. 7.4), requires the additional nodal planes of the octahedral potential function (eqn 7.4) in which the N—CH_3 groups lie in the derived complexes, **8, 9, 11** and **12** (figs. 7.6b, 7.7).

The problem of the ninth-order pseudoscalar potential (eqn 7.4) for the d-electron optical activity of chiral complexes containing an octahedral chromophore through the one-electron mechanism was circumvented by adding a second component of the ligand field to the original trigonal potential (eqn 7.2). The second component is *gerade*, with T_{2g} symmetry in the O_h group, so that the required A_{1u} perturbation arises from concurrent T_{2u} and T_{2g} components in the crystal field. A finite T_{2g} potential breaks the degeneracy of T_1 and T_2 octahedral d-electron states, so that the one-electron mechanism is precluded to second order in chiral complexes of O symmetry, and in complexes of lower symmetry where the component d-electron states with a common T_1 or T_2 octahedral parentage remain accidentally degenerate, such as Λ-(+)-$[Co(en)_3]^{3+}$, **1**, where the D_3 components of the $^1A_1 \rightarrow {}^1T_1$ octahedral transition in the visible-wavelength region have a common band origin. On carrying the treatment of concurrent T_{2u} and T_{2g} crystal-field potentials to all orders of perturbation theory, it is found, for a wide range of parameter sets, that the computed d → d rotational strengths of tris-chelate D_3 complexes are at variance, quantitatively and in qualitative form, with the corresponding observed CD spectra (Hilmes and Richardson, 1976; Richardson, 1979).

7.4 Dynamic polarization theory of d-electron optical activity

The dynamic coupling mechanism for the optical activity of chiral metal complexes takes into account the leading electric moment of a magnetic-dipole d → d transition, either a quadrupole or a hexadecapole, and the electric dipole induced in each ligand group by the dynamic field of the electric-multipole transition moment centred upon the metal ion. The particular component, $d_{xy} \rightarrow d_{x^2-y^2}$, of the octahedral $^1A_1 \rightarrow {}^1T_1$ transition of cobalt(III) has, as its leading moments, the z-component of a magnetic dipole and the $[xy(x^2-y^2)]$ component of an electric hexadecapole. The potential of the electric hexadecapole, $H_{xy(x^2-y^2)}$, produces a determinate correlation of the induced electric dipole in each ligand group which is not located in an octahedral symmetry plane of the $[Co^{(III)}N_6]$ chromophore (fig. 7.8). The resultant of the correlated dipoles induced in the ligand groups is generally non-zero in complexes without an inversion centre, constituting the first-order electric-dipole transition moment which takes energy from the radiation field and enhances the isotropic absorption.

In chiral complexes based upon the $[Co^{(III)}N_6]$ cluster, the resultant electric-dipole moment has a z-component, collinear with the zero-order magnetic-dipole moment of the $d_{xy} \rightarrow d_{x^2-y^2}$ transition component. The scalar product of these two moments gives the z-component of the rotational strength, R_{om}^z, of the octahedral $^1A_1 \rightarrow {}^1T_1$ transition of cobalt(III),

$$R_{om}^z = i \, m_{mo}^z \, H_{om}^{xy(x^2-y^2)} \sum_L \bar{\alpha}(L) \, G_{xyz(x^2-y^2)}^L \tag{7.6}$$

where $\bar{\alpha}(L)$ is the mean polarizability of the ligand group, L, at the $d \rightarrow d$ transition frequency. If the $d \rightarrow d$ transitional charge distribution is taken to be a point-hexadecapole at the position of the metal ion, and it is assumed that the moment induced in the ligand group is a point-dipole at the position, X,Y,Z, in

Fig. 7.8. The correlation of the electric-dipole moments induced in each ligand group of a N_1-methyl-(S)-1,2-propylenediamine chelate ring by the potential field of the electric-hexadecapole moment of the $d_{xy} \rightarrow d_{x^2-y^2}$ transition in the $[Co^{(III)}N_6]$ chromophore. The transition involves a clockwise d-electron charge rotation, viewed from the $+Z$-axis, so that the N-methyl group provides a positive contribution and the chelate-ring alkyl groups a negative contribution to the overall rotational strength.

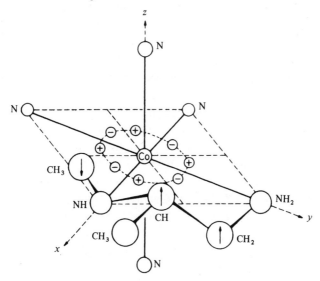

the octahedral chromophore frame at a distance R from the metal ion (fig. 7.8), the geometric tensor of eqn 7.6 has the form,

$$G^{L}_{xyz(x^2-y^2)} = 315[XYZ(Y^2 - X^2)/(2R^{11})] \tag{7.7}$$

The other two components of the octahedral $^1A_1 \rightarrow {}^1T_1$ transition of cobalt(III), due to the orbital promotions, $d_{yz} \rightarrow d_{y^2-z^2}$ and $d_{zx} \rightarrow d_{z^2-x^2}$, give rise analogously to the rotational strength components, R^x_{om} and R^y_{om}, respectively. The forms of the latter two components are obtained in the point-multipole approximation by the cyclic permutation of the coordinates in eqns 7.6 and 7.7, employing the C_3 rotation operation of the octahedron around the (1,1,1) axis. The higher-energy d-orbitals of the latter two transition components represent linear combinations of the basis d-orbitals of the octahedral e sub-shell,

$$d_{y^2-z^2} = -(1/2)[d_{x^2-y^2} + \sqrt{3d_{z^2}}] \tag{7.8}$$

and,

$$d_{z^2-x^2} = -(1/2)[d_{x^2-y^2} - \sqrt{3d_{z^2}}] \tag{7.9}$$

The orbital descriptions, $d_{x^2-y^2}$, $d_{y^2-z^2}$, and $d_{z^2-x^2}$, conform to the rotationa equivalence of the members of the set under the $C_3(1,1,1)$ operations, and express the isotropy of octahedral properties.

For a complex containing the octahedral $[Co^{(III)}N_6]$ chromophore and a chiral chelate ring spanning the X and the Y axes (fig. 7.8), the sum of R^x_{om} and R^y_{om} is equal in magnitude and opposite in sign to R^z_{om} in the point-multipole approximation, and the net first-order dynamic-polarization rotational strength goes to zero for a complex of O symmetry in this approximation. If the symmetry of the chromophore is reduced, so that the components of the octahedral $^1T_1$ state are separated in energy, as in the *trans*-$[Co^{(III)}N_4Cl_2]$ cluster, the oppositely-signed component rotational strengths no longer mutually cancel, appearing at different frequencies in the CD spectrum (fig. 7.5). The functional form of the geometric tensor (eqn 7.7) is isomorphous with the D_{4h} pseudoscalar function (eqn 7.5), and it provides the dynamic-polarization basis for the hexadecant sector rule relating the sign of the CD absorption in the visible-wavelength region of the *trans*-$[Co^{(III)}N_4Cl_2]$ complexes, 13, 14 and 15 to the location of the ligand groups in the tetragonal chromophore frame (fig. 7.6c). The notable contribution of N-methyl groups, opposite in sign to that of the chelate-ring alkyl groups, is accommodated by the dynamic polarization model (fig. 7.8).

The point-multipole approximation for a given potential is satisfactory when the distance R separating the charge distributions considered is large compared to the individual dimensions of those distributions. The radial factor for the hexadecapole moment of a 3d-electron transition has an extension in the range from 0.956 to 0.684 Å for neutral and tripositively charged cobalt, respectively. These extensions provide an upper and a lower bound for the radial maximum of the hexadecapole charge distribution, and neither value is negligible relative to the Co—N bond length (2.0 Å) or the metal–carbon distance (3.0 Å) in the tris-ethylenediamine complex, Λ-(+)-$[Co(en)_3]^{3+}$.

An allowance for the finite radial extension of the electric-multipole transition moment is afforded by summing the potential between each pole of the hexadecapole moment and the dipole induced in a ligand group, followed by summation over all ligand groups. The net first-order rotational strength, $R(T_1)$, of a chiral complex is found to be non-vanishing by this procedure. The dominant component of the octahedral $^1A_1 \rightarrow {}^1T_1$ transition of cobalt(III) in the contributions to $R(T_1)$ has its charge distribution in the mean plane of the chelate ring considered, e.g. the $d_{xy} \rightarrow d_{x^2-y^2}$ component for a ring spanning the X and the Y axis of the chromophore (fig. 7.8). The sign of the net first-order rotational strength, $R(T_1)$, due to a 1,2-diamine chelate ring is governed by the ring chirality, being positive for the λ-conformation and negative for the δ-conformation, as observed for $[Co(S\text{-}pn)(NH_3)_4]^{3+}$, **7** (fig. 7.4). The net rotational strength, $R(T_1)$, is expected to be additive over the number of chelate rings with a common conformation, in the absence of contributions of configurational origin, as is found for the *trans*-complex, **10**, and the pseudoracemate of **3** and **4** (fig. 7.4).

The d → d optical activity of configura... ...al origin, arising from the screw-dissymmetry of the mean planes of two or more chelate rings around the metal ion, is taken into account for the tris-chelate complexes by reducing the effective symmetry of the $[Co^{(III)}N_6]$ chromophore from O to D_3. The three components of the octahedral $^1T_1$ excited d-electron state of cobalt(III) are then no longer necessarily degenerate, and the D_3 component states, $^1A_2$ and ^1E, are each represented by symmetry-determined linear combinations of the excited configurations arising from the octahedral orbital promotion, $d_{xy} \rightarrow d_{x^2-y^2}$, and the two other orbital excitations, related by cyclic permutation of the electronic coordinates. The first-order dynamic polarization treatment of the D_3 component transitions, $^1A_1 \rightarrow {}^1A_2$ and $^1A_1 \rightarrow {}^1E$, of Λ-(+)-$[Co(en)_3]^{3+}$ and its analogues, gives the individual rotational strengths, $R(A_2)$ and $R(E)$, with the correct sign and magnitude, corresponding to the values measured from the axial single-crystal CD spectra. The overall value, $R(T_1)$, representing the sum of the individual component rotational strengths, is equal to the conformational rotational strength, due to the chiral-puckered conformation of each individual chelate ring, in the first order, whereas the CD absorption observed for the randomly-orientated complex ion requires a specific configurational contribution to $R(T_1)$, present when each chelate ring is planar, as in the tris-oxalato metal complexes, and independent of the ring-conformation optical activity.

The specific configurational contribution to $R(T_1)$ in the tris-chelate complexes, generally requiring the inequality, $|R(E)| > |R(A_2)|$, appears on extending the dynamic polarization to second-order, by taking into account the mixing of the D_3 components of the octahedral $^1A_1 \rightarrow {}^1T_1$ d-electron transition of cobalt(III) in the visible-wavelength region with the corresponding components of the high-intensity ligand–metal charge-transfer transition observed in the ultra-violet region. The absorption and CD spectra of Λ-(+)-$[Co(en)_3]^{3+}$ in the ultra-violet region, including the high-frequency tail of the axial single-crystal CD spectrum, characterise two strong electric-dipole transitions, the first at 208 nm to a ^1E state and the second, higher in frequency by $\sim 10000 \text{ cm}^{-1}$, to a $^1A_2$ state (fig. 7.2). These two high-energy transitions provide the main source from which electric-dipole strength is borrowed within the $[Co^{(III)}N_6]$ chromophore by the d-electron transitions of the corresponding symmetry at lower energy in the visible-wavelength region.

Table 7.2. *The observed and the calculated rotational strengths, R (10^{-2} Debye–Bohr magneton), and the dipole strength enhancement, ΔD (10^{-2} square Debye), relative to $[Co(NH_3)_6]^{3+}$ of the octahedral cobalt(III) $^1A_1 \rightarrow {}^1T_1$ d−d transition in an optical isomer of the tris-diamine complexes. Values are listed both for the net rotational strength, $R(T_1)$ from solution measurements, and the dihedral component, $R(E)$, measured from the axial single-crystal CD spectrum (Table 7.1). Calculated values refer to the absolute configuration and molecular structure determined by X-ray crystallography, from Mason and Seal (1976)*

Complex salt[a]		$R(E)$	$R(T_1)$	$\Delta D(T_1)$
(+)-$[Co(en)_3]Cl_3.H_2O$	obs.	+57	+4.7	3.5
$\Lambda(\delta\delta\delta)$ *lel*	calc.	+68	+4.3	4.0
(−)-$[Co(R\text{-}pn)_3]Br_3$	obs.	−45	−4.5	4.5
$\Delta(\lambda\lambda\lambda)$ *lel*	calc.	−70	−4.1	4.0
(−)-$[Co(tn)_3]Cl_3.H_2O$	obs.	−11.3	−1.47	1.0
Δ(tris-chair)	calc.	−11.0	−0.22	0.2

[a] Abbreviations: en = ethylenediamine;

R-pn = R-(−)-1,2,-propylenediamine; tn = trimethylenediamine.

According to the second-order dynamic polarization mechanism, the induced electric dipoles in the individual ligand groups, themselves correlated by the field of the electric-hexadecapole moment of either the $^1A_1 \rightarrow {}^1E$ or the $^1A_1 \rightarrow {}^1A_2$ d-electron transition in the visible region, produce in turn a correlation of the electric-dipole moment of the corresponding D_3 component in the high-intensity ultraviolet transition of the $[Co^{(III)}N_6]$ chromophore. The energy-interval between the d $\rightarrow$ d transition in the visible region and the charge-transfer transition in the ultraviolet region is the smaller for the ^1E components than for the $^1A_2$ components of the respective excitations (fig. 7.2), so that the second-order borrowing of an electric-dipole moment is the more effective for the $^1A_1 \rightarrow {}^1E$ than the $^1A_1 \rightarrow {}^1A_2$ d-electron transition, and the second-order augmentation of the d-electron optical activity is the larger in $R(E)$ than in $R(A_2)$. Computations based upon the second-order dynamic polarization mechanism account for the d $\rightarrow$ d rotational strengths, and the dipole-strength enhancement, $\Delta D(T_1)$, measuring the absorption-intensity increase relative to $[Co(NH_3)_6]^{3+}$, of the tris-diamine cobalt(III) complexes for which an X-ray crystal and molecular structure is available. Representative calculated values of $R(E)$, $R(T_1)$, and $\Delta D(T_1)$, are compared with the corresponding experimental values in table 7.2.

7.5 Chiral tetrahedral coordination

The study of the d-electron optical activity of chiral transition-metal complexes has long been based upon the six-coordinate chelate complexes, of the type first optically-resolved by Werner. Many of these enantiomers are well-characterised structurally, by X-ray crystallography, and spectroscopically, from solution and single-crystal CD studies over the readily-accessible visible and quartz ultraviolet region. More recently enantiomers in the six-coordinate series of the hexa-unidentate type have been isolated, such as (R)-$(+)$-*all-cis*-$[Co(NH_3)(H_2O)_2(CN)_2]^+$, (Ito and Shibata, 1977) (fig. 1.7), together with chiral chelate complexes of the four- and five-coordinate series, and the d-electron optical activity of these types has been investigated, involving the extension of circular dichroism spectroscopy to the infrared region for the latter two series.

The main four-coordinate quasi-tetrahedral chiral complexes at present characterised, by X-ray crystallography and CD spectroscopy, belong to the $[M^{(II)}(diamine)X_2]$ series, where M(II) is a transition-metal ion of the first long period, X is a halide or a pseudohalide ion, and the diamine is chiral and ditertiary. The diamines employed are the alkaloid, $(-)$-spartein (l-sp) 16, and its epimers, $(-)$-α-isospartein (l-α-isp), 17, and $(+)$-β-isospartein, 18, or a N-tetraalkyl derivative of an enantiomer of a synthetic chiral diamine, such as, (R)-$(+)$-N,N,N',N'-tetramethyl-1,2-propylenediamine (R-Me$_4$pn). The molecular structures of the complexes, $[Cu(l$-β-isp)Cl$_2]$, $[Co(l$-sp)Cl$_2]$, $[Co$-l-α-isp)Cl$_2]$,

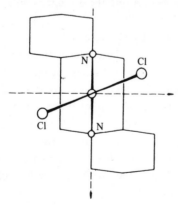

16 (6R, 11S)
17 (6R, 11R)
18 (6S, 11S)

19

and $[Co(R-Me_4 pn)Cl_2]$, determined by X-ray crystallography, show that the coordination is tetrahedral, although the molecular and the chromophoric $[MN_2 Cl_2]$ symmetry is C_2 at the highest (fig. 7.9).

The absorption and the CD spectrum of the complex $[Co(l-\alpha-isp)Cl_2]$, **19**, over the range 5.0 to 0.5 μm of the d → d manifold divide into three regions, characterised by the dissymmetry ratios, $(\Delta\epsilon/\epsilon)$, $g \sim 10^{-1}$ (2000 - 7000 cm^{-1}), $g \sim 10^{-2}$ (7000 - 12000 cm^{-1}), and $g \sim 10^{-3}$ (1500 - 20000 cm^{-1}) (fig. 7.10; table 3.1). The absorption and the CD in each of the three regions correspond to the low-symmetry components of the tetrahedral cobalt(II) d → d transitions, $^4A_2 \rightarrow {}^4T_2(F)$, $^4T_1(F)$, and $^4T_1(P)$, respectively, and characterise the lowest-energy group as a magnetic dipole-allowed set. The lowest-energy tetrahedral cobalt(II) transition, $^4A_2 \rightarrow {}^4T_2(F)$, is made up, together with two equivalent and orthogonal excitations, of the d-orbital promotion, $d_{x^2-y^2} \rightarrow d_{xy}$, which has

Fig. 7.9. A projection on a plane perpendicular to the C_2 rotation axis of the molecular structure of $[Co\{(-)-\alpha-isospartein\}Cl_2]$, **19**.

the z-component of a magnetic dipole and the $[xy(x^2-y^2)]$ component of an electric hexadecapole as its leading moments, like the converse d-orbital promotion of octahedral cobalt(III).

In a complex with regular tetrahedral coordination, the field of the electric-hexadecapole transition moment, $H_{xy(x^2-y^2)}$, while generally inducing an electric dipole in an anisotropic ligand group, produces no overall constructive correlation of those induced dipoles, and their resultant goes to zero. The dynamic polarization expression for the resultant of the dipoles induced in isotropic ligand groups is given by,

$$\mu_z^{dd'} = -H_{xy(x^2-y^2)}^{dd'} \sum_L \bar{\alpha}(L) \, G_{xyz(x^2-y^2)}^L \tag{7.10}$$

and the required geometric tensor, $G_{xyz(x^2-y^2)}^L$ (eqn 7.7), shows that each term

Fig. 7.10. The absorption spectrum (upper curve) and the circular dichroism (lower curve) of $[Co\{(-)-\alpha\text{-isospartein}\}Cl_2]$, **19**, in CDCl$_3$ solution over the frequency-region of the d $\rightarrow$ d transition manifold. The narrow band near 2900 cm^{-1} arises from the C—H stretching vibrational fundamentals of the organic ligand (fig. 7.13).

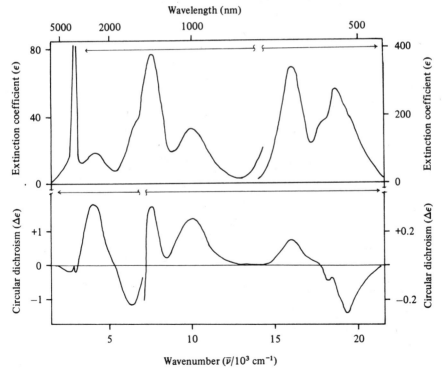

in the sum of eqn 7.10 vanishes for regular tetrahedral coordination. A D_2 or lower-symmetry distortion of a tetrahedral complex gives non-zero individual terms and a finite sum in eqn 7.10, affording the d → d transition a first-order electric-dipole moment with a z-polarization (fig. 7.11).

In the complex $[Co(l\text{-}\alpha\text{-isp})Cl_2]$, **19**, the coordinated atoms are displaced from the regular tetrahedral location, the $CoCl_2$ plane being torsionally displaced from orthogonality to the CoN_2 plane by 19° (fig. 7.9). The distortion gives the $d_{x^2-y^2} \to d_{xy}$ transition, and the two equivalent excitations producing the excited $^4T_2(F)$ tetrahedral cobalt(II) state, a small first-order electric-dipole transition moment (eqn 7.10) and a corresponding weak isotropic absorption. The first-order electric-dipole moment is collinear, however, with the zero-order

Fig. 7.11. The Coulombic correlation of the electric dipole induced in each of the coordinated ligand atoms by the potential field of the electric-hexadecapole moment, $H_{xy(x^2-y^2)}$, of the $d_{x^2-y^2} \to d_{xy}$ transition in a chirally-distorted tetrahedral metal complex of D_2 or lower symmetry. The transition involves an anticlockwise d-electron charge rotation, viewed from the $+Z$-axis, so that each coordinated atom makes a negative contribution to the rotational strength, for the absolute configuration illustrated.

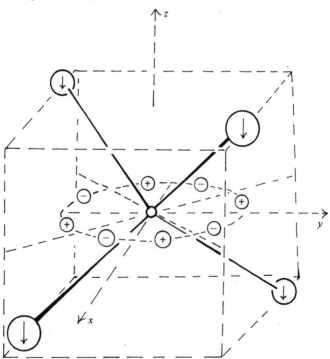

magnetic-dipole moment of the excitation, giving each of the three components of the tetrahedral $^4A_2 \rightarrow {}^4T_2(F)$ cobalt(II) transition a relatively large rotational strength, while the corresponding dipole strength is small. Thus the dissymmetry ratio, $(g = 4R/D)$, is large for each of the three CD bands of **19** observed over the lowest-energy range, 2000 - 7000 cm^{-1} (fig. 7.10).

The higher-energy d $\rightarrow$ d transitions of a tetrahedral cobalt(II) complex, $^4A_2 \rightarrow {}^4T_1(F), {}^4T_1(P)$, derive a non-zero transition probability from the one-electron promotion, $d_{z^2} \rightarrow d_{xy}$, and two other equivalent and orthogonal excitations. Both the $^4T_1(F)$ and the $^4T_1(P)$ excited states are represented by linear combinations of one-electron $(e^3 t_2{}^4)$ and two-electron $(e^2 t_2{}^5)$ excited configurations, and promotions from the ground configuration $(e^4 t_2{}^3)$ to the two-electron excited configuration have a vanishing probability in a radiation field. The leading moment of the one-electron, $d_{z^2} \rightarrow d_{xy}$, is the xy-component of an electric quadrupole, θ_{xy}, but there is no concurrent magnetic-dipole moment in this case, nor that of the two other equivalent excitations.

For a given metal–ligand bond distance, the constructive correlation of the electric dipoles induced in each coordinated ligand atom or group by the field of the d $\rightarrow$ d quadrupole transition moment, θ_{xy}, is an optimum, as a function of the bond angles, at regular tetrahedral coordination (fig. 7.12). The resultant first-order electric-dipole transition moment, located wholly in the ligands, is given by the expression,

$$\mu_z^{dd'} = -\theta_{xy}^{dd'} \sum_L \bar{\alpha}(L) \, G_{xyz}^L \qquad (7.11)$$

and the moment is relatively large as the appropriate geometric tensor, G_{xyz}^L, has a functional form (eqn 3.13) optimising for the regular tetrahedron.

Steric distortions from regular tetrahedral coordination reduce the resultant first-order electric-dipole transition moment induced in the ligands by the field of the d $\rightarrow$ d electric-quadrupole moment (eqn 7.11), but the moment remains substantial in complexes, such as **19**, where the angular distortions are not large. More significantly, the distortions limit the validity of the tetrahedral description, and the allowed and the forbidden magnetic-dipole d $\rightarrow$ d transitions in the complex **19**, which is reduced to C_2 symmetry, become mixed to a degree dependent, among other factors, upon the inverse of the energy-separation between the allowed and the forbidden magnetic-dipole transition types. The magnetic-dipole allowed d $\rightarrow$ d transitions of **19**, with $^4A_2 \rightarrow {}^4T_2(F)$ tetrahedral parentage, lie in the 2000 - 7000 cm^{-1} region with $g \sim 10^{-1}$, and the progressively smaller dissymmetry ratios of the higher-energy band systems, those with $^4A_2 \rightarrow {}^4T_1(F), {}^4T_1(P)$, tetrahedral parentage, in the 7000 - 12000 cm^{-1} $(g \sim 10^{-2})$ and 15000 - 20000 cm^{-1} $(g \sim 10^{-3})$ region, reflect the reduced mixing of allowed and forbidden magnetic-dipole transitions with increasing energy separation (fig. 7.10).

The lowest-energy component of the $^4A_2 \rightarrow {}^4T_2(F)$ tetrahedral d-electron transition of cobalt(II) in the complex **19** and its analogue, $[\text{Co}(l\text{-sp})\text{Cl}_2]$, overlaps on the frequency ordinate the fundamentals of the C—H stretching vibrational modes of the alkaloid ligand, **16** or **17**, as does the corresponding component of the $^3T_1(F) \rightarrow {}^3T_2(F)$ tetrahedral d $\rightarrow$ d transition of nickel(II) in the complex $[\text{Ni}(l\text{-sp})\text{Cl}_2]$ (fig. 7.13). The electronic d $\rightarrow$ d bands are intrinsically weak and broad, whereas the vibrational C—H stretching-mode absorptions are sharp and relatively strong, and such conditions are conducive to interference effects, producing resonance and antiresonance absorption line-shapes. While an interaction between the superimposed vibrational and electronic transitions is not evident in the isotropic absorption, there is a

Fig. 7.12. The Coulombic correlation of the electric dipole induced in each of the coordinated ligand atoms by the potential field of the electric-quadrupole moment, θ_{xy}, of the $d_{z^2} \rightarrow d_{xy}$ transition in a tetrahedral metal complex.

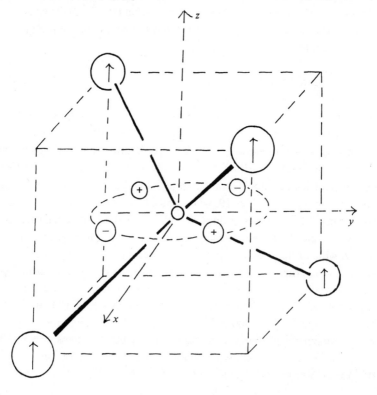

marked enhancement of the CD associated with the C—H stretching-mode fundamentals in the cobalt(II) and the nickel(II) complexes, relative to the purely vibrational CD absorption observed in the case of the corresponding zinc(II) complex, $[Zn(l\text{-}sp)Cl_2]$ (fig. 7.13).

Thus the primary interaction connects, not the electric-dipole moments

Fig. 7.13. The absorption spectra (upper curves) and the circular dichroism (lower curves) of the quasi-tetrahedral (–)-spartein complexes, $[Ni(l\text{-}sp)Cl_2]$ (dashed curve), $[Zn(l\text{-}sp)Cl_2]$ (dash-dot curves), and $[Co(l\text{-}sp)Cl_2]$ (full curves), in $CDCl_3$ solution over the frequency region of the C—H stretching vibration fundamentals of the organic ligand, and the lowest-energy d → d transitions of the open-shell metal complexes.

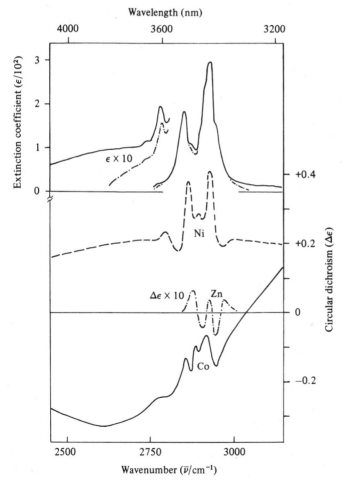

of the two superimposed transitions, but the magnetic moment, m_{dd}, of the electronic d → d excitation with the electric-dipole moment, μ_{01}, of the 0 → 1 fundamental of the C—H stretching-vibration mode, producing a new source of rotational strength in the cross-term, $[\mu_{01} \cdot m_{dd}]$. The additional rotational strength due to the cross-term is also dependent upon the energy-ratio, $[hc(\bar{\nu} - \bar{\nu}_{01})/V_{ev}]$, where $\bar{\nu}_{01}$ refers to the wavenumber of the C—H stretching vibration transition, $\bar{\nu}$ to the wavenumber of observation, and V_{ev} to the Coulombic coupling energy of the electronic and the vibrational transitional charge distributions, that is, the electric-hexadecapole moment of the former transition and the electric-dipole moment of the latter. The energy-ratio, and thus the additional rotational strength due to the cross-term, $[\mu_{01} \cdot m_{dd}]$, is zero when $\bar{\nu} = \bar{\nu}_{01}$, and changes sign, as a function of frequency, across the band-width, $\Delta\bar{\nu}_{01}$, of the C—H stretching-vibration fundamental. The resultant line-shape of the CD absorption arising from the additional rotational strength of the cross-term has a dispersion-curve form, which is evident in the CD spectra of **19** and its analogues over the 2750 – 3000 cm^{-1} region, particularly in the case of the [Ni(*l*-sp)Cl$_2$] complex (fig. 7.13) (Mason, 1981).

7.6 Chiral trigonal bipyramid coordination

The tetradentate tripod ligand, tris(2-aminoethyl)amine (tren), and its hexamethyl derivative, tris(2-dimethylamino)amine (Me$_6$ tren), form five-coordinate trigonal bipyramid coordination compounds with a range of metal ions. These compounds are chiral, due to the common λ- or δ-conformation of the three chelate rings in a given complex ion, although they are optically-

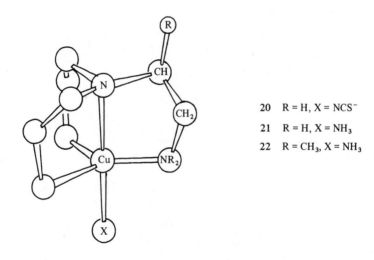

20   R = H, X = NCS$^-$

21   R = H, X = NH$_3$

22   R = CH$_3$, X = NH$_3$

labile in solution. In several cases, the racemic mixture spontaneously resolves on crystallisation, and all of the complex ions in a given single crystal have a common chirality.

The salt, [Cu(tren)(NCS)] (SCN), **20**, crystallises in the chiral orthorhombic space group, $P2_1 2_1 2_1$ (Jain and Lingafelter, 1967), while its analogue, [Cu(tren)(NH$_3$)] (ClO$_4$)$_2$, **21** (Duggan *et al.*, 1980), and each member of the series, [M$^{(II)}$(Me$_6$ tren)Br] Br, where the metal M is Mn, Fe, Co, Ni, Cu, or Zn, crystallise in the enantiomorphous cubic space group, $P2_1 3$ (Vaira and Orioli, 1967, 1968). Thus a single crystal of **21**, or a powdered single crystal of **20** incorporated into a potassium halide matrix, provides an isotropic homochiral assembly of the respective complex ion salts for spectroscopic study (fig. 7.14). A chiral derivative of the ligand Me$_6$ tren, (S)-2,4,8-trimethyl-5-(2-methyl-2-azabutyl)-2,5,8-triazanonane (S-tan), synthesised from (S)-(+)-alanine, gives an

Fig. 7.14. The absorption spectrum (upper curves) and the circular dichroism (lower curves) of (a) a single crystal of [Cu(tren)(NH$_3$)] (ClO$_4$)$_2$, **21**, and (b) of a powdered single crystal of [Cu(tren)(NCS)] (SCN), **20**, in a potassium bromide matrix.

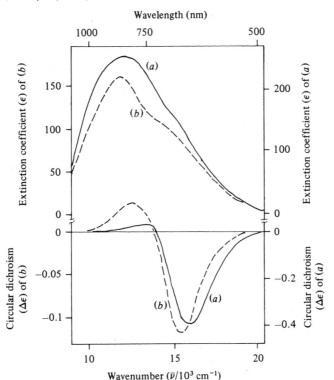

analogous series of trigonal bipyramid complexes, $[M^{(II)}(S\text{-tan})(NH_3)]^{2+}$ and $[M^{(II)}(S\text{-tan})Br]^+$, which are optically stable in solution (Endo, Horikoshi and Utsuno, 1981). The solution absorption and CD spectrum of a copper(II) member of the S-tan series, $[Cu(S\text{-tan})(NH_3)](ClO_4)_2$, **22**, are similar to the corresponding single-crystal spectra of **21** (fig. 7.14).

The five-coordinate trigonal bipyramid complexes have C_3 symmetry at the highest, as in the case of **21**, although the coordination cluster $[Cu^{(II)}N_5]$ approximates to D_{3h} point-symmetry. The relative energies of the d-electrons of a transition-metal ion in a D_{3h} five-coordinate ligand field give the copper(II) ion a $^2A_1'$ ground state, corresponding to a singly-occupied d_{z^2} orbital in an otherwise-filled 3d-shell (fig. 7.15). The two doubly-degenerate d-electron excitations correspond to the hole-promotions, $a_1' \rightarrow e'$, e'', which remain pure in D_{3h} symmetry, but intermix in the C_3 five-coordinate complexes (Furlani, 1968; Wood, 1972).

The leading transition moment of the lower-energy hole-promotion, $d_{z^2} \rightarrow d_{xy}$, $d_{x^2-y^2}$, is a component of an electric quadrupole, θ_{xy} and $\theta_{x^2-y^2}$, respectively. Neither of these two hole-promotions has a magnetic-dipole moment. In contrast, each of the paired higher-energy hole-promotions, $d_{z^2} \rightarrow d_{yz}$, d_{xz}, has a magnetic-dipole moment, m_x and m_y, respectively, together with an electric-quadrupole moment, θ_{yz} and θ_{xz}, respectively. If the coordination cluster, $[Cu^{(II)}N_5]$, has exact D_{3h} symmetry, only the electric

Fig. 7.15. The relative energies of a set of d-electrons, degenerate in the gaseous metal ion, in the ligand field of a trigonal bipyramid five-coordinate complex with D_{3h} symmetry.

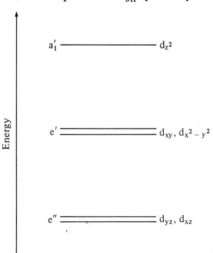

Fig. 7.16. The Coulombic correlation of the y-component of the electric dipole induced in each coordinated ligand group by the field of the xy-component of the electric-quadrupole d → d transition moment, centred on the metal ion, in a five-coordinate trigonal bipyramid complex with C_3 or higher symmetry.

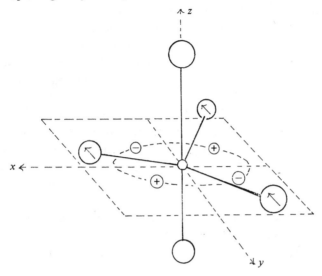

Fig. 7.17. The Coulombic correlation of the x-component of the electric dipole induced in each coordinated ligand group by the field of the xz-component of the electric-quadrupole d → d transition moment, centred on the metal ion, in a five-coordinate trigonal bipyramid complex with C_3 or higher symmetry.

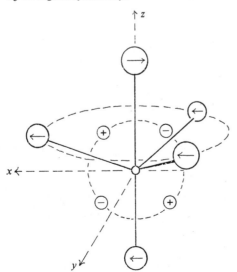

quadrupole components, θ_{xy} and $\theta_{x^2-y^2}$, of the lower-energy hole-promotion produce a construction correlation of the electric dipoles induced in each ligand, giving a non-vanishing resultant first-order electric-dipole transition moment (fig. 7.16). The resultant electric-dipole moment remains substantial for lower chromophoric symmetry, C_{3v} or C_3.

The higher-energy hole-promotion of five-coordinate copper(II) acquires a non-vanishing first-order electric-dipole transition moment through the correlation of the dipoles induced in the ligands by the electric quadrupole, θ_{yz} or θ_{xz}, only if the chromophoric symmetry is reduced from D_{3h} to C_{3v} or lower. With inequivalent axial ligands, or equatorial ligands equivalently displaced from the molecular XY plane (fig. 7.17), the electric-quadrupole transition moment, θ_{xz}, at the metal ion correlates the x component of the dipole induced in each coordinated ligand atom, to give a non-zero resultant first-order electric-dipole transition moment. Similarly the y component of the dipole induced in each coordinated ligand group is correlated by the isoenergetic-quadrupole transition moment at the metal ion, θ_{yz}.

The resultant electric-dipole transition moment of the higher-energy hole-promotion, dependent upon distortions of the $[Cu^{(II)}N_5]$ cluster from D_{3h} symmetry, is small relative to the corresponding dipole moment of the lower-energy hole-promotion, substantial for strict D_{3h} symmetry. Thus the lower-frequency d → d absorption band of the five-coordination copper(II) complexes, **20** and **21**, has the major intensity, the higher-frequency transition appearing only as a shoulder absorption (fig. 7.14).

Although the higher-energy hole-promotion, $d_{z^2} \rightarrow d_{xz}$, has a zero-order magnetic-dipole moment, $\mathbf{m}_y$, the first-order electric-dipole moment produced by the correlation of the dipoles induced in the coordinated ligand atoms is x-polarized. No rotational strength arises by this mechanism, as expected from the effective C_{3v} symmetry of the $[Cu^{(II)}N_5]$ coordination cluster (fig. 7.17). The chirality of the complexes, **20**, **21** and **22**, is due to the common δ-conformation of three chelate rings, giving each of these coordination compounds an actual or an effective C_3 symmetry. The groups and bonds of the tripod ligand which are not directly coordinated to the copper(II) ion are polarized by the electric-quadrupole transition moment of the metal ion in the same manner as the coordinated atoms.

The C—C bond has a relatively large polarizability anisotropy (Le Fèvre, 1965). The electric dipole induced in the C—C bond of each of the three chelate rings of **20**, **21** and **22** by the electric-quadrupole moment of the higher-energy copper(II) hole-promotion, θ_{xz} and θ_{yz}, gives a resultant first-order electric-dipole transition moment collinear with the corresponding zero-order magnetic-dipole moment, $\mathbf{m}_y$ and $\mathbf{m}_x$, respectively. If the δ-conformation is common to the three chelate rings, as in **20, 21** or **22**, the resultant of the induced electric dipole directed along the C—C bond of the rings is antiparallel to the magnetic-dipole moment of each component of the higher-energy copper(II) hole-promotion.

As the lower- and the higher-energy hole-promotions of five-coordinate copper(II) are intermixed in C_3 symmetry, the ligand-polarization expectations for the complexes **20** and **21** with the δ-conformation of the chelate rings are a weak positive and a strong negative CD from lower to higher frequency, associated with strong and weak isotropic absorption, respectively (fig. 7.14) (Mason, 1980; Kuroda *et al.*, 1981). The expectations are supported by the absorption and CD spectra of the complex **22** in which the δ-conformation of the chelate rings is adopted on account of the equatorial preference of the exocyclic methyl group (Endo, *et al.*, 1981).

8 VIBRATIONAL OPTICAL ACTIVITY

8.1 The fixed partial charge model

A concern with optical activity of vibrational origin arose initially from analyses of the optical rotation of molecules that owe their chirality to isotopic substitution. The optical rotation may be relatively large, considering the generally small magnitude of isotope effects. The 2-axial- and the 2-equatorial-deuteroadamantan-4-one of table 4.1, for example, have a specific rotation of $-1.2°$ and $-2.7°$, respectively, at the mercury yellow line (578 nm). While isotopic substitution generally gives rise to only minor modifications of electronic transition energies and probabilities, the changes produced in the corresponding vibrational transition properties are large. Accordingly an origin of the optical rotation of isotopically-chiral molecules was sought in the vibrational rather than the electronic rotational strengths.

A general treatment originally derived for an estimate of the vibrational rotational strengths of CHDBrCl and its isotopic analogues was subsequently extended to other chiral systems. The treatment is based upon a fixed partial charge model, in which each atom of the molecule is assigned a static residual charge, determined by the nuclear charge screened by the electronic distribution in the equilibrium nuclear configuration of the molecule in its electronic ground state. A normal coordinate analysis affords the nuclear displacements from the equilibrium configuration of the molecule in a radiation field at a normal-mode vibration frequency. The displacements of the fixed residual charges generate local dipoles which couple to give a resultant electric-dipole and magnetic-dipole transition moment. The scalar product of the two moments gives the vibrational rotatory strength of the particular fundamental mode, and the vibrational contribution to the optical rotation at a given frequency follows from Rosenfeld's relation (eqn 2.2) summed over the corresponding strengths of all the fundamentals. For an enantiomer of CHDBrCl the estimated vibrational contribution to the specific rotation at the sodium D-line is only $\sim 10^{-4}$ degrees, compared with 0.1 degree at 3000 cm^{-1} close to the C—H stretching mode frequency (Cohan and Hameka, 1966).

Extensions of the fixed partial charge model to helical macromolecules, notably polypropylene oxide and the polypeptide α-helix, indicated that dissymmetry ratios ($\Delta\epsilon/\epsilon$) of the order of $g \sim 10^{-4}$ are expected in vibrational Cotton effects. Vibrational g-ratios of the expected order were observed when circular dichroism measurements of the required precision became feasible, first in the H-stretching region, and subsequently over the major part of the rocksalt infrared range (table 3.1). The vibrational g-ratios are found to be of a comparable order whether the enantiomer owes its chirality to isotopic substitution, as in the case of (R)-(−)-neopentyl-1-d-chloride, 1, where $g = 2 \times 10^{-5}$ for the

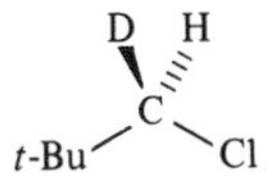

C—D stretching mode, or not, as for (R)-(−)-2,2,2-trifluoro-1-phenylethanol, 2, where $g = 6 \times 10^{-5}$ for the aliphatic C—H stretching mode (Faulkner *et al.*, 1974; Holzwarth *et al.*, 1974).

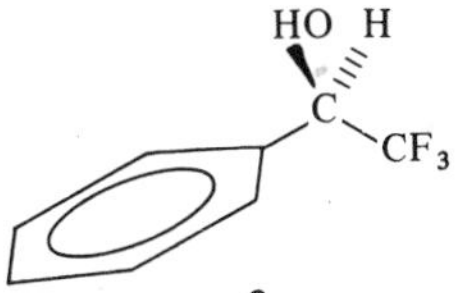

The *ab initio* calculation of the residual charges on the atoms of molecules as complex as 1 and 2 is not readily feasible, and for an analysis of the vibrational optical activity of these molecules, in terms of the fixed partial charge model and a normal coordinate calculation, the residual atomic charges are based upon initial empirical estimates obtained from bond dipole-moment data. The set of charges so derived are found, however, to give vibrational dipole strengths as much as an order of magnitude smaller than the corresponding observed values. A subsequent set of charges, scaled up to accommodate the observed dipole strengths, are employed to calculate the corresponding vibrational rotational strengths, although the latter charges are no longer consistent with the empirical bond-dipole moments, nor with the overall molecular-dipole moment. The principal limitation of the fixed partial charge model is the neglect of the response of the electrons to the nuclear vibrational motion. In general the nuclear displacements of a given vibrational mode mix the different electronic states of an enantiomer. The form and the extent of the mixing is changed between the initial and the final state of a vibrational transition, so that there are generally dynamic contributions from the electrons to the vibrational dipole and rotational strengths of a chiral molecule (Faulkner, Marcott, Moscowitz and Overend, 1977).

8.2 Dynamic coupling models

Each of the general mechanisms for optical activity (chapter 3) apply with appropriate modification to both electronic and vibrational optical activity. The two-group electric-dipole coupling mechanisms (§ 3.4) are of general application whether the two transition dipoles are both electronic, or both vibrational, or even one of each kind. The isoenergetic two-group coupled electric-dipole mechanism for the rotational strengths of a dissymmetric dimer is possibly applicable to the H—O and the C=O stretching vibrations of dimethyl-(+)-tartrate, **3**, which gives a bisignate (CD) absorption, characteristic of the mech-

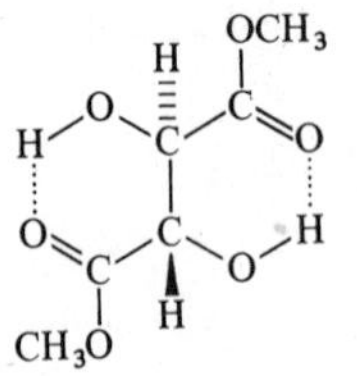

3

anism, in both the H—O and the C=O stretching vibration regions (Keiderling and Stephens, 1977; Su and Keiderling, 1980). The interpretation is questioned, however, on the grounds that the two C—H groups of (+)-tartaric acid are not analogously coupled, although separated by only one bond, and give rise to a single negative CD absorption in the C—H stretching vibration region, while the CD spectrum of (R)-(−)-methyl mandelate **4** and its antipode in the H—O

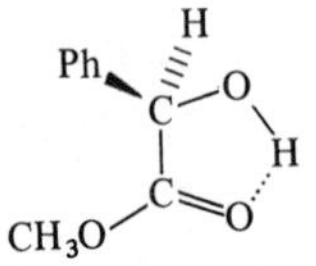

4

stretching vibration region suggests that the bisignate CD absorption of dimethyl-(+)-tartrate in that region is due to two different absorbing species, one with six-membered hydrogen-bonded rings, **3**, and the other with five-membered rings, analogous to **4** (Marcott *et al.*, 1978).

It is probable that the isoenergetic two-group electric-dipole coupling mechanism is not generally important in the vibrational CD spectra of dissymmetric dimers on account of the relatively small energy splitting (eqn 3.17) between the oppositely-signed rotational strengths of equal magnitude produced by the mechanism (eqn 3.18). The frequency interval between the rotational strengths is proportional to the corresponding dipole strength, measured by the associated isotropic absorption intensity, which is typically $\sim 10^3$ times larger for electronic than vibrational transitions. In the case of the alkaloid dimer, calycanthine, **5**, the vibrational CD absorption is essentially monosignate for the N—H stretching mode and for each of the C—H stretching fundamentals (fig. 8.1), whereas the corresponding electronic CD spectrum shows the bisignate CD absorption,

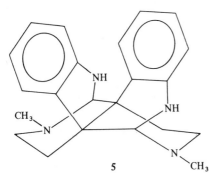

5

associated with a given isotropic absorption band system, characteristic of the isoenergetic electric-dipole coupling mechanism (fig. 6.6).

The N—H stretching mode of calycanthine, **5**, gives an isotropic and CD absorption isolated on the frequency ordinate at 3448 cm^{-1}, free from overlap by adjacent absorptions (fig. 8.1). The dipole strength of the isotropic N—H absorption gives from eqn 3.17 a frequency interval of only 0.7 cm^{-1} between the oppositely-signed rotational strengths produced by the coupling of the electric-dipole transition moments of the two N—H stretching fundamentals (eqn 3.18). The calculated frequency interval is equivalent to only 2% of the

Fig. 8.1. The absorption spectrum (upper curve) and the circular dichroism (lower curve) of calycanthine, **5**, in CDCl$_3$ solution over the frequency region of the C—H and the N—H stretching vibration fundamentals. The molecular coordinate frame is chosen for the application of eqn 3.20 to the estimation of the rotational strengths of a N—H stretching mode (Barnett, Drake and Mason, 1979).

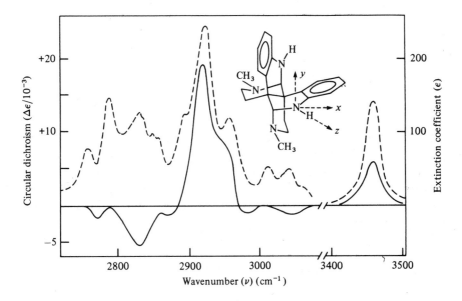

observed N—H stretching-mode band-width, whereas the corresponding
frequency interval calculated from the dipole-strength of the lowest-energy
electronic absorption band system of calycanthine at 308 nm, associated with a
bisignate CD absorption, represents 22% of the width of that band system. Thus
the mutual cancellation of the isoenergetic-coupling rotational strengths (eqn
3.18) is more substantial for the N—H stretching mode vibrational transition
of 5 than for the corresponding lowest-energy electronic transition. The resulting
bisignate CD absorption, apparent for the latter transition (fig. 6.6), is too weak
to become apparent in the presence of a stronger monosignate CD absorption,
with an origin other than isoenergetic coupling, over the N—H stretching vibrational mode region (fig. 8.1).

Infrared CD spectroscopy covers at present mainly the higher-frequency range
of the characteristic group frequencies, where the absorption bands arise from an
electric-dipole vibrational transition largely localised to a particular bond or set
of bonds to a common centre, constituting a vibrational chromophore. In general
the vibrational chromophore is symmetric, and located in a chiral molecular
environment of anisotropic substituent groups. The non-degenerate coupling
mechanism is effective for relatively weak electric-dipole transitions, and in its
limiting form (eqn 3.20) the mechanism accounts for the rotational strength of a
number of monosignate CD bands due to H-stretching vibrational modes.

In the example of calycanthine, 5, the field of the electric-dipole moment of
the N—H stretching fundamental at 3448 cm^{-1} is taken to induce an electric
dipole in each of the groups in the molecular environment of the vibrational
chromophore. The groups with the largest polarizability anisotropy, $\beta = (\alpha_{\|} - \alpha_{\perp})$,
are the benzene rings, for which $\beta = -5.62$ Å^3 at zero frequency (Bridge and
Buckingham, 1966), compared with the corresponding values of -0.12,
$+0.72$ Å^3, and zero, for the C—N, C—C, and C—H group, respectively (Le
Fèvre, 1965). The main interaction, accounting for the sign and some 80% of
the observed rotational strength of the N—H stretching mode in calycanthine
(fig. 8.1), consists in the Coulombic coupling of the transition moment of the
N—H group substituted into one benzene ring with the dipole induced in the
other benzene nucleus (Barnett *et al.*, 1979).

The polarization dipole-coupling mechanism of eqn 3.20 allows for the response of the electrons in the substituent groups to the nuclear motions in the
vibrational chromophore. The dissymmetry ratio of the non-degenerate mech-
anism is independent of the dipole strength and the ratio, $(\Delta\epsilon/\epsilon)$, remains observable for weak transitions, other factors being equal. In contrast, the observable
ratio, $(\Delta\epsilon/\epsilon)$, resulting from the corresponding isoenergetic coupling mechanism
is proportional to the dipole strength, since the ratio in this case is dependent
upon the energy separation (eqn 3.17) between the rotational strengths produced

as well as their magnitudes (eqn 3.18), so that the mechanism is important mainly for transitions of high intensity. Both the isoenergetic and the non-degenerate mechanism are subject to the constraint of the dimension ratio, (d/λ), which predicates smaller g-ratios for the infrared than the ultraviolet region.

The polarizability anisotropy of the substituent groups, which enters the expression for the rotational strength due to non-degenerate coupling (eqn 3.20), is not a significant variable with respect to the frequency of the chromophore transition over the range from the H-stretching infrared region to the beginning of the quartz ultraviolet region. The high-intensity transitions of the vacuum ultraviolet region dominate in the expression for each component of the molecular polarizability tensor (eqn 3.11). In the case of benzene, the vibrational contribution to the mean polarizability, estimated from the infrared absorption band areas, is 0.29 Å^3, a value which represents some 3% of the total mean polarizability at zero frequency, 10.4 Å^3. At H-stretching vibration frequencies, the corrections to the polarizability tensor components for benzene from zero frequency are negligible, a 0.2% increase from the electronic contributions being offset by an equally minor decrease from the lower-frequency vibrational contributions (Barnett *et al.*, 1980).

8.3 Raman optical activity

Vibrational optical activity is spectroscopically accessible, not only by the infrared absorption method but also through the Raman emission technique (Barron and Buckingham, 1975; Barron, 1980). The two methods are complementary in a number of respects but especially in the frequency range covered. While infrared CD absorption measurements to frequencies as low as $\sim 1250 \text{ cm}^{-1}$ are feasible with a 3kW carbon rod radiation source (Su *et al.*, 1980), the majority of Raman circular intensity differential determinations cover the lower frequencies, in the 100 to 1700 cm^{-1} range (Barron, 1980; Hug *et al.*, 1980).

A laser-Raman spectrometer is modified for CID studies by placing an electro-optic modulator in the optical train between the source and the sample, to give LCP and RCP excitation periodically in square-wave modulation (fig. 8.2). The radiation scattered by the sample is collected in a direction perpendicular to the excitation propagation direction, and transmitted through a linear polarizer to the frequency-scanning monochromator, equipped with a photomultiplier detector. The polarizer is orientated to pass radiation polarized parallel to the scattering plane, in order to minimise the optical artefacts which are the more prominent in the perpendicularly-polarized scattered radiation (fig. 8.2).

At a given spectrometer frequency-setting, the photons scattered during each period of RCP excitation are fed to one channel of a two-channel photon counter, and the photons collected during the corresponding periods of LCP excitation are monitored by the other channel. After the accumulation of sufficient counts, the sum and the signed difference of the counts in the two channels are recorded, and the counting process is repeated at incrementally successive frequencies to obtain the Raman CID spectrum of $(I_\alpha^L - I_\alpha^R)$ and $(I_\alpha^L + I_\alpha^R)$ as a function of the frequency separation from the Rayleigh-scattering line. The subscript α refers to the parallel or the perpendicular polarization of the scattered radiation transmitted by the linear polarizer (fig. 8.2) (Barron and Buckingham, 1975).

The incremental frequency scanning of a Raman CID spectrum in time requires an extended period during which instrumental drift and sample degradation may not be negligible. These hazards are minimized by the use of a Raman spectrometer equipped with an image-intensifier, followed by a linear array of diode detectors, each receiving a different scattered radiation frequency, in successively larger increments from the Rayleigh scattering line. Each of the 500 or more

Fig. 8.2. The Raman circular intensity differential technique. A laser Raman spectrometer is modified by placing an electric-optic modulator (EOM) in the excitation optical train to provide LCP and RCP monochromatic radiation periodically in square-wave modulation. The scattered radiation is collected in a direction perpendicular to the excitation direction through a linear polarizer (P) orientated, in general, to transmit the radiation polarized parallel to the scattering plane to the spectrograph monochromator (M). The Raman spectrum is recorded either (*a*) by a frequency-scan in time, employing a photomultiplier detector (PM) and a two-channel photon counter, or (*b*) by a spatial frequency-scan, using an image-intensifier (Ii) and a diode array (Da) as a multi-frequency detector (Barron, 1980).

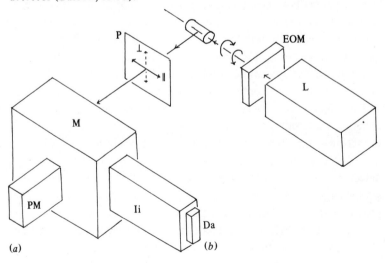

diode detectors is monitored by a single channel of a multi-channel analyser, the signals collected during periods of RCP excitation being segregated from those received during the corresponding periods of LCP excitation. Each set of signals is stored and, after accumulation, the sum and the difference of the two sets are recorded for each scattered frequency monitored. A 350 cm^{-1} region of the Raman CID spectrum is covered at any one time in the instrument of Brocki, Moskovits and Bosnich (1980), or of some 2100 cm^{-1} in the instrument of Hug and Surbeck (1979). The collection time for a Raman CID spectrum is reduced from days to hours with a laser excitation power level as low as 50 mW.

The determination of a small differential circular polarization in absorption or emission is dependent upon a relatively large photon flux throughout the optical train from the source to the detector. The limitations of CD spectroscopy in the vacuum ultraviolet and the infrared regions derive primarily from the radiation power levels attainable. Even so, a dissymmetry ratio, $g = \Delta\epsilon/\epsilon$, of the order of 10^{-6} is both required and routinely achieved in infrared CD spectroscopy. The Raman scattering, in contrast, is weaker than the Rayleigh scattering, the entire Raman spectrum containing only some 1% of the total scattering intensity which, in turn, may be of the order of 10^{-5} of the incident intensity (Woodward, 1967). Accordingly, the attainable dissymmetry ratio in a Raman CID spectrum is more limited, the current level of g_{Ram} being $\sim 10^{-4}$, where,

$$g_{Ram} = 2(I_\alpha^L - I_\alpha^R)/(I_\alpha^L + I_\alpha^R) \tag{8.1}$$

Despite the disparity between the currently attainable g-ratios, comparable levels of vibrational optical activity are accessible by Raman CID and infrared CD absorption spectroscopy, since the Boltzmann–Kuhn ratio of molecular dimensions to radiation wavelength, (d/λ), favours the former technique. The typical argon ion laser excitation source used in Raman CID spectroscopy has a wavelength (488 nm) smaller by one to two magnitude orders than the infrared wavelengths of molecular vibrations directly investigated by the corresponding CD absorption technique.

The vibrational Raman scattering intensity is dependent upon the variation of the molecular electric-dipole polarizability tensor components (eqn 3.11) with nuclear displacement, through the square of the derivative $(\partial\alpha/\partial Q)$ with respect to the normal coordinate, Q. The optical-activity polarizability tensor, which has components given by the analogue of eqn 3.11 with the molecular dipole strengths replaced by the corresponding rotational strengths, has non-zero derivatives with respect to the normal coordinate nuclear displacements. These derivatives and those of the dipole–quadrupole polarizability tensor govern the Raman CID scattering intensity (Barron and Buckingham, 1971).

For a molecule composed of two groups with a chiral mutual orientation (fig. 8.3) the dissymmetry ratio of the Raman CID scattering, and that of the corresponding Rayleigh scattering, are expressed by a simple relation between the dimer dimension, R_{21}, and the wavelength of the laser excitation, λ_{ex} (Barron and Buckingham, 1974). With the collected radiation polarized parallel to the scattering plane, the Raman CID dissymmetry ratio is given by the expression,

$$g_{Ram} = (4\pi/3)\,[\sin 2\theta/(1 - \cos 2\theta)]\,(R_{21}/\lambda_{ex}) \tag{8.2}$$

and the corresponding Rayleigh scattering ratio takes the form,

$$g_{Ray} = -\,[4\pi\sin 2\theta/(5 + 3\cos 2\theta)]\,(R_{21}/\lambda_{ex}) \tag{8.3}$$

The infrared CD dissymmetry ratio for the chiral two-group structure (fig. 8.3), where one group absorbs at the characteristic group vibrational wavelength, λ_{vib}, and the other has the polarizability anisotropy, $(\alpha_{\parallel} - \alpha_{\perp}) = \beta(L)$ at that wavelength, is given by eqn 3.20 as,

$$g_{IR} = -\,2\pi\sin 2\theta\,[\,\beta(L)/(\lambda_{vib}R_{21}^2)\,] \tag{8.4}$$

Fig. 8.3. The two-group model for a comparison of the dissymmetry ratio of Rayleigh and Raman scattering optical activity with infrared absorption optical activity.

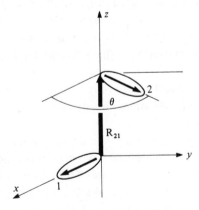

In a favourable case, such as a twisted biphenyl, the group polarizability may be so large that, $\beta(L) \sim R_{21}^3$, but the Raman dissymmetry ratio, g_{Ram}, is larger than the corresponding infrared CD absorption ratio, g_{IR}, by the approximate factor of ($\lambda_{vib}/\lambda_{ex}$) nonetheless. The same approximate factor favours g_{Ram} over g_{IR} for the particular case of the torsional vibration mode of the methyl group ($100 - 300$ cm^{-1}) in chiral organic molecules (Barron and Buckingham, 1979). The methyl torsion mode gives a strong CID scattering, with $g_{Ram} \sim 10^{-3}$, although the corresponding CD absorption remains as yet undetermined.

Both the infrared CD absorption and the Raman CID scattering of the methyl deformation mode near 1450 cm^{-1} have been measured for the series of (R)-(+)-isomers of the α-phenylethyl derivatives, Ph-CHX-CH$_3$, 6, where

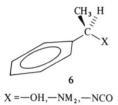

6

X = —OH, —NM$_2$, —NCO

X = —OH, —NH$_2$, or —NCO (Su and Keiderling, 1981). While the infrared CD absorption consists of a single band at 1450 cm^{-1}, negative for the (R)-(+)-isomers, with $g_{IR} \sim 10^{-4}$, the corresponding Raman CID scattering comprises a bisignate couplet, positive at 1460 cm^{-1} and negative at 1440 cm^{-1}, each with $g_{Ram} \sim 4 \times 10^{-4}$. Initially the Raman bisignate CD couplet centred on 1450 cm^{-1} had been ascribed to the split components of the doubly-degenerate antisymmetric methyl deformation mode, but Raman CID studies of aryl- and N-substituted α-phenylethylamine enantiomers (Barron, 1977, 1978) and of a series of deuterated (R)-(+)-1-methylindanes (Hug *et al.*, 1980) suggest that the feature arises from the coupling of the methyl group modes with the benzene ring modes. The interpretation is supported by the infrared CD absorption studies of the (R)-(+)-isomers, 6, in the 1450 cm^{-1} region (Su and Keiderling, 1981).

9 THE CHARACTERISATION OF CHIRAL STRUCTURES

9.1 Pasteur on dissymmetry in nature

From his discovery of the optical resolution of (±) -tartaric acid by salt formation with quinine and related alkaloids, Pasteur (1853) drew the significant and general conclusion that 'the absolute identity of the physical and chemical properties of left and right non-superposable substances ceases to exist in the presence of another active substance'. With the discovery of the chemical and physical differentiation of compounds containing two chiral units, (+)-A(+)-B and (-)-A(+)-B, one common to both and the other antipodal between the two substances, the term 'physical isomers' used to describe dextrorotatory and laevorotatory analogues, and the corresponding racemate, progressively became outmoded.

The term 'physical isomers' had arisen from the apparent identity, apart from the presence or absence of optical rotation, or the sign of the rotation, in substances such as (+) -tartaric acid and the corresponding racemic or paratartaric acid. The term survived with an added morphological significance after the earlier discovery by Pasteur (1848) of the hemihedral mirror-image facets in the two crystal types, corresponding to the (+)- and the (-) -isomers, formed by aqueous solutions of the sodium ammonium salt of (±) -tartaric acid at ambient temperature. The different morphologies of the two crystal types afforded an additional physical distinction between optical isomers. More significantly, the particular mirror-image form of the crystal-morphology difference provided Pasteur with the concept that optical isomers have non-superposable left- and right-handed structural forms, exhibiting at the molecular level the macroscopic dissymmetry of the corresponding crystals.

During his period at Strasbourg, 1850—54, Pasteur generalised the concept of dissymmetry to the forces of nature and the structure of the physical world as a whole. The earth, Pasteur argued, is inherently dissymmetric, on account of the diurnal rotation and the polarity of terrestrial magnetism. The same holds for the system of planets at large, since on 'placing before a mirror the group of bodies which compose the solar system, with their proper movements, we obtain in the mirror an image not superposable on the reality' (Vallery-Radot, 1885).

The combination of a rotation with a linear motion is expected to generate dissymmetry, Pasteur held, as appeared to be already exemplified by Faraday's discovery in 1845 of optical activity magnetically-induced in otherwise inactive media by the flow of an electric current through a helically-wound conductor (Faraday, 1846).

Pasteur investigated the postulated dissymmetry of the forces of nature, taken in appropriate combinations, by growing crystals, normally holohedral, in a magnetic field with the expectation that the magnetically-induced dissymmetry would be manifest in hemihedral crystal forms. Plants were grown under conditions of constant rotation, driven by a clockwork mechanism, and a motorised mirror-system was employed to present to plants grown under standard conditions of cultivation the aspect of a sun which rose in the west and set in the east. The plants, however, continued to produce their customary optical isomers, and the crystals grown in magnetic fields retained their usual holohedral forms.

From these experiments, Pasteur arrived at the view that natural optical activity provides a dividing line between synthetic and natural products and sets a bound to synthetic chemistry. Initially Pasteur held that synthetic chemistry could not proceed beyond *détordu meso*-forms, having isolated *meso*-tartaric acid in 1853 from the residues remaining after heating cinchonine tartrate at 170°C for several hours. After (±)-tartaric acid had been synthesised, indirectly from ethylene, by Perkin and Duppa in 1860, Pasteur adjusted his limit of synthetic chemistry to racemic and *meso*-forms.

Optical activity as the demarcation criterion between animate and laboratory chemistry was strengthened for Pasteur by his discovery in 1858 of the differential metabolism of the optical isomers in a racemic mixture of micro-organisms. The common mould, *Penicillium glaucum*, was found to employ specifically (+)-tartrate as a carbon source, in a solution of ammonium hydrogen racemate containing a little phosphate, leaving the (−)-isomer. The procedure provided Pasteur and his successors with a third method of optical resolution, although it was not without its hazards. In 1892 Le Bel attempted the optical resolution of citraconic, 1, and mesaconic acid, 2, employing micro-organisms, in order to

1   2

investigate his view that the valencies of carbon are not exactly regular-tetrahedral in orientation, so that the atoms of ethylene and its simple derivatives are not coplanar. Obtaining optically-active products, Le Bel initially believed that his view was established, but subsequently he found in 1894 that the apparently-positive result was spurious, presumably due to an enzymatic stereospecific addition of the elements of water across the double bond of the substrate acids.

The extension of organic stereochemistry to three dimensions by Le Bel and van't Hoff in 1874 gave an added emphasis to the chemical significance of optical isomerism, and brought to the fore Pasteur's distinction, based upon optical activity, between the chemistry of the laboratory and of organic metabolism. The vitalist school of thought (Japp, 1898) was strengthened by the development of organic stereochemistry, regarding optical activity as the product of a force unique to living organisms. According to the vitalist view, the primary expression of the unique force was the asymmetric carbon atom, and all optically-active molecules were expected to contain carbon. The vitalist challenge was countered by Werner (1914) who, in a *tour de force*, synthesised and optically resolved a purely inorganic tris-chelate complex, the dodecammine hexa-μ-hydroxotetra-cobalt (III) ion $[Co\{(OH)_2 Co(NH_3)_4\}_3]^{6+}$, 3, containing no carbon.

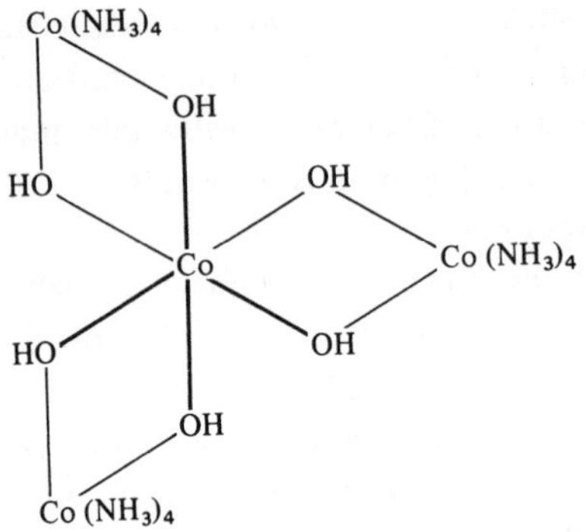

3

Pasteur too, in his later years, was antipathetic to the use made of his earlier views by the vitalist organic chemists. 'Not only have I not set up as absolute the existence of a barrier between the products of the laboratory and those of life', Pasteur wrote in 1884, 'but I was the first to prove that it was merely an artificial barrier, and I indicated the general procedure necessary to remove it, by recourse to those forces of dissymmetry never before employed in the laboratory'.

Forces of dissymmetry additional to those considered by Pasteur in the 1850s had been proposed during the intervening period. In his classic paper on the connection between optical activity and chiral organic structures, Le Bel (1874) suggested the use of left- or right-handed circularly-polarized radiation in photo-chemical reactions involving racemates, and of chiral catalysts in the corresponding thermal reactions, in order to produce an excess of a particular enantiomeric product. Subsequently van't Hoff indicated the possibility of dissymmetric photosynthesis employing circularly-polarized radiation and, on discovering the circular dichroism of chiral molecules in solution, Cotton (1909) attempted the photoresolution of the (±) -tartrate complex of copper (II) by circular irradiation into the absorption band in the red region of the visible spectrum. The attempt was not successful and the first unambiguous photoresolutions were achieved by Kuhn (1929) and coworkers, employing circular ultraviolet radiation for the differential photolysis of the enantiomers in racemic α-bromo- and α-azido-proprionate esters, **4**.

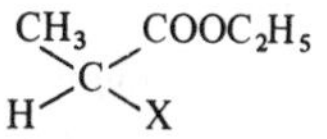

$$\begin{array}{c} CH_3 \quad COOC_2H_5 \\ \diagdown C \diagup \\ H \diagup \quad \diagdown X \end{array}$$

(X = Br or N₃)

4

The production of an enantiomeric excess in a racemic product by means of the dissymmetric agencies suggested by Le Bel, circularly-polarized radiation or a chiral catalyst, is well established (Izumi and Tai, 1977). The possibility of generating chiral molecules through the application of a dissymmetric force field, corresponding to the appropriate combination of the rotation and translation of a charge or a mass, as envisaged by Pasteur, remains conjectural however. The reaction of isophorone, **5**, with hydrogen peroxide under basic conditions in the

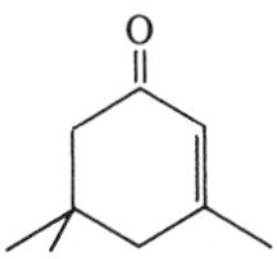

5

presence of an electric and magnetic field has been reported to yield, unsystematically, an enantiomeric excess in the product isophorone oxide, **6**. The same reaction, carried out in a centrifuge with the axis of rotation collinear to the earth's gravitational field, has been reported to yield an excess of the (+)-isomer of **6** if the rotation, viewed from above, is clockwise, or of the (−)-isomer for

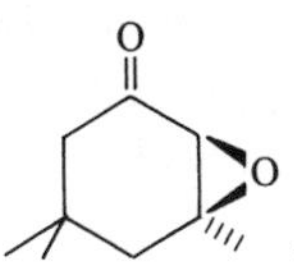

6(a) 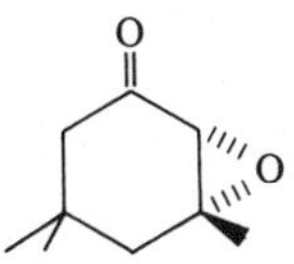   6(b)

an anticlockwise rotation sense (Dougherty, 1980; Dougherty, Edwards and Cooper, 1980). The positive result of both the electric and magnetic field experiment, and the centrifuge-gravitational field experiment, have been questioned on theoretical grounds, and on the practical consideration that the possible order of magnitude of the dissymmetric forces in either experiment is far too small to account for the extent of the enantiomeric excess reported (Mead and Moscowitz, 1980).

The view of Pasteur, that the spherical earth rotating in the reference frame of the solar system expresses a macrocosmic dissymmetry, has a more recent microcosmic atomic and molecular analogue. Two molecules in the gas phase, even those of the high-symmetry spherical top category, such as methane, form two types of chiral pair, distinguished by the rotation of the molecules in the same sense or in contrary senses. The polarizability of one molecule is sensitive to the rotating field due to its neighbour, and an assembly of co-rotating molecules is expected to have an optical activity differing from that of an assembly of counter-rotating pairs (Atkins, 1980). The role of the molecular rotation is taken over by that of electron spin in gaseous atoms containing unpaired electrons, such as the alkali metals. In a non-bonded, but proximate, pair of alkali-metal atoms, the polarizability of one atom is dependent upon the electron-spin state of its neighbour. The atom pair has one doubly-degenerate and two non-degenerate spin-orbital states, which are distinguished by the dispersion energy between the two atoms, dependent upon the atomic polarizability. The ratio of the discrimination energy to the average dispersion energy is estimated at 1.3×10^{-4} for a pair of caesium atoms, compared to an estimate of $\sim 10^{-9}$ for the corresponding discrimination ratio between co-rotating and counter-rotating methane molecules (Buckingham and Joslin, 1981).

9.2   Enantiomer–racemate phase equilibria

The investigation of the phase relationships between a racemate and the corresponding enantiomers was initiated by van't Hoff (van't Hoff and van Deventer, 1887) and extended by his successor at the University of Amsterdam, Roozeboom (1899). From measurements of the aqueous solubility of active and racemic tartrate salts as a function of temperature, van't Hoff found a number of cases of spontaneous optical resolution on crystallisation in the series, additional to the instance of the sodium ammonium salt discovered by Pasteur. The temperature of the transition from the formation of racemic crystals to a conglomerate of enantiomorphous active crystals is given by the intersection point of the curves relating the solubility to the temperature of the active and the corresponding racemic salts (fig. 9.1).

In all cases, the salt crystallising above the transition temperature proves to be less hydrated than the crystal formed below that temperature, exemplifying the more general rule of van't Hoff (1899) that the less-solvated salt is the more stable at the higher temperature. The active sodium ammonium and sodium potassium tartrates form tetrahydrates, whereas the corresponding racemates are a monohydrate and a trihydrate, respectively, and a conglomerate of active crystals is afforded by crystallising the racemate below the respective transition temperatures of +27° and −6°C. In contrast, the active dipotassium and dirubidium tartrate crystals are a hemihydrate and anhydrous, respectively, whereas the corresponding racemates are both dihydrates, and solutions of the latter crystallise as a conglomerate of active crystals above the respective transition temperatures of +71.8° and +40.4°C (fig. 9.1).

Roozeboom investigated the solid–liquid equilibria of enantiomeric mixtures as a function of temperature, after his attention had been drawn to the phase rule of Willard Gibbs (1876) by his colleague, van der Waals, in 1886. Three principal types of solid–liquid phase diagram for enantiomeric mixtures were distinguished by Roozeboom (1899). The first type (fig. 9.2) covers the cases where active crystals are formed from the melt whatever the enantiomeric ratio in the liquid phase, so that the eutectic, with the racemic composition of a 0.5 mole fraction in each enantiomer, forms a conglomerate of active crystals, each

Fig. 9.1. The relation between the temperature (°C) and the solubility in water (mole per 100 mole of solvent) of dirubidium tartrate for the racemic dihydrate (R) and the anhydrous optically-active salt (A). After van't Hoff (1899).

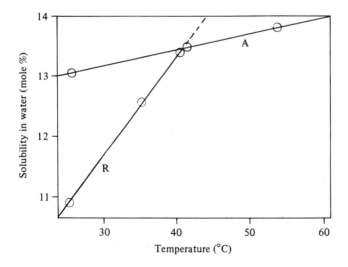

Fig. 9.2. The phase diagram relating the enantiomer composition (mole fraction, x) to the onset of fusion (lower line) at the melting point of the eutectic racemic mixture, T_E equal to T_R, and the termination of fusion (upper relations), for an enantiomeric mixture which is spontaneously resolved into a conglomerate of active crystals by crystallisation from the melt. In the case of [6]-helicene, $T_A = 270°$ and $T_E = T_R = 240°C$ (Newman, Darlak and Tsai, 1967).

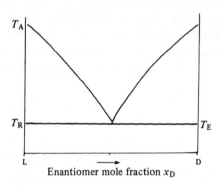

Enantiomer mole fraction x_D

Fig. 9.3. The phase diagram relating the enantiomer composition (mole fraction, x) to the onset of fusion at the eutectic temperature (lower line, T_E), and the termination of fusion (upper relations) for an enantiomeric mixture forming a true racemate on crystallisation from the melt. The sub-cases are distinguished by (a) a racemate with a higher melting point than the corresponding enantiomers, e.g. dimethyl tartrate, $T_R = 89.4°$ and $T_A = 43.3°C$, and (b) a racemate with a lower melting point than the corresponding enantiomers, e.g. mandelic acid, $T_R = 118.0°$ and $T_A = 132.8°C$ (Findlay, 1951).

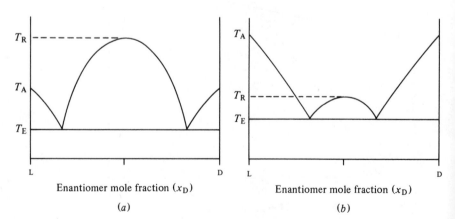

Enantiomer mole fraction (x_D)

(a)

Enantiomer mole fraction (x_D)

(b)

crystal containing a single optical isomer. The second type refers to the majority of racemates. Here the melt with a racemic composition deposits only a 'racemic compound', forming crystals which contain the two enantiomers in an equimolecular ratio (fig. 9.3). The racemic compound may have a melting point, T_R, higher than that of the corresponding single enantiomer, T_A (fig 9.3a) or a lower melting point (fig. 9.3b), so that the condition, $T_A > T_R$, for spontaneous optical resolution by crystallisation from the melt, while necessary (fig. 9.2), is not sufficient. The conclusion is reinforced by the third main type of solid–liquid phase diagram, which covers the minority of cases where the two enantiomers form solid solutions continuously as a function of composition (fig. 9.4). The subdivisions of the third type refer to ideal solid solutions (fig. 9.4a) and to positive (fig. 9.4b) and negative (fig. 9.4c) deviations from ideality.

The general phase diagram of the first type (fig. 9.2) was analysed thermodynamically by Schröder (1893), and the analysis was extended by van Laar

Fig. 9.4. The phase diagram relating the enantiomer composition (mole fraction, x) to the onset and the termination (lower and upper relations, respectively) for enantiomeric mixtures forming solid solutions. The sub-cases refer to (a) ideal behaviour, e.g. camphor, $T_A = T_R = 178°C$, (b) positive deviation, e.g. carvoxime, $T_A = 72°$ and $T_R = 91.5°C$, and (c) negative deviation from ideality, e.g. o-chlorophenyliminocamphor, $T_A = 123°$ and $T_R = 113°C$ (Singh and Perti, 1963).

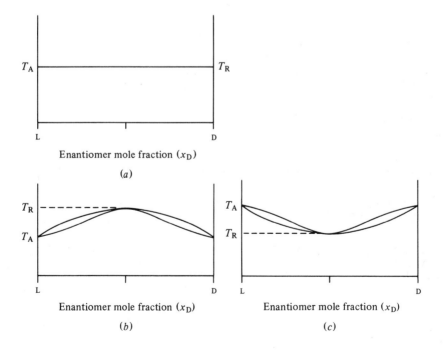

(1908), to show that an enantiomeric mixture with a mole fraction, x, of the major isomer terminates fusion at a temperature T (K) governed by the relation,

$$ln\ x = \Delta H_A\ [(1/T_A) - (1/T)]/R \tag{9.1}$$

where ΔH_A and T_A (K) are the enthalpy and temperature of fusion of a pure enantiomer, and R is the ideal gas constant. An additional term enters eqn 9.1 if the heat capacity of the liquid and the solid differ appreciably, but the additional term is generally negligible in relation to the uncertainties introduced by the assumptions upon which the equation is based. These assumptions are, firstly, the ideal behaviour of the enantiomer mixture in the liquid state and, secondly, the immiscibility of the enantiomers in the solid state. The application of eqn 9.1 to the third general type of enantiomer phase diagram (fig. 9.4) covering solid solutions is thus excluded.

The Schröder-van Laar equation (eqn 9.1) refers additionally to each branch of the liquidus curve in the second type of phase diagram (fig. 9.3) between the pure enantiomer and the adjacent eutectic composition. For the intermediate region between the two eutectic compositions an analogous expression (Prigogine and Defay, 1967) governs the relation between the enantiomer composition, in terms of the mole fraction, x, and the temperature, T (K), of the termination of fusion,

$$ln\ 4x(1 - x) = 2\Delta H_R\ [(1/T_R) - (1/T)]/R \tag{9.2}$$

where ΔH_R and T_R (K) are the enthalpy and temperature of fusion of the racemic compound, respectively. The assumptions underlying eqn 9.2 are again the ideality of the liquid phase, the immiscibility of the solid phases and the negligible contribution of heat capacity differences.

The analysis of the enantiomer mixture phase diagrams of the first and second principal type (figs. 9.2 and 9.3) in terms of eqns 9.1 and 9.2 afford values of ΔH_A and ΔH_R which are generally in good accord with the corresponding calorimetric values. The assumption that the enantiomers in the liquid phase exhibit ideal behaviour, upon which eqn 9.1 is based, implies that the enthalpy of mixing goes to zero and that the entropy of mixing, ΔS_m, is purely statistical,

$$\Delta S_m = -\ R[x_D ln x_D + x_L ln x_L] \tag{9.3}$$

where x_D and x_L refer to the respective mole fractions of the dextrorotatory and the laevorotatory isomers. The entropy of mixing of equal mole fractions of the liquid isomers at the racemic composition accordingly is expected to have the value $R \ln 2$ (5.76 J mol^{-1} K^{-1}). The entropy of mixing for liquid enantiomers forming a conglomerate of active crystals (fig. 9.2) is found to approximate to $R \ln 2$ from measurements of the enthalpy and the temperature of fusion of a single enantiomer and of the racemic mixture (Leclerq, Collet and Jacques, 1976). The analogous conclusion for liquid isomers forming a racemic compound on crystallisation follows less directly from the congruence between the corresponding phase diagram (fig. 9.3) and eqn 9.2, which is based upon the assumption of an ideal liquid phase.

The enthalpy of mixing is generally small ($\sim$ J mol^{-1}) for liquid enantiomers forming either a conglomerate of active crystals or a racemic compound on crystallisation. In the latter case, the enthalpy and the entropy of fusion of both the racemic compound and one of the constituent enantiomers are accessible directly from calorimetric measurements. The thermochemical measurements give the free energy of formation of the racemic compound crystals from an equimolecular mixture of the single-enantiomer crystals, $\Delta G^{\phi}(T_m)$, at the melting point, T_m, of the active crystal, T_A, or the racemic crystal, T_R, whichever is the lower. For an extensive set of enantiomers belonging to the second main class (fig. 9.3), the free energy of formation of the racemic from the active crystals, $\Delta G^{\phi}(T_m)$, is found to lie in a range from zero to -8.8 kJ mol^{-1}, with a mean value of -3.2 kJ mol^{-1} over 37 cases (Leclerq *et al.*, 1976).

The free energy of formation, $\Delta G^{\phi}(T_m)$, is proportional to the difference between the melting point of the racemic and the corresponding active crystals, $(T_R - T_A)$, following the approximate relation,

$$\Delta G^{\phi}(T_m) = \Delta S_m(T_R - T_A) - T_m R \ln 2 \tag{9.4}$$

where ΔS_m is the constant of Walden (1908) for the entropy of fusion of organic molecular crystals ($\sim$ 70 J mol^{-1} K^{-1}). The empirical relation between the free energy, $\Delta G^{\phi}(T_m)$, and the melting-point difference, $(T_R - T_A)$, in eqn 9.4, suggests the probability of spontaneous optical resolution on the crystallisation of a racemate from the melt if the melting point of the enantiomer lies $\geqslant 30°$C above that of the corresponding racemate (Leclerq *et al.*, 1976).

The crystallisation of enantiomers from solution, rather than from the melt, introduces an added degree of variability, on account of the additional component, the solvent. At a given pressure, the thermal behaviour of the solid-liquid equilibrium in a melt is uniquely determined, but temperature becomes a variable of choice in the corresponding crystallisation from solution. While the phase diagrams for an enantiomeric mixture and solvent fall into one of three principal classes, corresponding to the three main types of melt-diagram, it is occasionally possible to transfer a given solution system from one main class to another by changing the temperature, as van't Hoff showed for a number of tartrate salts.

The first main type of ternary phase diagram (fig. 9.5) covers the optical isomers which crystallise from solution as a conglomerate of active crystals over the whole enantiomer-composition range. At a given temperature, the solubility of the racemic and other mixed compositions is larger than that of the individual enantiomers. Neutral and charge-compensated enantiomers which are spontaneously resolved on crystallisation approximate to the double-solubility rule of Meyerhoffer (1904), according to which the solubility of a racemate is twice as large as that of the corresponding enantiomers under the same conditions. The double solubility rule refers to ideal solubility behaviour and to effectively neutral species, where the solubility and supersaturation are related to the mole fraction composition by lines parallel to the solvent–enantiomer sides of the triangular ternary phase diagram (fig. 9.5a).

The solubility ratio is smaller for the salts formed by optically-active acids or bases with achiral counter-ions, due to the common-ion effect of the counter-ion. For a uni-univalent electrolyte, which is completely dissociated in solution and forms a conglomerate of active crystals, the ratio of the solubility of the racemate to that of an enantiomer is $2^{1/2}$ and, in the general case where the achiral counter-ion carries n-charges, the corresponding ratio becomes $2^{1/(n+1)}$ (Yamanari, Hidaka and Shimura, 1973).

Fig. 9.5. The ternary phase diagram for the solid–liquid equilibrium at a given temperature between a solvent (S) and a mixture of enantiomers (L and D), which crystallise from the solution as a conglomerate of active crystals. The full lines refer to the thermodynamic equilibrium and the broken lines to the metastable supersaturation in the case (a) of neutral or charge-compensated species and (b) of a uni-univalent electrolyte composed of a chiral ion and an achiral counter-ion. A single enantiomer, (L) or (D), has the same mole-fraction solubility in the two cases. After Collet *et al.*, (1980).

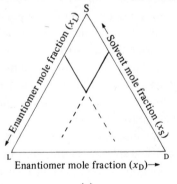
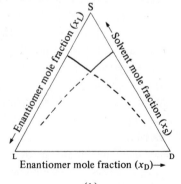

(*a*)   (*b*)

In the ternary phase diagram, the relations between the solubility and the supersaturation of the chiral salts and the enantiomer–solvent composition are no longer linear and parallel to the solvent–enantiomer edges of the ternary diagram (fig. 9.5*b*). For a neutral enantiomer and a comparable enantiomeric salt, with the same solubility in a given solvent at a particular temperature, the area of the phase diagram referring to supersaturation is the larger for the salt than the neutral species (figs 9.5*a*, *b*). More particularly, the region of the super-saturation area of the phase diagram relating to the crystallisation of a single enantiomer, from a solution with a racemic or near-racemic composition following seeding with a particular enantiomer crystal, is the more extensive for an enantiomer salt than the corresponding neutral species. Thus the optical resol-ution of a racemate giving a conglomerate of active crystals, by seeding a super-saturated solution of the racemate with a crystal of one of the optical isomers, is the more facile for an enantiomeric salt than the corresponding neutral species (Collet, Brienne and Jacques, 1980).

The second main class of ternary phase diagram refers to the majority of ra-cemic substances, which crystallise from solution as racemic compounds, each crystal being equimolecular in the isomer composition (fig. 9.6). In general, the ratio of the solubility of the racemic compound to that of the corresponding enantiomer in a given solvent at a particular temperature is less than the theor-

Fig. 9.6. The ternary phase diagram for the solid–liquid equilibrium at a given temperature between a solvent (S) and a mixture of enantiomers (L and D), which crystallise from the solution as a true racemate (R). The tie-lines (broken) relate a particular solution to the solid with which it is in equilibrium. The sub-classes refer to the case where the racemate is (*a*) less soluble and, (*b*) more soluble, than the correspond-ing enantiomer. The sub-classes are the respective analogues of the cases (*a*) and (*b*) or the corresponding melt phase diagram (fig. 9.3).

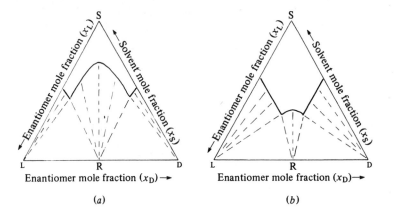

Enantiomer mole fraction (x_D) →

(*a*)

Enantiomer mole fraction (x_D) →

(*b*)

etical value for the conglomerate case, $2^{1/(n+1)}$, where n is the number of charges carried by any achiral counter-ion, or zero for a neutral species. The solubility ratio becomes substantially less than unity in the cases where the formation of racemic crystals from the corresponding enantiomeric crystals is energetically favoured to the degree, $T_R > T_A$ (eqn 9.4).

For neutral species, an approximate measure of the free energy of formation of the racemic crystals from the corresponding active crystals may be obtained from the solubility ratio, (S_R/S_A), through the relation,

$$\Delta G^{\phi}(T) = RT[ln(S_R/S_A) - ln2] \tag{9.5}$$

In the case of a α-naphthyl-2-propionic acid, the enantiomer is some 50 times more soluble in benzene at 25°C than the racemate, corresponding to a free energy of -11.4 kJ mol^{-1} (eqn 9.5), compared with the value of -7.1 kJ mol^{-1} at the melting point of the enantiomer (69°C), determined thermochemically, or the corresponding value of -7.7 kJ mol^{-1}, given by eqn 9.4 (Leclerq *et al.*, 1976).

The third main class of ternary phase diagram covers the co-crystallisation of optical antipodes from solution over part or the whole of the composition range to give solid solutions (fig. 9.7). For ideal solid solutions of optical antipodes the solubility is independent of the enantiomer composition and the ratio of the solubility of the racemate to that of an individual isomer is unity (fig. 9.7a), with a smaller or a larger ratio for negative or positive deviations, respectively, from ideal behaviour (fig. 9.7b, c).

Fig. 9.7. The ternary phase diagram for the solid–liquid equilibrium at a given temperature between a solvent (S) and a mixture of enantiomers (L and D), which crystallise as solid solutions. The tie-lines (broken) relate a particular ternary solution to the corresponding solid binary solution with which it is in equilibrium. The sub-classes refer to (*a*) ideal behaviour, (*b*) positive deviation, and (*c*) negative deviation from ideality. These are the respective analogues of the cases (*a*), (*b*) and (*c*) of the corresponding melt phase diagram (fig. 9.4).

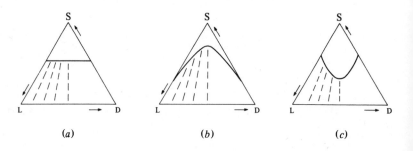

(*a*)   (*b*)   (*c*)

Optical isomers forming solid solutions over a part or the whole of the composition range are relatively uncommon, but they are well represented among the chiral molecules which give plastic or rotationally-disordered crystals, such as camphor. The general criteria for the formation of a solid solution are, firstly, the isomorphism of the crystal structures of the two components and, secondly, the isosterism, or similarity of shape and volume, of the molecules of the two species. Optical antipodes which rotate at a lattice site in a crystal, or are rotationally disordered and switch from one orientation in the lattice to another, have a virtually complete isosterism. The molecular rotation confers an effective spheroidal shape upon each of the enantiomeric molecules and ensures a crystallographic isomorphism, based upon the close-packing of spheres, over the whole composition range.

The enantiomers of camphor derivatives and of other chiral cage-molecules, frequently form solid solutions consisting of a rotationally-disordered phase on crystallising from the melt or from solution, and the corresponding phase diagrams often approximate to ideal behaviour (figs 9.4a and 9.7a). Mixtures of (+)- and (-)- camphoric anhydride, 7, have a melting point (222°C) invariant with respect to the enantiomeric composition (fig. 9.8). Between the melting

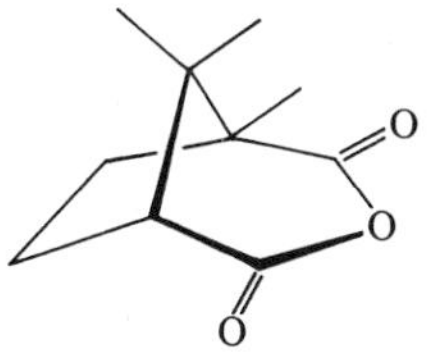

7

point and the transition temperature from the rotator phase to the ordered-solid phase, the solid solution has a hexagonally close-packed structure with a volume of 268 Å^3 per molecule, independent of the enantiomeric composition. Below the transition temperature to the ordered solid phase, (+)-camphoric anhydride crystallises in the orthorhombic system (space group $P2_1 2_1 2_1$), while the racemate crystallises in the monoclinic class (space group $P2_1/c$), the volume per molecule being 230 and 236 Å^3, respectively (Mjojo, 1979).

The enantiomer is slightly more closely packed than the racemate in the ordered phase, where both are rather more economically packed than in the hexagonal rotator phase. The equilibrium between the plastic rotator phase and the ordered solid phase of 7 as a function of enantiomeric composition (fig. 9.8) has a form analogous to the limiting or intermediate case between the first two Roozeboom types (figs 9.2 and 9.3b), while the corresponding solid–liquid equilibrium relation of 7 exemplifies the ideal third type (fig. 9.4a).

Fig. 9.8. The binary phase diagram of D- and L-camphoric anhydride, relating the enantiomer composition (mole fraction, x) to the melting point of the hexagonal close-pack rotator phase, invariant at 495 K (upper line), and to the equilibria between the rotationally disordered solid phase and the active orthorhombic and racemic monoclinic ordered phase (lower relations). After Mjojo (1979).

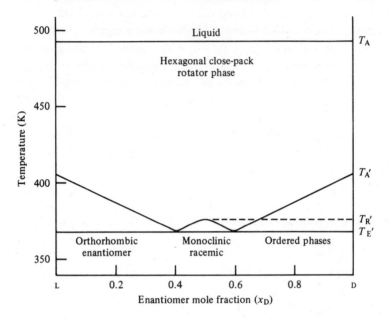

9.3 Phase-equilibria methods for relative configuration

An important qualitative application of the investigation of phase equi-libria is the correlation of stereochemical configuration. Following the proposal of Fischer (1891) that chiral substances related by chemical interconversions to (+)- or (−)-glyceraldehyde have the D- or the L-configuration, respectively, Winther (1895) formulated the solubility rule that 'two acids precipitated by the same base have the same configuration', and, equally, that two chiral bases forming a less-soluble diastereomer with a common chiral acid are configur-ationally related in the Fischer convention. The configurational correlations based upon the solubility rule proved to be consistent with those based upon chemical interconversions, or upon the optical activity, for many chiral organic acids and bases, but a number of exceptions and ambiguous cases were discovered. Crystals of apparently-related less-soluble diastereomers may contain different numbers of solvent molecules in the lattice, and a diastereomeric salt less soluble in one solvent may become the more-soluble salt in another.

Physical methods of correlating stereochemical configuration were of particu-

lar importance in the series of the chiral coordination compounds of the transition-metal ions. Compared to the organic field, the range of reactions relating
one chiral metal complex to another stereochemically is limited, and some of
these involve a configurational inversion. Werner (1912a) suspected that a Walden
inversion occurred in the reaction of the bis-ethylenediamine complex, $(-)$-*cis*-
$[Co(en)_2 Cl_2] Cl$, with potassium carbonate to give the $(+)$-isomer of the
carbanato-product, $[Co(en)_2 CO_3] Cl$. The conjecture was confirmed by Bailar
and Auten (1934) who, carrying out the corresponding reaction with silver
carbonate, obtained the $(-)$-isomer of the product.

   Following Winther, Werner (1912b) proposed the rule that, in a series of
chiral complex ions, the enantiomers forming the less-soluble diastereomeric
salt with a common chiral counter-ion have the same stereochemical configuration. However, he himself obtained each enantiomer of the trisoxalato complex,
$[Cr(ox)_3]^{3-}$, as a less-soluble diastereomer with $(-)$-strychnine under different
conditions (Werner, 1912c). A more restricted and reliable form of the solubility
rule for the correlation of stereochemical configuration in the transition-metal
complex series limited comparisons to isomorphous less-soluble diastereomers,
determined by the crystal morphology (Jaeger, 1919, 1930).

   The related method of active racemates, introduced by Delépine (1921), was
based similarly upon the comparison of a series of less-soluble isomorphous crystals. Two different chiral complex ions of the same charge-type and similar in
structure, such as those of the tris-ethylenediamine series, $[M(en)_3]^{3+}$, or the
tris-oxalato series, $[M(ox)_3]^{3-}$, form mixed crystals with an achiral counter-ion.
The heterochiral double salt, such as $(+)$-$[Cr(en)_3](+)$-$[Rh(en)_3] Cl_6 .6H_2 O$, is
less soluble than the corresponding homochiral salt and is isomorphous with the
corresponding racemates, $(\pm)$-$[Cr(en)_3] Cl_3 .3H_2O$ and $(\pm)$-$[Rh(en)_3] Cl_3 .3H_2O$.

   In the heterochiral active racemate, the crystal lattice of the corresponding
racemate is preserved, apart from the inversion centre and other secondary
symmetry elements, and the complex ions of one metal are replaced by complex
ions with the same configuration of the other metal. Delépine (1934),
Delépine and Charonnat (1930) and Jaeger (1937) established the isomorphism of
a heterochiral active racemate and the corresponding racemates by comparing the
X-ray diffraction powder patterns. The comparisons gave a common stereochemical configuration for $(+)$-$[Co(en)_3]^{3+}$, $(+)$-$[Cr(en)_3]^{3+}$ and $(-)$-$[Rh(en)_3]^{3+}$
in the ethylenediamine series, and for $(+)$-$[Co(ox)_3]^{3-}$, $(-)_{546}$-$[Rh(ox)_3]^{3-}$,
and $(-)$-$[Ir(ox)_3]^{3-}$ in the oxalato series. More recently, $(-)_{350}$-$[Ru(en)_3]^{3+}$
has been added to the former series by the method of active racemates (Elsbernd
and Beattie, 1969).

   Applied to chiral organic substances, the method of active racemates, or quasi-
racemates as they were termed by Fredga (1944), was extended to the determi-

nation of the phase diagrams of corresponding homochiral and heterochiral mixtures as a function of temperature (Timmermans, 1929). Two related chiral compounds give a relatively-stable quasi-racemic association if the stereochemical configurations are enantiomeric, and the solid–liquid phase diagram connecting composition with temperature has the Roozeboom type IIa form (fig. 9.3*a*) with $T_R > T_A$. If the chiral compounds have a common configuration, the intermolecular association formed at the equimolecular composition is relatively unstable and the corresponding phase diagram has the form of type IIb (fig. 9.3*b*) with $T_A > T_R$, or the association is so weak that a eutectic or solid solution results, with a phase diagram of the first (fig. 9.2) or third (fig. 9.4) type. The configurations of a wide range of chiral dicarboxylic acids have been correlated by the quasi-racemate method (Fredga, 1960). The method is of particular utility for chiral species devoid of acidic or basic groups, where the criterion of the less-soluble diastereomeric salt is inapplicable, such as the *trans*-1, 2-diaryloxirans, for which the quasi-racemate procedure gives relative configurations consistent with the absolute configurations determined by CD spectroscopic analysis (Gottarelli and Samori, 1972).

9.4 Enantiomer and racemate crystal structures

The racemates that crystallise from the melt or from solution as a conglomerate of active crystals, while not common, are more frequent than those which form solid solutions of the enantiomers. Some 250 cases of the spontaneous resolution of racemates on crystallisation have been listed (Collet *et al.*, 1980), compared with the ~ 60 cases identified of chiral solid solutions (Chion *et al.*, 1978). The frequency of spontaneous optical resolution on crystallisation is larger, by a factor of two or more, for racemic salts than neutral racemates. A survey of 1308 neutral racemates, taken from ten volumes of Beilstein, reveals 83 cases of crystallisation into a conglomerate of active crystals, placing the frequency, F, of spontaneous resolution within the percentage bounds, $5.0 < F < 7.6$, at the 95% confidence level. In contrast, the corresponding percentage bounds for the spontaneous resolution of racemic salts on crystallisation are, $7 < F < 23$, based upon the identification of 14 conglomerates of active crystals in a set of 94 crystalline salt racemates (Jacques, Leclerq and Brienne, 1981).

Only a minority of the crystallographic lattice symmetries are available for the homochiral assemblies of single enantiomers. Of the 230 space groups, there are 66 which accommodate homochiral sets of optical isomers, or 55 with different X-ray diffraction patterns, if each of the enantiomorphous pairs is counted singly. Racemates normally crystallise in one of the remaining 164 space groups but they are unrestricted and, like achiral molecules, racemates occasionally crystallise in one of the enantiomorphous lattice systems. The range of available space groups is not, however, an important factor governing the low incidence of the spontaneous optical resolution of racemates on crystallisation.

A survey of some 5000 X-ray structure determinations of homomolecular crystals carried out up to 1975 (Belsky and Zorkii, 1977) shows that neutral molecules crystallise predominantly in the systems of low symmetry, mainly monoclinic and orthorhombic (table 9.1). Of the 219 space groups with distinctive X-ray diffraction patterns, only 89 are represented in the collection, and over one third of all the structures belong to the monoclinic space group, $P2_1/c$. The six most common space groups cover 83% of all the crystal structures studies (table 9.2).

A chirality classification of crystal structures distinguishes between (A) homochiral, (B) heterochiral and (C) achiral lattice types (Zorkii, Razumaeva and Belsky, 1977). In the type A structure, the molecules occupy a homochiral system or system of equivalent lattice positions. Secondary symmetry elements are precluded in the type A lattices, i.e., inversion centres, mirror or glide planes, or the higher-order inversion axes. In the racemic type B lattice, the molecules

Table 9.1. *The distribution of homomolecular crystals over the crystal systems, from Belsky and Zorkii (1977)*

Crystal system	Number of Structures	%
Monoclinic	2794	55.9
Orthorhombic	1562	31.2
Triclinic	478	9.6
Tetragonal	92	1.8
Hexagonal and trigonal	68	1.4
Cubic	8	0.1

Table 9.2. *The most common space groups of molecular crystals, based upon a survey of some 5000 crystal structure determinations, from Belsky and Zorkii (1977). The chiral crystal structure types are (A) homochiral, (B) heterochiral and (C) achiral systems of equivalent molecular positions in the lattice*

Space group	Type	Number	%
$P2_1/c$	B,C	1897	37.9
$P2_1 2_1 2_1$	A	839	16.8
$P\bar{1}$	B,C	449	9.0
$P2_1$	A	418	8.4
$C2/c$	B,C	310	6.2
Pbca	B,C	247	4.7
$Pna2_1$	B	120	2.4
Pnma	C	94	1.9

occupy heterochiral systems of equivalent positions, and antipodal optical isomers are related by secondary lattice symmetry operations. In type C, the molecules occupy achiral systems of equivalent positions, and each molecule is located either on an inversion centre, or on a mirror plane, or on a special position of a higher-order inversion axis. If there are two or more independent sets of equivalent positions in a crystal lattice, a fourth type, D, becomes feasible, with one set of type B and another of type C. The type D lattice is, however, rare. Of the 5000 crystal structures studied, the distribution by chiral lattice type is: A 28.4%, B 55.6%, C 15.7%, and D 0.3%. A large number of racemates (51.4%) crystallise in the type B structural class, $P2_1/c$, with four molecules in the unit cell ($Z = 4$), and many enantiomers (56.1%) in the type A lattice, $P2_1 2_1 2_1$, also with $Z = 4$ (Belsky and Zorkii, 1977). Another survey of the distribution of racemic and single-enantiomer crystals over the space groups gives a similar distribution (Cesario *et al.*, 1978).

The prevalence of low-symmetry crystal systems and space groups among molecular crystals is an expression of a substantial loss of point-group molecular symmetry at a lattice site on crystallisation. Centrosymmetric molecules generally occupy a lattice site with inversion symmetry in the crystal, e.g. 92% of 387 cases studied (Belsky, 1974). Molecules with a twofold rotation symmetry axis occupy a lattice site with the same symmetry in the crystal at the 43% level, but dihedral molecules with D_p point-group symmetry lose the principal rotation axis C_p and all but one of the p twofold axes at the lattice site in 80% of the crystals investigated. Only 10% of molecules with planar symmetry, belonging to the point-group C_s, lie on lattice mirror planes in the corresponding crystal. Molecules with C_{2v} point-symmetry retain at a lattice site one of the two mirror planes in 12% of the cases, and the two-fold rotation axis in 18%, but the site symmetry corresponds to the full point symmetry in only five of the 305 crystals studied (Belsky, 1974).

The basis of the high frequency of some space groups in molecular crystals and the virtual absence of others has been investigated by Kitaigorodsky (1973) in terms of the close packing of arbitrarily-shaped molecules into a unit cell of minimum volume. The model is essentially geometrical, and each molecule is taken to have a hard boundary surface with the crystal structure sustained by orientation-insensitive intermolecular forces. Kitaigorodsky defines for molecular crystals a packing coefficient, or compacity, given by the ratio of the volume occupied by the molecules in a unit cell to the unit-cell volume. The observed packing coefficient is generally quite large (0.65 – 0.77) in organic crystals, comparable to that obtained by the close packing of spheres and ellipsoids. The packing efficiency is due to the mutual fitting of complementary irregularities in the boundary surfaces of nearest-neighbour molecules in the crystal. An analysis of the symmetry relations between irregularly-shaped molecules for close packing, first into layers, and then into three-dimensional lattices, results in the conclusions of table 9.3.

Close-packed layers of molecules, with an arbitrary shape in a flatland projection, are not possible in the two-dimensional plane groups which contain mirror-symmetry or rotation axes of order higher than two. In a close-packed plane layer, each molecular projection is in contact with six of its neighbours. The coordination number of six is generally feasible in the plane group, p1, where the projected molecular shapes are related by two primitive translations in the layer, and in the plane groups, p2, pg, and pgg, where the projected shapes are additionally related by a twofold rotation axis, 2, and by one and by two glide-lines, g, respectively (fig. 9.9).

The stacking of layers, with a coordination number of six to achieve close packing in three dimensions with a coordination number of twelve, requires a translation at an oblique angle to the layer plane, or a twofold screw-axis, a glide-plane, or an inversion centre, relating the neighbouring layers. A mirror-plane relation between the layers, like the corresponding relation in a layer, does not allow economic packing. The only molecular symmetry element preserved in a close-packed crystal structure is a centre of inversion. The retention of other elements of molecular symmetry in the crystal lattice results in less economic packing and it reduces the packing coefficient by some 0.02 to 0.03. The efficient packing of molecules with symmetry elements other than an inversion centre generally entails a reduction in the intrinsic molecular symmetry, the site symmetry in the crystal being lower than that of the molecular point group. With the restriction that molecular symmetry is retained at the crystal lattice site, Kitaigorodsky has derived those space groups permitting the optimal packing of homochiral and racemic assemblies of chiral molecules with C_2 ($\equiv 2$) and D_2 (a) After Kitaigorodsky (1973).

Table 9.3. *Closest-packed space groups, and the space groups of maximum density for homochiral and racemic molecular crystals*[a]

Molecular site symmetry		Closest-packed space groups	Optimally-packed space groups
Homochiral	1	$P2_1$, $P2_12_12_1$	
	2		$P2_12_12$
	222		$C222$, $F222$, $I222$
Racemic	1	$P\bar{1}$, $P2_1/c$, $Pca2_1$, $Pna2_1$	
	2		$Pbcn$, $C2/c$
	222		$Ccca$

(a) After Kitaigorodsky (1973).

The molecules of the majority of organic crystals are of low intrinsic symmetry and they often have a degree of conformational lability which allows the mutual accommodation of the molecules into a close-packed crystal lattice. Some 85% of the molecular crystals with a known structure have a closest-packed lattice, including conformationally-constrained molecules of relatively high-point sym-

Fig. 9.9. Two-dimensional close packing in the plane groups (a) p1, (b), p2, (c) pg and (d) pgg, and the limitation to close packing imposed by mirror symmetry in the plane group (e) pm. After Kitaigorodsky (1973).

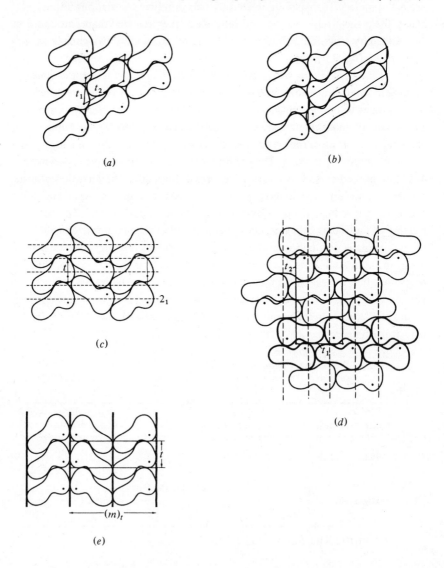

metry, as well as labile molecules with low symmetry. In the former case the higher-symmetry lattices become less infrequent, particularly in the ionic crystals.

The crystal symmetry elements forbidden in homochiral lattices but allowed in racemic structures are significant, particularly taken in conjunction with the close-packing criterion, for the prevalence of true racemates, rather than con-glomerates of active crystals, among the products from the crystallisation of enantiomeric mixtures. The inversion centre and the glide plane, both conducive to close-packing, are forbidden in homochiral crystal lattices, together with mirror planes and inversion-axial symmetry. The presence of these crystallographic symmetry elements implies that a molecule at a special position in the lattice contains the corresponding element in its point group, and is necessarily a *meso* form. If the molecule lies at a general position in the lattice, these crystallographic elements require an antipodal molecule related by a glide or mirror plane, or an inversion axis, giving thereby a racemic structure. The twofold screw-axis is the principal crystallographic symmetry element available for a close-packed homo-chiral lattice, and the addition of the inversion centre and the glide plane allows the corresponding racemate to take up the more compact structure.

The close-pack criterion for crystal stability, generally favouring the racemic over the homochiral lattice, has an empirical expression in the rule of Wallach (1895) which states that the combination of two optical antipodes to form a racemate is accompanied by a volume contraction. The rule has many exceptions (table 9.4) and expresses only a general trend. The modification due to Walden (1896b), while the more restricted, is the more generally valid, namely, that if an enantiomer has a lower melting point than the corresponding racemate, the crys-tals of the latter have the higher density.

The chiral molecules with the highest point symmetry at present readily accessible are the dihedral (D_3) tris-chelate coordination compounds with the molecular morphology of a three-bladed propeller. The neutral tris-chelate com-plexes, such as those formed by β-dicarbonyl ligands with trivalent metal ions, generally give molecular crystals with a close-pack low symmetry lattice con-taining no elements of the molecular point-symmetry of the complex. The tris (pentane-2, 4-dionato) complexes, $[M(pd)_3]$, crystallise mainly in the monoclinic system, and both the racemates (M = Al, Co, Cr, Mn, Rh) and the homochiral crystals (M = Co, Cr) form isomorphous series with the respective space group, $P2_1/c$ and $P2_1$. A crystal of the enantiomer, Δ-(-)-$[Cr(pd)_3]$, 8, is less densely packed than the corresponding racemate by some 6% (table 9.4).

The general geometric model of Kitaigorodsky applied to the particular class of neutral D_3 tris-chelate complexes suggests that the packing of homochiral molecules with a three-bladed propeller shape is inevitably inefficient, relative to the corresponding heterochiral packing of the racemate. The tris-chelate com-

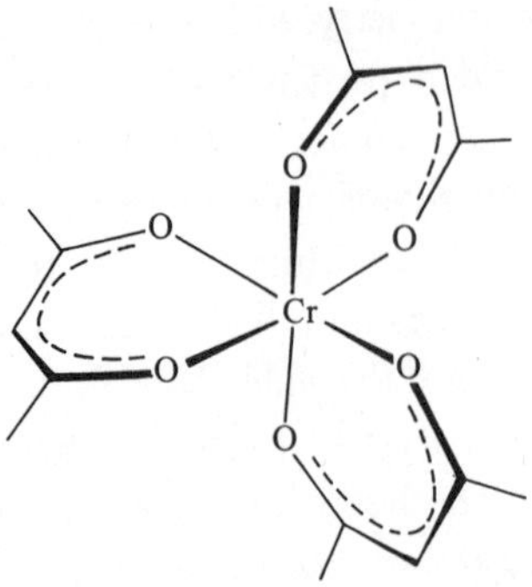

8

plexes with D_3 point-symmetry stack economically in homochiral columns with the principal C_3 rotational axis of each molecule parallel, or at a small angle, to the column axis. Along the direction of a molecular C_2 rotational axis, nearest-neighbour molecules pack efficiently, with interleaving chelate rings and with the principal C_3 axes of the molecules maintained parallel to one another, only if the two molecules have antipodal configurations, one with a left-handed and the other a right-handed propeller form (fig. 9.10). Thus the racemic $[M(pd)_3]$ complexes which crystallise in the monoclinic system, $P2_1/c$ (M = Al, Co, Cr,

Fig. 9.10. The interleaving of the chelate rings of two tris-chelate molecules along the direction of a common two-fold rotation axis. Molecules (a) and (b) have antipodal stereochemical configurations and pack economically with parallel principal (C_3) molecular axes, allowing the formation of parallel molecular stacks along the respective C_3 axial directions. Molecules (b) and (c) have the same stereochemical configuration and do not pack efficiently, with interleaved chelate rings, unless they mutually rotate about the line of centres, the common C_2 axis, so that the dihedral angle between the C_3 axes of the two molecules has the tetrahedral value.

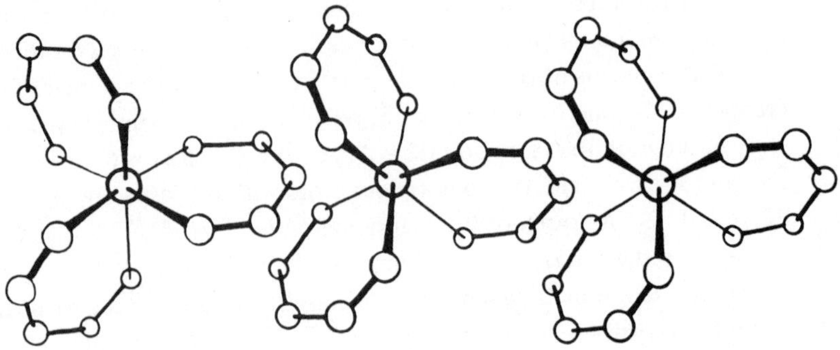

(a)   (b)   (c)

Table 9.4. *The density (d, g cm⁻³) and the space group of corresponding racemic (R) and homochiral active (A) crystals, and the percentage density increment, $d_A \to d_R$, (% A → R)*

Substance	System (A)	System (R)	d_A	d_R	%(A → R)	Ref.
camphoric anhydride[a]	P2₁2₁2₁	P2₁/c	1.315	1.282	−2.5	a
mandelic acid	P2₁	Pbca	1.35	1.32	−2.2	b
p-fluorophenylhydracrylic acid	P2₁	Pna2₁	1.417	1.390	−1.9	b
1-(α-naphthyl)-N-acetylethylamine	P2₁	Pca2₁	1.20	1.18	−1.7	c
trans-1,2-dibromoacenaphthene	P2₁2₁2	Pna2₁	1.973	1.945	−1.4	b
trans-cyclopentane-1,2-dicarboxylic acid	P2₁	C2/c	1.340	1.341	0	b
9,10-dihydro-dibenzo [c,g] phenanthrene	P3₁ 1	P4₂bc	1.240	1.250	+0.8	d
alanine	P2₁2₁2₁	Pna2₁	1.374	1.393	+1.4	b
tyrosine	P2₁2₁2₁	Pna2₁	1.414	1.436	+1.5	b
trans-cyclohexane-1,2-dicarboxylic acid	C2₁	C2/c	1.38	1.43	+3.6	b
valine	P2₁	P2₁/c	1.261	1.326	+5.0	b
tris(pentane-2,4-dionato)chromium (III)	P2₁	P2₁/c	1.285	1.360	+5.8	e
1,1'-binaphthyl	P4₁2₁2	C2/c	1.180	1.297	+9.9	d
1-phenyl-N-pivaloyl ethylamine	P4₁2₁2	P2₁/c	1.03	1.17	+13	c

(a) Data for the ordered phase; the hexagonal close-pack rotator phase has a density of 1.131; Miojo (1979).
(b) Cesario *et al.*, (1978). (c) Weinstein and Leiserowitz (1980). (d) Kuroda and Mason (1981a,b).
(e) Kuroda and Mason (1979).

Mn, Rh), and those in the orthorhombic class, Pbca (M = Sc, V, Fe), have lattices made up of pairs of parallel homochiral columns, with the molecules of one column antipodal to those of the adjacent parallel column.

Neighbouring tris-chelate molecules with the same chirality pack economically along the direction of a common C_2 rotational axis, with interleaving chelate rings, only if the molecules are mutually rotated around the line of centres so that the dihedral angle between the two principal (C_3) axes has the tetrahedral value, or its supplement (70.5°). Such a condition is less favourable for the formation of homochiral columns in a direction parallel, or at a small angle, to the principal molecular C_3 axes. In the crystal of Δ-(−)-[Cr(pd)$_3$], three of the four independent angular relationships between the molecular C_3 axes in the unit-cell are close to the tetrahedral value, and the fourth is a near-parallel relation (Kuroda and Mason, 1979).

The close-packing model goes some way to account for the higher incidence of the spontaneous optical resolution of racemic salts on crystallisation, relative to corresponding neutral or charge-compensated racemic mixtures. With two or three species of different shape and size in the crystal structure, the chiral ion, the counter-ion, and frequently water of hydration, the range of packing modes is the wider for the chiral salt. The wider latitude implies that the compacity-discrimination between a racemic and the corresponding active structure is likely to be smaller for the chiral salt than the neutral chiral compound.

An expression of the wider scope for economic packing in chiral salts is the relatively-common high crystallographic symmetry of both homochiral and heterochiral ionic crystals (table 9.5). The space groups have been listed for 116 chiral coordination compounds for which the absolute stereochemical configuration has been determined by the anomalous X-ray scattering method (Saito, 1977). The crystals of 45% of the ionic tris-chelate complexes listed are either isotropic (cubic) or uniaxial (trigonal, tetragonal or hexagonal), whereas examples of neutral tris-chelate complexes with a higher crystallographic symmetry than orthorhombic are rare. In the ionic tris-chelate crystals, the counter-ions and the water molecules occupy the voids in the lattice created by a high crystallographic symmetry based upon elements of the D_3 point-symmetry of the chiral molecular ion.

Some ionic tris-chelate coordination compounds with D_3 point-symmetry share common geometric packing features with the corresponding neutral complexes in the respective crystal lattices. The homochiral salt, K{Λ-(−)-[Co(ox)$_3$] Λ-(+)-[Ni(phen)$_3$]}2H$_2$O, forms cubic crystals, belonging to the space group, P2$_1$3, with Z = 4. Each of the four pairs of dihedral complex ions in a unit-cell is a member of a homochiral column of alternating complex anions and cations extending throughout the lattice. In a given column, the molecular threefold rotation axis of each complex ion is directed along the column axis, satisfying the geometric condition for the efficient packing of molecules with the shape of a three-bladed propeller along the direction of the principal molecular axis, as in the case of the neutral $[M(pd)_3]$ complexes. The four homochiral columns of the ionic diastereomer are made up of two parallel pairs, and the column direction of one pair lies at the tetrahedral angle to that of the other pair. The latter angular relationship satisfies the geometric condition for the economic interleaving of the chelate rings of two tris-chelate molecules with the same chirality along the direction of a common twofold rotation axis, the angular relation being dominant in the crystal structure of the neutral Δ-(-)-$[Cr(pd)_3]$ complex.

Other ionic tris-chelate coordination compounds form heterochiral columns in the crystal, with the threefold rotation axis of each complex ion parallel to the column axis. In the crystal lattice of the racemates, $(\pm)$-$[Co(en)_3]Cl_3.2.8H_2O$ and $(\pm)$-$[Cr(en)_3]Cl_3.3H_2O$, or of the pseudo-racemate, Λ-(+)-$[Co(en)_3]$ Δ-(-)-$[Cr(en)_3]Cl_6.6.1H_2O$, and its analogues, the antipodal complex cations are stacked alternately along the column axis, which is parallel to the trigonal crystal axis. Each cation is separated from its enantiomeric neighbour by a chloride anion, so that the stereochemical constraints favouring the homochiral packing of D_3 tris-chelate complexes in contact with one another into a column are no longer significant (Spinat, Brouty and Whuler, 1980).

Table 9.5. *The space group of homochiral and related heterochiral ionic crystals*

Crystal[a]	Space group	Ref.
$(Na(NH_4)$-(+)-tartrate.$4H_2O$	$P2_12_12$	b
$(Na(NH_4)$-($\pm$)-tartrate.H_2O	$P2_1/a$	b
Λ-(+)-$[Co(en)_3]Cl_3.H_2O$	$P4_32_12$	c
$(\pm)$-$[Co(en)_3]Cl_3.3H_2O$	$P\bar{3}cl$	d
(+)-$[Co(en)_3]$-(-)-$[Cr(en_3]Cl_6.6 \cdot 1H_2O$	$P3\ 2\ 1$	d
(-)-$K_3[Rh(ox)_3].2H_2O$	$P3_1 2\ 1$	e
$(\pm)$-$K_3[Rh(ox)_3].4 \cdot 5H_2O$	$P\bar{1}$	e
Λ-(+)-$[Co(ox)(en)_2]H$-(+)-tartrate.H_2O	$P2_1$	f
Δ-(-)-$[Co(ox)(en)_2]H$-(+)-tartrate.$2H_2O$	$P2_12_12$	f

(a) The abbreviations are: ethylenediamine (en); oxalate (ox).

(b) Kuroda and Mason (1981d).

(c) Saito (1977).

(d) Spinat, Brouty and Whuler, (1980).

(e) Herpin (1958).

(f) Kuramoto, Kushi and Yoneda, (1980).

The geometric-packing model has a limitation in that $\sim 15\%$ of known crystal structures are not close packed. In addition, the model provides no sure guide to an expectation of the spontaneous resolution of a racemic mixture on crystallisation, or the formation of a true racemate. The racemic crystal of 1,1'-binaphthyl, **9**, is some 10% more densely packed than the corresponding active crystal (table 9.4), yet the conglomerate of active crystals is more stable thermodynamically

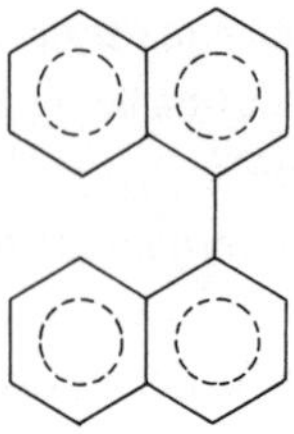

9

than the racemic crystals above the transition temperature of 76°C, and a metastable conglomerate of active crystals is readily obtained by crystallisation at ambient temperature.

The space group of active 1,1'-binaphthyl is exceptional to the close-packing criterion, and so too are those of both the active and the racemic crystals of the 2,2'-bridged analogue, 9,10-dihydrodibenzo[c,g] phenanthrene, **10**, (table 9.4). Orientation-dependent intermolecular attractions in the crystal lattice, notably,

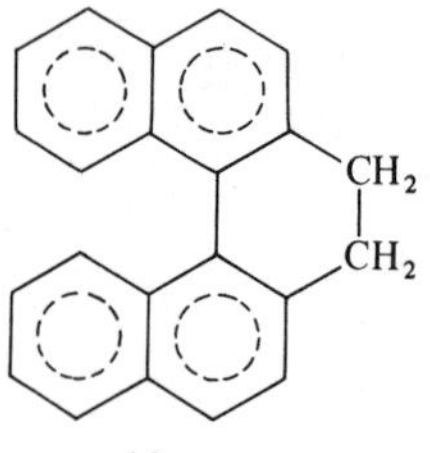

10

those arising from the dispersion energy, become significant in chiral π-systems, particularly those which are conformationally constrained, and contain groups with a substantial anisotropy between the components of the electric-dipole polarizability tensor, such as the naphthalene nuclei of **9** and **10**. The dihedral angle between the planes of the naphthalene nuclei is constrained to the value of 49° in both the active and the racemic crystal of the 2,2'-bridged 1,1,-binaphthyl, **10** but the parent molecule, **9**, is conformationally labile and the corresponding dihedral angle changes from the *transoid* value of 103° in the active crystal to the *cisoid* value of 68° in the racemic crystal, where the packing is optimal (table 9.4 and 9.5).

9.5 Absolute stereochemical configuration

Attempts to place the Fischer–Rosanoff convention for relative stereo-chemical configuration upon an absolute basis were founded initially, in the 1930s, on the classical and the early quantum theory of optical activity. The sodium D-line optical rotation of simple organic substances of central importance to the convention were calculated but with little agreement between the different procedures. The molecules chosen were often conformationally labile, such as (R)-(-)-(butan-2-ol), **11**, for which Kuhn (1935) and Gorin, Walter and Eyring

HO   H

C

CH$_3$ C$_2$H$_5$

11

(1938) calculated a D-line laevorotation, while Boys (1934) and Kirkwood (1937) obtained a D-line dextrorotation. Subsequently, Kirkwood, Wood and Fickett (1952) discovered an error in the original (1937) calculation requiring a change in the sign of the optical rotation. Although it was appreciated that the optical rotation at a given wavelength represents a sum of oppositely-signed contributions from all of the spectroscopic transitions in a chiral molecule (eqns 2.1 and 2.2), the D-line rotations were then the principal optical data available for most chiral molecules. The determination of the CD spectrum of a butan-2-ol enantiomer, requiring vacuum ultraviolet instrumentation, is relatively recent (Snyder and Johnson, 1973).

The assignment of the absolute configuration of a chiral molecule from an analysis of the CD spectrum due to an individual electronic transition, or a given set of transitions, avoids the summation problem inherent in the calculation of an optical rotation at a particular wavelength, although other limitations remain. A CD assignment of absolute configuration requires a conformationally-con-strained chiral molecule with an established structure, apart from configuration. For a conformationally-labile chiral molecule, CD spectroscopy may provide an indication of the preferred conformation adopted in a given solvent at a particu-lar temperature, and even a measure of the conformational equilibrium, but only if the absolute configuration is already established.

The more direct methods for determining the absolute configuration of a chiral molecule by an analysis of the CD spectrum are based upon the two-group electric-dipole mechanisms, particularly the isoenergetic two-group mechanism appropriate for the dissymmetric dimer or oligomer (§ 3.4). The assignment of the absolute configuration of a chiral oligomer from the frequency-order of the two adjacent and oppositely-signed CD bands of comparable magnitude, charac-teristic of the two-group electric-dipole mechanism, generally requires a knowledge of the polarization direction of the electric transition moment, and even of its mean location, in each individual chromophore. In the biaryl series (§ 6.1) the polarization directions of the monomer transition moments are evident from the CD spectrum, from the presence or absence of the bisignate CD couplet, and in the case of calycanthine and its analogues (§ 6.3) the sign and frequency order of the CD couplet bands depend only on the absolute configuration, irrespective of the polarization direction of the in-plane $\pi \to \pi^*$ transitions of the aniline chromophores. Such examples are uncommon, however, and generally a CD assignment of absolute configuration, based upon the two-group electric-dipole mechanisms, requires a prior linear dichroism study of the monomer chromophore orientated in a single crystal, a nematic mesophase, or a strain-orientated polymer film.

The CD determination of the absolute configuration of an inherently-dissymmetric molecule, such as a helicene (§ 5.3) where the electronic transitions are both electric- and magnetic-dipole allowed, is similarly more direct than the corresponding assignment from the CD due to an allowed magnetic-dipole transition in a molecule consisting of a chirally-substituted symmetric chromophore. In the latter case, the analysis becomes the more extended and less precise through the necessity of invoking a mechanism for the origin of the first-order electric-dipole transition moment collinear with the zero-order magnetic moment, e.g. the static field or the dynamic-polarization mechanism (§ 3.2 and 3.3). The need, in all cases, for a conformationally-constrained molecule with an established structure, and a chromophore or chromophores with well-characterised optical transitions, bounds the scope of determining absolute configuration by electronic or vibrational CD absorption or CID emission methods.

The view of Pasteur, that the macroscopic dissymmetry of a crystal implies the microscopic dissymmetry of the constituent molecules, carried forward to provide, in principle, a method for determining absolute configuration. On the basis of an early crystal structure of (+)-tartaric acid, Waser (1949) investigated from models of the D- and the L-acid in the Fischer convention the probable rate of growth of the dihedral (011) and (01$\bar{1}$) faces of the (+)-crystal, which are enantiomorphous to the (0$\bar{1}$1) and the (0$\bar{1}\bar{1}$) faces of the (−)-crystal (fig. 9.11). The (0$\bar{1}$1) and (0$\bar{1}\bar{1}$) faces do not appear in the (+)-crystal, as the rate of crystal growth normal to these faces is fast, and they develop to extinction, whereas the rate of growth is slow normal to the (011) and (01$\bar{1}$) faces, which become prominent. The converse growth rates obtain for the (−)-crystal, where the (0$\bar{1}$1) and (0$\bar{1}\bar{1}$) faces become prominent, and the (011) and (01$\bar{1}$) faces develop to extinction. The models suggested that the steric and hydrogen-bonding requirements for crystal growth normal to a (011) or a (01$\bar{1}$) face are stringent for the L-configuration of tartaric acid, the (−)-acid of the Fischer convention. It was concluded that the postulated absolute basis of the Fischer convention is incorrect in sign (Waser, 1949), although knowledge of the factors governing crystal growth was then too incomplete to allow a firm decision (Turner and Lonsdale, 1950).

The anomalous-scattering X-ray diffraction method for the determination of absolute configuration, proposed by Bijvoet (1949), substantiated the absolute basis adopted in the Fischer convention. The initial application of the method, employing the crystal of sodium rubidium (+)-tartrate tetrahydrate (Bijvoet *et al.*, 1951), indicated that (+)-tartaric acid has the (2R, 3R)-configuration of the Cahn, Ingold, and Prelog convention (1966), with the absolute stereochemistry assumed by Fischer (1891) and Rosanoff (1906). A firm absolute basis was provided for the extensive sets of correlated configurations of chiral organic molecules, established mainly by chemical interconversions, aided by the phase-equilibria methods, and by comparisons of the sodium D-line optical rotations of the substances or, less frequently, the corresponding ORD curves.

Subsequently ORD comparisons became more common (Djerassi, 1960), followed by CD correlations (Velluz, Legrand and Grosjean, 1965) and CD methods for absolute configuration (Mason, 1967). Reference lists of the absolute configuration of organic structures determined by the Bijvoet X-ray

Fig. 9.11. The crystal morphology of (*a*) (−)-tartaric acid and (*b*) (+)-tartaric acid.

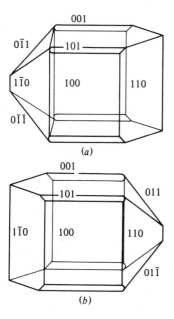

(a)

(b)

method, initiated with 54 entries (Allen and Rodgers, 1966), soon extended to 186 cases (Allen, Neidle and Rodgers, 1970), while by 1980 the Cambridge Crystallographic Data Centre had over 900 X-ray determinations of absolute configuration on file. The accumulated correlations and absolute determinations provided the basis for comprehensive surveys of the absolute configurations of organic molecules, combining the results of the several physical and chemical methods (Jacques *et al.*, 1977; Klyne and Buckingham, 1978).

The application of the anomalous X-ray scattering method to the series of chiral transition-metal coordination compounds similarly provided an absolute reference standard for the sets of related configurations established by the isomorphous less-soluble diastereomer and the active-racemate methods, or the configurational correlations based upon the CD spectra, introduced by Mathieu (1936). In the initial application, by Saito and coworkers (1954), it was found that the tris(ethylenediamine) cobalt(III) enantiomer, $(+)$-$[Co(en)_3]^{3+}$, has the stereochemical form of a left-handed three-bladed propeller, subsequently termed the Λ-configuration (IUPAC, 1970).

Owing to the restricted range of chemical reactions relating stereochemical configuration in the transition-metal complex series, and the configurational inversion involved in some reactions, the anomalous X-ray scattering method has been important in providing, through the determination of absolute configuration, a stereochemical cross-correlation of metal complexes differing in charge-type, in the denticity or the coordinating atoms of the ligands, or in the particular group or period to which the metal ion belongs. The crystal data, molecular structure, and the absolute configuration determined by X-ray methods, have been listed (Saito, 1977) for some 116 transition-metal coordination compounds differing in these various respects.

The anomalous X-ray scattering method for the determination of absolute configuration is based upon the violation of the law of Friedel (1913) if the radiation is partly absorbed by a set of like atoms in a crystal. With negligible absorption, the X-ray diffraction pattern of a crystal is centrosymmetric according to Friedel's law, whether the crystal structure is centrosymmetric or not, and whether the molecules composing the lattice are a homochiral, heterochiral, or achiral set. In the case of 1,1′-binaphthyl, the (R)-$(-)$-isomer crystallises in the tetragonal space group, $P4_1 2_1 2$, whereas the crystals of the (S)-$(+)$-isomer belong to the enantiomorphous group, $P4_3 2_1 2$. The normal X-ray diffraction patterns of these two crystals are identical, as are those of crystals of enantiomers belonging to each of the other pairs of enantiomorphous space groups.

Friedel's law is dependent upon the assumption that the phase differences between the electromagnetic waves scattered at different centres in a crystal arise solely from the path differences between the centres, that is, any intrinsic phase change due to the scattering process is uniform over the scattering centres. The assumption is a good approximation and, in normal diffraction, the intensities of the reflection from an array of crystal planes (hkl) and of the corresponding counter-reflection ($\overline{hkl}$) are equivalent, giving a normal Friedel pair. The assumption breaks down if the X-radiation frequency lies close to one of the absorption-edge frequencies of an inner-shell electron, the K, L, or M shell, in a set of like-atoms in the crystal.

According to the classical model, electromagnetic radiation incident upon an assembly of atoms gives rise to forced oscillations of the elastically-bound electrons, and the oscillating electric dipoles produce secondary radiation of the same frequency. The electron oscillations are in phase with those of the incident electromagnetic wave if the frequency of the latter is much smaller than that of the electronic resonance absorption, or out of phase by π if the frequency of the incident wave is much larger. Throughout the absorption frequency region the phase difference between the primary incident wave and the secondary scattered wave changes from zero to π continuously, with the particular value of $\pi/2$ at the frequency of the absorption maximum. The phase angle difference, β, between the incident and the scattered wave is given, for small angles, by the ratio of the imaginary, α', to the real, α, part of the atomic or molecular polarizability, represented macroscopically by the ratio of the absorption index to the refractive index. More generally, $\tan \beta = (\alpha'/\alpha)$.

In the X-ray frequency region the counterparts of the atomic polarizability components are the real ($f_0 + \Delta f'$) and the imaginary ($\Delta f''$) atomic structure factors. The real atomic structure factor is broken down into f_0, the scattering factor at zero frequency, corresponding to the static field polarizability, α_0, and $\Delta f'$, the correction to the real part required at a particular radiation frequency, analogous to the increment between the static and the dynamic polarizability of an atom. As in other regions of the electromagnetic spectrum, the angle β of the phase shift due to the anomalous scattering of X-rays near to an absorption frequency corresponds to the ratio of the imaginary to the real part of the atomic scattering factor, $\tan \beta = [\Delta f''/(f_0 + \Delta f')]$.

On account of the anomalous scattering, the relative intensities of the two components in some Friedel-paired reflection become inequivalent in the X-ray diffraction pattern of a crystal near to an absorption wavelength. The effect was first detected by Nishikawa and Matsukawa (1928) and by Coster, Knol and Prins (1930), employing a crystal of zinc-blende (sphalerite). The cubic crystal of zinc-blende has a truncated tetrahedron form (fig. 9.12). Each vertex of the regular tetrahedron is replaced by a triangular face. The small facet and the parallel larger face correspond to an atomic plane, (111) and ($\overline{111}$) or their equivalents. One of these planes is made up of zinc atoms and the other of sulphur atoms (fig. 9.13).

The normal X-ray diffraction pattern does not indicate which of the two types of atomic plane corresponds to the small facet or the larger opposed face. The two faces are inequivalent in growth rate, and in optical and piezo-electrical properties. Generally, one face is reflective and the other matt in appearance and the former becomes positively charged when the crystal is compressed along the normal to the two faces.

The polarity sense of the zinc-blende crystal was determined from the anomalous scattering observed with X-rays close in wavelength to the K-shell absorption edge of zinc at 1.281 Å. Using the gold (Au)Lα_1 ($\lambda = 1.2738$ Å) or the tungsten (W) Lβ_1 ($\lambda = 1.2792$ Å) radiation, it was found that the reflective face of zinc-blende corresponds to a sulphur layer and the matt face to a zinc layer.

Fig. 9.12. The crystal morphology of zinc-blende.

Fig. 9.13. The crystal structure of zinc-blende; (*a*) the atomic arrangement and (b) the relative spacing of the planes of zinc and of sulphur atoms parallel to (111).

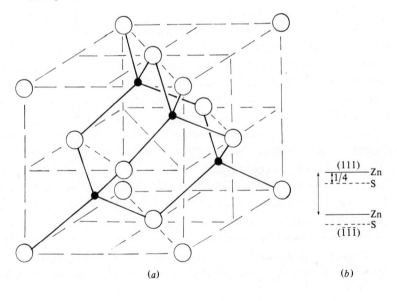

(*a*)   (*b*)

With Au Lα_1 radiation, the wave scattered from the zinc atoms is advanced in phase by 10.5° relative to the wave scattered from the sulphur atoms. The phase shift results in a differential interference of the two scattered waves for some of the Friedel-paired reflections. In particular, the intensity of the reflection of Au Lα_1 radiation from the ($\overline{1}\overline{1}\overline{1}$) face of zinc-blende becomes larger than that from the (111) face (fig. 9.14).

Fig. 9.14. The X-ray reflections from (a) the (111) planes and (b) the ($\overline{1}\overline{1}\overline{1}$) planes of zinc-blende, and the amplitudes of the waves reflected from the zinc and from the sulphur planes, with the resultant amplitude (Zn + S) for (c) normal diffraction, following Friedel's law, and (d) anomalous diffraction, giving inequivalent (111) and ($\overline{1}\overline{1}\overline{1}$) resultant amplitudes.

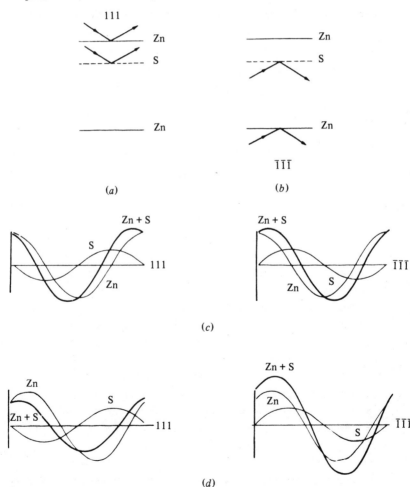

The problem of determining the absolute stereochemical configuration of a molecule or a polyatomic ion in a crystal, while more complex, is analogous to that of assigning the absolute polarity sense in a polar structure. The anomalous scattering of X-ray photons and of particles with a non-zero rest-mass, notably neutrons, provides not only the path difference between two types of scattering centre, i.e. a bond length, but also the particular polar orientation of a hetero-atomic bond. For a chiral structure, the set of polar-bond orientations defines the absolute stereochemical configuration.

The particular choice of absolute molecular configuration, and the absolute polarity sense in a polar crystal, adopted in the anomalous scattering X-ray diffraction method rests upon the classical result, supported by quantum field theory, that the imaginary component of the atomic scattering factor, $\Delta f''$, is positive in an absorption region. On the basis of a new quantum field model, Tanaka (1972) suggested that $\Delta f''$ is, in fact, negative, so that all previous absolute configurations determined by X-ray analysis should be reversed.

The suggestion appeared to be supported by an analysis of the CD spectrum of (−)-1,5-diamino-9,10-dihydro-9,10-ethenoanthracene, **12**, and other chiral

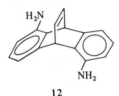

12

molecules containing two aniline chromophores (Tanaka *et al*, 1972, 1973). The CD spectrum of **12** consists of two bisignate CD couplets associated with the 285 and the 220 nm aniline absorptions, each couplet consisting of a negative CD band at lower frequency followed by a positive CD absorption at higher energy. The analysis of the CD spectrum of **12** employing the two group iso-energetic electric-dipole mechanism (§ 3.4), with the location of the electric-dipole $\pi \rightarrow \pi^*$ transition moment in each aniline chromophore evaluated by the dipole-length method (§ 5.2), gave the absolute configuration of the (−)-isomer as (9R, 10R), in agreement with the anomalous scattering X-ray assignment on the postulated basis that $\Delta f''$ is negative (Tanaka *et al* , 1973)

It was well established, however, that rotational strengths evaluated by the dipole-length method are, in general, dependent upon the choice of an origin for the coordinate frame in molecules of low symmetry (Gorin *et al.*, 1938), and Moffitt (1956*b*) had shown that origin-independent rotational strengths are ensured by working consistently in the dipole-velocity representation (§ 5.2). In a re-examination of the CD analysis of **12**, it was found that the electric moment of the lowest-energy π-transition of the aniline chromophore is displaced from the centre of the benzene ring towards the amino-substituent if estimated by the dipole-length method, but shifted towards the *para*-carbon atom if evaluated by the dipole-velocity procedure. The latter, more-reliable, procedure requires the (−)-isomer, 12, to have the (9S, 10S)-configuration (fig. 9.15), in conformity with the X-ray assignment for $\Delta f''$ positive, and with all previous cases where absolute configuration had been determined by both CD and X-ray methods (Mason, 1973a). Absolute configurations assigned by the CD method alone, such as that of calycanthine (§ 6.3), were found to be supported by the X-ray method with $\Delta f''$ positive (Beecham, Hurley, Mathieson and Lamberton, 1973).

The problem of the sign of $\Delta f''$, the imaginary component of the atomic scattering factor, was solved by an independent determination of the polarity sense of the zinc-blende and analogous crystals. Brongersma and Mul (1973) reflected a beam of Ne^+ ions with an energy of 1000 eV from the (111) and ($\overline{1}\overline{1}\overline{1}$) faces of a ZnS crystal at an overall angle of 45° and analysed the energy of the reflected ions in a mass spectrometer. The conservation of energy and momentum in the collision of the Ne^+ ions with the zinc or the sulphur surface atomic layer of the crystal identified, through the measured energy-loss, the particular atoms of the crystal face investigated. It was found that the (111) consists of zinc atoms and the ($\overline{1}\overline{1}\overline{1}$) face of sulphur atoms, supporting the earlier anomalous X-ray scattering result for $\Delta f''$ positive.

The anomalous X-ray scattering method is the most general procedure available for the determination of absolute configuration and it has few limitations. Chiral structures containing two or more types of atom which scatter X-rays anomalously do not violate Friedel's law, and the absolute configuration of those structures cannot be established by the technique (Iwasaki, 1974). A practical limitation, apart from the requirement of a crystalline solid, is the small X-ray scattering produced by the lighter atoms. The determination of the absolute configuration of a molecule owing its chirality to the replacement of a hydrogen by a deuterium atom is scarcely feasible by the anomalous scattering X-ray method. The corresponding determination for a chiral hydrocarbon presents problems, since the K-shell absorption edge of the carbon atom lies at some remove (43 Å) from the longest wavelengths commonly employed in X-ray diffraction studies, e.g. [24]Cr Kα (λ = 2.291 Å)

The extension of the anomalous scattering method for absolute configuration to neutron diffraction may alleviate the light-atom problem of the corresponding X-ray technique. The anomalous scattering of neutrons in the thermal energy region by some nucleides is large, due to absorption, e.g. ^{113}Cd + n → ^{114}Cd. The resonance scattering is so large that the anomalous components are generally an order of magnitude greater than the normal scattering amplitudes. The absolute

Fig. 9.15. The origin sensitivity of the electric-dipole moment of a π-electron transition estimated by the dipole-length procedure. The longest-wavelength CD band of (9S,10S)-1,5-diamino-9,10-dihydro-9,10-ethenoanthracene is negative, in accord with the negative rotational strength of the lower-frequency coupling mode (a) of the π-excitation moments of the two aniline chromophores calculated by the dipole velocity method. The highest-occupied (b) and lowest-unoccupied (c) π-orbital AO coefficients of the carbanion corresponding to the aniline chromophore give, (d) the atomic transition charge densities, $C_{\nu i}C_{\nu j}$, required for the dipole-length calculation (eqn 5.4), and (e) the transitional bond-order changes, $P_{\mu\nu}^{ij}$, required for the dipole-velocity calculation (eqns 5.5 and 5.6). The dipole-length estimate gives a positive rotational strength for the lower-frequency coupling mode of the aniline π-excitations in the enantiomer (a).

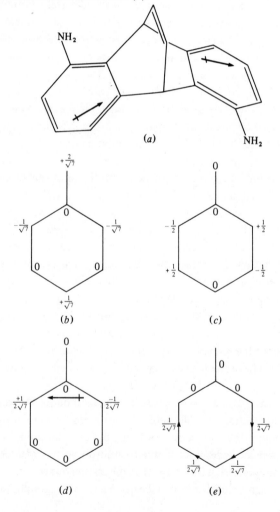

configuration of the (S)-glutamate complex of cadmium(II), $[Cd(S\text{-}glu)(H_2O)]$ .H_2O, has been determined by neutron diffraction, employing [113]Cd in natural abundance (12.3%) as the anomalous scattering nucleide, with neutrons of 0.981 Å wavelength (Flook, Freeman and Scudder, 1977).

10 ENANTIOMERIC DISCRIMINATION

10.1 Optical resolution procedures

Two classical methods for the optical resolution of racemates developed from Pasteur's studies of molecular dissymmetry. The first involves the direct crystallisation of enantiomer mixtures and the second concerns the fractional crystallisation of the diastereomers formed by a racemic mixture with a chiral reagent. While the second procedure is the more important for small-scale laboratory separations of enantiomeric mixtures (Wilen, Collet and Jacques, 1977), the first method is the more significant for the large-scale optical resolution of racemates. Some 13 000 tons of (S)-glutamic acid were obtained annually over the 1963–73 period by the direct crystallisation of the racemic mixture synthesised from acrylonitrile (Collet *et al.*, 1980).

10.1.1 Entrainment

As originally practised by Pasteur (1848), i.e. the hand-sorting of enantiomorphous crystals, the first method is limited, not only in expedition, but also by the absence of distinguishing hemihedral facets from the crystals formed by some optical isomers, particularly those crystallising in the higher-symmetry crystal classes. The tetragonal crystals of (R)-(–)- and (S)-(+)-1,1′-binaphthyl lack distinctive facets and visually appear to be identical. Owing to the high 422 crystal symmetry, the lowest-order crystal face which would not be accompanied by a parallel counterpart, and so give enantiomorphism, is the 121 face in the active 1,1′-binaphthyl crystal. The higher-order crystal faces form with a smaller probability than the lower-order faces and, in general, hemihedral facets are not expected for the enantiomers crystallising in the higher-symmetry crystal classes (Kress *et al.*, 1980).

The lack of a visible distinction between the crystals of two enantiomers is not a limitation to more recent developments of Pasteur's first method for enantiomer separation. His one-time student, Gernez (1866), found that a supersaturated solution of racemic sodium ammonium tartrate gave crystals only of the enantiomeric salt used to seed the solution, and Jungfleisch (1882), employing a seed of each antipode, obtained a large single crystal of the individual active

salts from a similar supersaturated solution. Subsequently, Werner (1914) discovered that each enantiomer of the (oxalato)bis(ethylenediamine)cobalt(III)
salt, $[Co(en)_2(ox)]$ Br, is precipitated step-wise from a saturated aqueous solution
of the racemate, initially containing a small excess of one isomer, by the successive addition of aliquots of ethanol. The first addition precipitates a relatively-
large quantity of the isomer initially present in excess, leaving the mother liquor
enriched in the antipodal isomer. In turn, the latter is precipitated by a second
addition of ethanol, again in a quantity larger than its excess, leaving the mother
liquor now enriched with the original isomer.

   The method of optical resolution by the direct crystallisation of a racemic
mixture with the appropriate seeding, termed entrainment by Amiard (1956), is
dependent upon the condition that the solubility of the enantiomer is less than
that of the corresponding racemate (Werner, 1914). More specifically, the ratio
of the solubility of the racemate to that of the corresponding enantiomer,
(S_R/S_A), is required to equal or exceed the factor, $2^{1/(n+1)}$, where n is zero for a
neutral species or the number of charges carried by an achiral counter-ion in a
salt (§ 9.2). The condition is not sufficient, however. In cases where the solubility ratio exceeds the theoretical factor, the formation of metastable crystals
of the racemic compound may be the preferred crystallisation mode (Collet
et al., 1980).

   The solubility ratio (S_R/S_A) is temperature-dependent, as van't Hoff had
found in his studies of the relative solubilities of corresponding active and racemic tartrate salts (§ 9.2). Following the rule of van't Hoff (1899), that the less-
solvated salt is the more stable at the higher temperature, Duschinsky (1934)
showed that histidine hydrochloride is optically resolved by direct crystallisation from water above ~ 40 °C, but not at lower temperatures. At 20 °C the
racemic salt crystallises as a dihydrate, whereas the salt of an enantiomer forms
a monohydrate, so that a conglomerate of active crystals is the expected crystallisation mode of the racemic mixture at more elevated temperatures (Duschinsky,
1934).

   The scope of the entrainment procedure for enantiomer separation is wider
for racemic salts than for the corresponding neutral acid or base, as the incidence
of spontaneous optical resolution on crystallisation is some two to three times
larger for ionic than for covalent racemates (Jacques, Leclerq and Brienne, 1981).
In addition, the 'entrainment area' of the ternary phase diagram (fig. 9.5) is the
larger for a racemic salt than the corresponding neutral racemate (Collet *et al.*,
1980). For the optical resolution of a neutral racemate by entrainment, a solubility ratio (S_R/S_A) of at least 2 is required, whereas the corresponding factor is
$\sqrt{2}$ for a racemic salt containing a singly-charged achiral ion. The larger ratio for
the neutral racemate implies that the antipode of the seeded isomer attains a

progressively larger supersaturation during the crystallisation of the isomer selected, and ultimately the antipode spontaneously nucleates and crystallises.

For glutamic acid and its N-acetyl derivative, with a solubility ratio (S_R/S_A) of 2.35 and 2.18, respectively, relatively large seed crystals of the (S)-isomer are employed and, after crystallisation, the small crystals of the (R)-isomer are separated with a sieve. In contrast, ammonium glutamate and glutamic acid hydrochloride, with a (S_R/S_A) ratio of 1.4 and 1.5, respectively, give virtually all of the supersaturated fraction as the isomer selected by seeding in an optically-pure form (Collet *et al.*, 1980).

Where seed crystals for optical resolution by the entrainment method are not available, the use of a chiral solvent may produce an initial discrimination. Kipping and Pope (1898) found that racemic sodium ammonium tartrate crystallised from aqueous solutions of D-glucose gives initially crystals of the dextro-rotatory salt, although the salts of each isomer have the same solubility in the glucose solution. The use of a chiral solvent for the resolution of a racemate involves diastereomer formation, however weak the association. The partial resolution of a [7]-heterohelicene by crystallisation from (−)-α-pinene, where no diastereomer formation was detected by NMR spectroscopy (Groen, Schadenberg and Wynberg, 1971), does not differ in principle from the corresponding resolution of [6]-helicene by crystallisation from a benzene solution of an enantiomer of α-(2,4,5,7-tetranitro-9-fluorenylideneaminoöxy)-proprionic acid (TAPA), **1**, where diastereoisomeric charge-transfer complex formation is evident from the

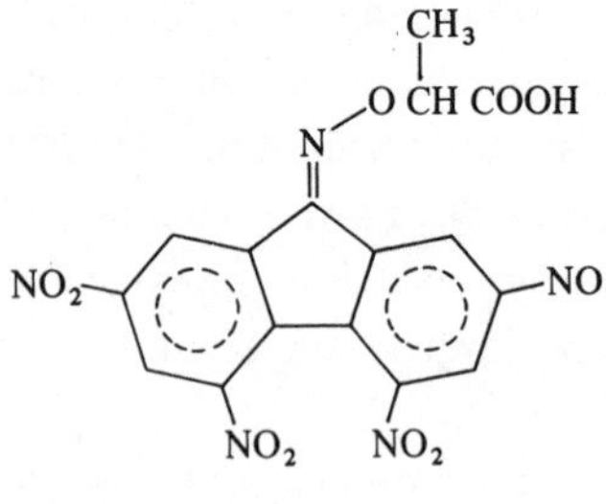

1

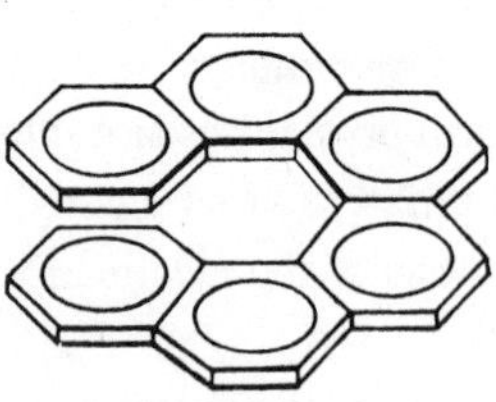

2

marked colour change (Newman and Lednicer, 1956). With (+)-TAPA, the isomer with the P-configuration, (+)-[6]-helicene, **2**, preferentially crystallises from the solution, as the (−)-isomer forms the stronger complex and remains dissolved. The (−)-isomer, with the M-configuration, similarly forms the stronger complex with (+)-TAPA in the case of [5]-helicene where the purple-red diastereomer, [(−)-[5]-helicene-(+)-TAPA], crystallises from the solution (Goedicke and Stegemeyer, 1970).

10.1.2 *Diastereomer separation*

The classical form of Pasteur's (1853) second method of optical resolution, the fractional crystallisation of diastereomeric salts, has been described variously as a matter of trial and error (Eliel, 1962), or as largely an art (Wilen, 1971). The chiral natural products introduced as resolving agents during the nineteenth century, notably the hydroxyacids for the resolution of bases, and the alkaloids for the separation of acidic enantiomers, remain the most generally used reagents (Boyle, 1971). Over one third of the resolutions of some 230 racemic acids reported during the decade 1960–70 were accomplished with quinine or brucine (Wilen, 1971). For the resolution of a new racemate, comprehensive tables of resolving agents and techniques of optical resolution are available (Wilen, 1972), and an analogous case may be found in a collection of resolution recipes extending back over a century (Newman, 1978).

Marckwald (1896) advocated the use of synthetic resolving agents for enantiomer separation by the fractional crystallisation of the diastereomers formed, on the basis that natural products generally afforded only one of the isomers in a racemate as the less-soluble salt. In correlating stereochemical configuration, Winther (1895) had shown, however, that while quinine, brucine and strychnine form the less-soluble salt with (+)-tartaric acid, the (–)-tartrate of cinchonine, cinchonidine, or morphine is the less-soluble salt. The enantiomers of lactic, malic, mandelic and other acids were separated by the two sets of alkaloids, and so were related configurationally to the isomers of tartaric acid. Both enantiomers of these acids became available for optical resolutions and each optical isomer of tartaric acid had been available since Pasteur's original resolution. In turn the resolved acids provided the enantiomers of synthetic chiral bases, notably 1-phenylethylamine and its analogues, introduced by Theilacker and Winkler (1954).

In an endeavour to improve the efficiency of the less-soluble diastereomer method of optical resolution, Marckwald (1896) proposed the use of a half-quantity of the resolving reagent, so that only the less-soluble salt is formed with the racemate, leaving the antipodal isomer as a free acid or base. The procedure was taken up by Pope and Peachey (1899) and developed further by the use of a half-quantity of an achiral acid or base, in addition to the half-quantity of the resolving reagent, for each equivalent of the racemate. Pope and Peachey (1899) observed that the solubilities of the diastereomers, (+)-A(+)-B and (+)-A(–)-B, formed by an active acid with a racemic base, often do not differ substantially, resulting in an optically-impure product. A discrimination in favour of the less-soluble diastereomer is introduced by ensuring salt formation by both isomers of a racemic base with a half-quantity of an achiral acid and of the active acid in order to reduce the competition for the latter by the enan-

tiomers of the base. Leclerq and Jacques (1979) have investigated and systematised the equilibria involved in the method of half-quantities, demonstrating the
importance of pH control for an optimisation of the method.

An approximate measure of the probable success of an optical resolution by
the procedure of diastereomeric salt formation may be obtained from a survey
of 144 salts formed by 14 acids and 9 bases commonly used in the method
(Jacques, Leclerq and Brienne, 1981). A racemic acid and a racemic base form
four diastereomers which are classified as positive, *p*, if the D-line rotations of
the acid and base enantiomers forming the salt have the same sign, both dextrorotatory or both laevorotatory, or negative, *n*, if the optical rotations of the
acid and the base isomers in the salt are opposed, one dextrorotatory the other
laevorotatory. The salts, (+)-*p*, [(+)-A(+)-B] , and (-)-*p*, [(-)-A(-)-B] , are enantiomeric, as are the two *n* salts. The recrystallisation of the mixture of salts
formed by a racemic acid with a racemic base results either in a conglomerate of
doubly-active crystals, each crystal being of the (+)-*p* or the (-)-*p* type or, equally,
of the (+)-*n* or the (-)-*n* type, or in racemic crystals, which contain all four of
the chiral species [(±)-A(±)-B] . Alternatively, no crystals may form, as was
found for 30 of the 144 salts studied. The formation of a *p*- or *n*-type conglomerate of active crystals suggests that the racemic acid or the racemic base used to
prepare the crystals may be resolvable by means of an enantiomer of the respective base or acid in the salt. Of the 114 salts obtained in crystalline form, 25
proved to be conglomerates of crystals of the (+)-*p* and (-)-*p* type, or of the
(+)-*n* and (-)-*n* type.

In a parallel survey of 94 crystalline salts formed by a racemic acid with an
achiral base, or a racemic base and an achiral acid, 14 of the salts examined
proved to be conglomerates of active crystals, spontaneously resolved during
crystallisation. A statistical analysis of the data obtained indicates that, at the
95% confidence level, the incidence of spontaneous optical resolution on crystallisation lies in the region of 15 ± 8% for the salts of a racemic acid or base with
an achiral counter-ion, rising to 22 ± 8% for the salts formed by a racemic acid
with a racemic base, compared with 6.3 ± 1.3% for neutral molecular racemates
(Jacques, Leclerq and Brienne, 1981).

The higher incidence of optical resolution on crystallisation among salts
suggests that the traditional method of separating the enantiomers of a neutral
racemate, by conversion to an acidic or basic derivative and the isolation of a
less-soluble diastereomeric salt, may be more generally reliable than the fractional crystallisation of a covalently-bonded diastereomer. The additional step
required makes the traditional method less expeditious, however, as in the case
of the conversion of a racemic alcohol with phthalic anhydride to the phthalate
half-ester, followed by the fractional crystallisation of the diastereomeric salts

formed with an alkaloid (Pickard and Kenyon, 1911). The direct formation of the diastereomeric esters from a racemic alcohol with a chiral acid may be the more convenient for a recrystallisation separation, and it is the more appropriate for a chromatographic resolution using an achiral stationary and mobile phase (Gerlach, 1968).

10.1.3 Chromatographic methods

Some of the more recently developed chromatographic techniques for optical resolution eliminate the step of diastereomer formation and allow the direct separation of an enantiomer mixture by the use of either a chiral stationary phase or a chiral mobile phase. Such techniques are particularly useful where the methods involving diastereomer formation are not readily applicable, as in the case of uncharged chiral transition-metal complexes, or chiral hydrocarbons, or substances which racemise in the presence of normal resolving agents, e.g. Tröger's base, **3**.

3

Tsuchida and coworkers (1936) reported that the neutral bis(dimethylglyoximato) complex, $[Co(NH_3)(dmg)_2Cl]$, is partly resolved by adsorption from aqueous solution on to powdered active quartz, the use of (+)- and (-)-quartz giving optical rotations of opposite sign and approximately equal magnitude. The method was extended to the salts of charged complexes, notably, $[Cr(en)_3]^{3+}$, by Karagounis and Coumoulos (1938).

Prelog and Wieland (1944) reported the chromatographic resolution of Tröger's base on a large column of (+)-lactose, and similar columns were used by Moeller and Gulyas (1958) and others (Collman, Blair, Slade and Marshall, 1963; Fay, Girgis and Klabunde, 1970) for the separation of the enantiomers of the neutral tris(pentane-2,4-dionato) complexes, $[M(pd)_3]$, containing chromium-(III), cobalt(III), rhodium(III) or ruthenium(III). Such resolutions occupied several days but, by the use of smaller columns and elevated pressures above the mobile phase, the time required for the resolution of the $[M(pd)_3]$ complexes has been reduced to less than an hour (Nordén and Jonás, 1976).

The native polysaccharides and their derivatives provide useful chiral stationary-phase materials for enantiomer separations. Krebs and Rasche (1954) partly resolved the neutral *trans*-tris(glycinato) complex, α-$[Co(gly)_3]$, on a starch column, and the complete resolution of 6,6′-dinitrodiphenic acid and *m*-poly-phenyl derivatives on similar columns has been described (Hess, Burger and Musso, 1978). A versatile chiral column material for liquid-chromatographic optical resolutions is microcrystalline triacetylcellulose, MCTC, introduced by Hesse and Hagel (1973). Chiral saturated hydrocarbons, such as 2-cyclohexyl-norbornane, **4**, and racemic olefins and biaryls, as well as chiral ketones, alcohols, esters, acids, and salts have been optically-resolved on a MCTC column (Hesse

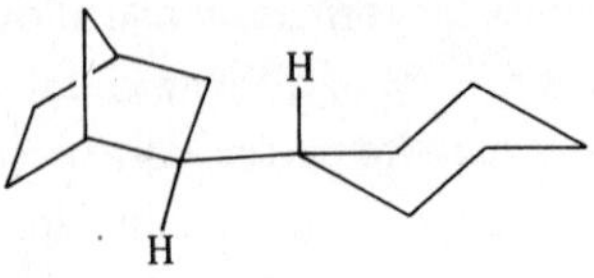

4

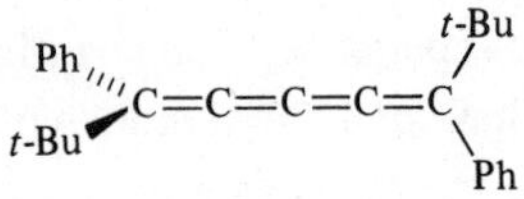

5

and Hagel, 1976). The chiral pentatetraene, **5**, which is optically labile with a half-life of an hour at 258 K, is resolved on elution from a MCTC column with pentane (Bertsch and Jochims, 1977). Liquid-chromatographic resolutions using MCTC columns become more expeditious at moderately elevated pressures of 2–10 atmospheres, the enantiomers of Tröger's base, **3**, and of *trans*-1,2-di-phenylcyclopropane, **6**, being separated in a few minutes (Linder and Mannschreck, 1980).

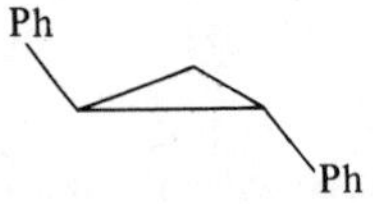

6

A range of chiral polyacrylate adsorbents for the chromatographic resolution of racemates have been developed, the more successful containing a chiral poly-mer chain. The restriction of chirality to the polymer sidechain substituents, as is produced by the reaction of polyacrylic acid chloride with a chiral amine, gives a less effective stationary phase for enantiomer separation than that resulting from the polymerisation of the corresponding chiral acrylamide monomer (Blaschke, 1980). Polymers of the latter type are particularly effective for the preparative optical resolution of racemic pharmaceutical substances with amide and imide groups (Blaschke, 1980). The achiral monomer triphenylmethyl methylacrylate is polymerised catalytically by (−)-sparteine butyllithium to an isotactic chiral polymer which, as a stationary phase, affords a chromatographic separation of the enantiomers of *trans*-bicyclo[8.8.0]octadeca-1(10)-ene, **7**, and

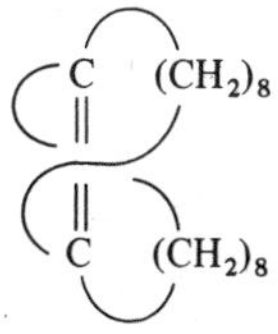

C (CH$_2$)$_8$
C (CH$_2$)$_8$

7

other chiral hydrocarbons, as well as racemates containing functional groups (Yuki, Okamoto and Okamoto, 1980).

Alternative procedures for the preparation of a chiral stationary phase are the adsorption or the covalent bonding of a resolving agent to an achiral chromatographic material. Columns of (+)-tartaric acid adsorbed on alumina have been employed to separate the enantiomers of neutral chiral metal complexes (Piper, 1961; Nordén and Jonás, 1976). The charge-transfer resolving agent, (+)- or (-)-TAPA, 1, adsorbed (Klemm and Reed, 1960) or covalently bonded to silica (Mikes, Boshart and Gil-Av, 1976; Mikes and Boshart, 1978) resolves chiral aromatic hydrocarbons including the helicenes. More recent resolving agents which have been bonded to silica or organic polymers for chromatographic enantiomer separation include the chiral crown ethers (Cram and Cram, 1978), and a series of aromatic fluoro-alcohols, the latter providing a broad spectrum of optical resolutions by means of high-performance liquid chromatography (Pirkle, House and Finn, 1980).

In order to follow the progress of an optical resolution by liquid chromatography it is desirable to monitor simultaneously the amount and the optical activity of the chiral solute in a given volume of the mobile phase emerging from the column. An attachment to a standard spectrophotometer allows the simultaneous recording at a given monitoring wavelength of the mean light absorption, as the absorbance (A), and the CD as the differential absorbance of left- and right-circularly polarized light, ($\Delta A = A_L - A_R$), of a chiral solute in the mobile phase from the chromatographic column passing through a flow-cell. A third simultaneous recording of the dissymmetry ratio, $g = \Delta A/A$), provides a measure of the optical purity (Drake, Gould and Mason, 1980). A non-zero g-ratio which is constant with respect to elution volume over a given range indicates that the chiral solute is optically pure over that range, as in the case of Tröger's base, 3, resolved on a MCTC column, where base-line separation of the enantiomers is attained (fig. 10.1a). If the g-ratio changes continuously with respect to elution volume, the resolution is incomplete, as is found for the mid-portion of the chromatogram given by pavine, 8, on a MCTC column (fig. 10.1b).

With a chiral stationary phase and an achiral mobile phase, the CD band-areas of the chromatogram, given by the individual enantiomers of an initially racemic

MeO / MeO — NH — OMe / OMe

8

substance, are equal in area (fig. 10.1). If a chiral mobile phase is employed, with
a chiral or an achiral stationary phase, the corresponding CD band areas of the
elution chromatogram are not necessarily equal, owing to discriminatory interactions between the chiral solvent and the individual enantiomers, and a resultant differential change in the CD absorption at the monitoring wavelength. The
effect is observed on monitoring over the d → d absorption wavelengths the
products from the chromatographic resolution of racemic transition-metal coordination compounds with a chiral mobile phase.

More than one hundred cationic chelate metal complexes have been optically
resolved by liquid chromatography on a column of a cation-exchange Sephadex
with an aqueous solution of a (+)-tartrate or an antimonyl (+)-tartrate salt as the

Fig. 10.1. Chromatograms of the optical resolution on a microcrystalline
triacetylcellulose column (400 × 15 mm I.D.) of (*a*) Tröger's base (**3**)
and (*b*) pavine (**8**) with a mobile phase of ethanol-water (9:1). The
chromatograms record the absorbance (*A*), the differential absorbance
of left- and right-circularly polarized radiation (Δ*A*), and the dissymmetry ratio (*g* = Δ*A*/*A*), as a function of the elution volume (*V*) at the
analytical wavelength of 280 nm in (*a*) and 295 nm in (*b*). In (*b*) the
fractions of pavine eluted between the vertical broken lines are incompletely resolved (Drake, Gould and Mason, 1980).

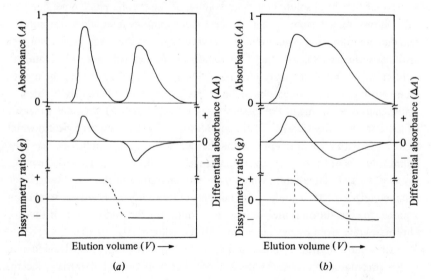

(*a*) (*b*)

eluent (Yoshikawa and Yamasaki, 1979). Sephadex, which is composed of the D-glucose polysaccharide, dextran, cross-linked with epichlorohydrin, shows itself some discrimination between the enantiomers of chiral metal complexes, but the optical resolution is generally the more complete with a chiral mobile phase. In the cation-exchange forms, alkylsulphate or carboxymethyl groups are linked by ether bonds to the dextran chains, and columns of the latter type, CM-Sephadex, effectively separate the enantiomers of the neutral β-alaninato complex, *trans*-[Co(β-ala)$_3$], with a 1:1 ethanol–water eluent (Yoneda, 1979).

In the chromatographic resolution of the tris(diamine) metal complexes on a standard Sephadex column with aqueous (+)-tartrate solutions, the Λ-isomer is first eluted, whereas the Δ-enantiomer of the complex appears in the early fractions if (+)-tartrate is covalently linked to the Sephadex through ether or ester bonds and an achiral mobile phase is employed (Yoshikawa and Yamasaki, 1979). The ion association constants of the tris(ethylenediamine) complex, [Co(en)$_3$]$^{3+}$, with either (+)-tartrate or antimonyl (+)-tartrate are larger for the Λ- than the Δ-isomer of the complex ion in aqueous solution, and the less-soluble diastereomeric salts of (+)-tartrate are generally formed with the Λ-enantiomer of a tris(diamine) metal complex.

The stereochemical basis of the chromatographic and fractional crystallisation methods of optical resolution connected with the preferred ion association has been investigated by X-ray crystal structure analyses of the salt, Λ-(+)-[Co(en)$_3$]. [(+)-tartrate].Br.5H$_2$O, and of analogous less-soluble diastereomers formed by (+)-tartrate with tris(diamine) complex cations (Yoneda, 1979). In the crystal

Fig. 10.2. The ion-pair relationship between the Λ-(+)-[Co(en)$_3$]$^{3+}$ complex ion and the (2R,3R)-(+)-tartrate ion in the crystal lattice of the salt, (+)-[Co(en)$_3$].Br.(+)-tartrate.5H$_2$O. The broken lines denote hydrogen bonding between the oxygen atoms of the anion and the N—H groups of the cation (Yoneda, 1979).

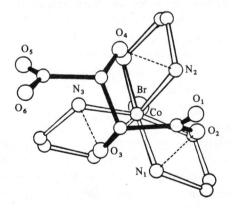

structure, the Λ-(+)-[Co(en)$_3$]$^{3+}$ complex cation and the (+)-tartrate anion are associated through three hydrogen bonds, involving the three N—H groups, one from each ligand, nearly parallel to the threefold rotation axis of the complex ion, and sharing a common octahedral face (fig. 10.2). These three N—H groups of the complex cation hydrogen bond to the two hydroxyl and one of the carboxyl groups of the (+)-tartrate ion, while the second carboxyl group of the anion acts as a steric discriminator, economically accommodated by the Λ-(+)-cation but not by the Δ-(−)-isomer. Extensions of the model account for the discriminatory associations observed between the enantiomers of other types of chelated cobalt(III) complex and both the (+)-tartrate and the antimonyl (+)-tartrate ion (Yoneda, 1979).

The resolution of optical isomers by gas-liquid chromatography was carried out initially with an achiral stationary phase by converting the racemate to a diastereomeric mixture (Gil-Av and Nurok, 1974). Subsequently the enantiomers of a number of chiral amines, amino-acid esters, and α-methyl- and α-phenyl-carboxyamides were separated, as their N-trifluoroacetyl-derivatives, by gas chromatography using a chiral secondary amide as the stationary phase (Weinstein, et al., 1976). A stereochemical model for the chiral recognition involved in these optical resolutions is based upon a series of crystal structure determinations, by X-ray diffraction, of an enantiomer and the corresponding racemate of the secondary amides, RCONHCH(CH$_3$)Ar, employed as a stationary phase (Weinstein, Leiserowitz and Gil-Av, 1980).

The dominant molecular-packing mode of the chiral secondary amides in the crystal consists of a 5 Å translation axis which is based upon linear intermolecular hydrogen bonding, and a close packing of the aromatic residues, with a separation of 3.4 Å. The 5 Å translation axis is partly retained in the melt of the larger molecules, notably, (R)-N-lauroyl-α-(1-naphthyl)ethylamine. As the resolution is particularly good for racemates of the same structural type, RCONHCH(CH$_3$)Ar, it is proposed that the isomers of the volatile racemate intercalate into the hydrogen-bonded stack of molecules forming the stationary phase. The chiral guest molecules with the same absolute configuration as the host molecules are sterically the more compatible in the hydrogen-bonded stack, accounting for the selective retention of the volatile enantiomers with the same configuration as the stationary phase substance (Weinstein et al., 1980).

10.2   Chiral photodiscrimination

The relatively small differential absorption of left- and right-circularly polarized light by an enantiomer limits the degree of photochemical discrimination between the two isomers of the corresponding racemate produced by direct irradiation with one of the circular components. Although enantiomeric photodiscrimination by circularly-polarized irradiation was proposed by Le Bel in 1874, authentic cases were established only some fifty years later by Kuhn and coworkers, on account of the small magnitude of the effect.

10.2.1 Circular photolysis

Kuhn and Knopf (1930) investigated the photolysis of racemic α-azido-N,N-dimethylpropionamide, **9**, with circularly-polarized radiation in the wavelength region of the weak n → π* azido group absorption (λ_{max} 280 nm, ϵ_{max}

$$CH_3CHCON(CH_3)_2$$
$$|$$
$$N_3$$

9

25). The (+)-isomer of the substrate, **9**, has a CD maximum, $(\epsilon_L - \epsilon_R)_{max}$, of +0.45 at 280 nm, giving a dissymmetry ratio, $g = \Delta\epsilon/\epsilon$, of 2.4% at this wavelength. Irradiation of the racemate, **9**, to the extent of 40% decomposition left the residual substrate with an enantiomeric excess, $([R] - [S])/([R] + [S])$, of 0.5%, estimated from the specific rotation of the residue, $[\alpha]_D$ 1.04° (neat), relative to that of a pure enantiomer.

In an analysis of the differential photolysis of a racemate with circular radiation, Kuhn and Knopf (1930) showed that the optical rotation of the residual substrate rises to an optimum value, proportional to the dissymmetry ratio, g, and subsequently falls during the course of the photodecomposition, and that the enantiomeric excess increases towards the limit of unity as the fraction of residual substrate goes to zero. The differential photolysis experiments and the analysis of Kuhn and Knopf (1930) have been reproduced and extended by Kagan and coworkers (1974), who show that the circular irradiation of racemic camphor, **10**, at 313 nm, where the g-ratio of an enantiomer has a value of 9%, gives after 90% photodecomposition a residual substrate with an enantiomeric excess of 20%.

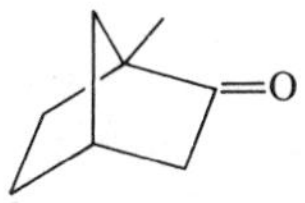

10

10.2.2 Circular photoequilibration

In addition to the differential photolysis of a racemic mixture, two other types of photodiscrimination by a circularly-polarized radiation field have been distinguished (Buchardt, 1974; Kagan *et al.*, 1974). Enantiomers which are chemically stable but optically labile in the photoexcited state re-equilibrate from the racemic-mixture composition on irradiation with circularly-polarized light to give an excess of the enantiomer with the smaller extinction coefficient for the particular circular component employed, ϵ_L or ϵ_R. In the photostationary state, the enantiomeric excess equals the corresponding extinction coefficient ratio, that is, one-half of the dissymmetry ratio $(g/2)$ at the wavelength of irradiation. The photostationary equilibria of several tris-chelate complexes of chromium(III) have been investigated (Stevenson and Verdieck, 1969). In the case of the pentane-2,4-dionato complex, $[Cr(pd)_3]$, for which the g-ratio has the value of 7.8% at 546 nm, the CD spectrum of an enantiomer was deduced from the results of photoequilibration measurements prior to the isolation of the individual enantiomers (Stevenson, 1972).

10.2.3 *Circular photosynthesis*

A third type of direct chiral photodiscrimination results if the enantiomers of a racemic mixture irradiated with circularly-polarized light form a photostable product at a faster rate than they racemise in the photoexcited state. The enantiomeric excess in the photostable product, $\{([R] - [S])/([R] + [S])\}_{(P)}$, is now independent of the extent to which the photoreaction proceeds to completion, and the excess equals in magnitude the extinction coefficient ratio of the enantiomers in the racemic substrate precursor for the circular radiation component employed. The magnitude of the extinction coefficient ratio again corresponds to one-half of the dissymmetry-ratio of a substrate enantiomer, *l-S* or *d-S*, e.g., for left-circular irradiation,

$$[\epsilon_L(l\text{-}S) - \epsilon_L(d\text{-}S)] / [\epsilon_L(l\text{-}S) + \epsilon_L(d\text{-}S)] = [(\epsilon_L - \epsilon_R)/(\epsilon_L + \epsilon_R)] (l\text{-}S) \qquad (10.1)$$

The third type, asymmetric or dissymmetric photosynthesis, is exemplified by the photocyclisation of annelated *cis*-stilbene derivatives with circularly-polarized light to give the corresponding dihydrophenanthrenes, which readily oxidise, thermally or photochemically, to the fully aromatic analogue. The photocyclisation, in contrast to analogous thermal reactions, generally produces a member of the helicene series (Martin, 1974), and the use of circularly-polarized radiation for the photoreaction gives an enantiomeric excess of up to 0.3% for [6]-helicene and its halogen-substituted derivatives, rising to some 2% for [8]-helicene (Buchardt, 1974). The dissymmetric photosynthesis of [5]-helicene (Goedicke and Stegemeyer, 1972), and of the analogues from [6]- to [9]-helicene (Kagan *et al.*, 1971; Bernstein, Calvin and Buchardt, 1972, 1973) have been investigated from a mechanistic viewpoint.

The possibility that the dissymmetric photosynthesis arises from the first type of chiral photodiscrimination, enantioselective photolysis of the helicene product, is ruled out by the observation that prolonged irradiation of a racemic helicene with a given circular component selectively decomposes the particular enantiomer preferentially produced by that circular component through the photocyclisation of the corresponding annelated *cis*-stilbene (Kagan *et al.*, 1971). Moreover, the relationship between the optical yield of the helicene and the wavelength of the circular irradiation for two different precursors of [8]-helicene differ one·from the other and from the spectroscopic curve relating the wavelength to the dissymmetry ratio of an enantiomer of [8]-helicene (Bernstein *et al.*, 1972, 1973). The dissymmetry ratio of the photoselected species and the optical yield of the product are expected to show a similar relationship to the wavelength of the circular radiation, so that the differential photolysis of the enantiomers in the helicene product is not a probable mechanism.

The photocyclisation of *cis*-1,2-(β-naphthyl)-ethylene, 11, formed by the photoisomerisation of the corresponding *trans*-isomer, yields the relatively stable dihydro-[5]-helicene, 12. The (4a,4b)-hydrogen atoms in the dihydro-[5]-helicene are *trans* to one another, as expected by the rules of Hoffmann and Woodward (1968) for the cyclisation of a *cis*-stilbene analogue from the lowest singlet photoexcited state. The *trans*-relation of the (4a,4b)-hydrogen atoms follows from

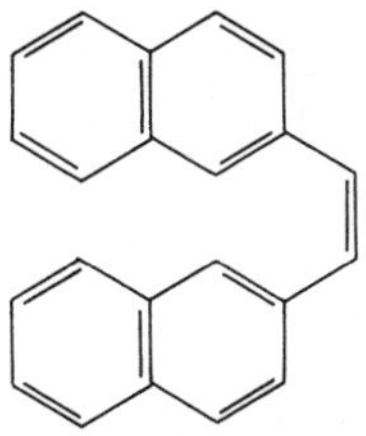

11 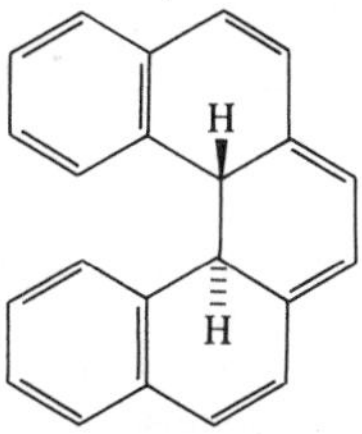   12 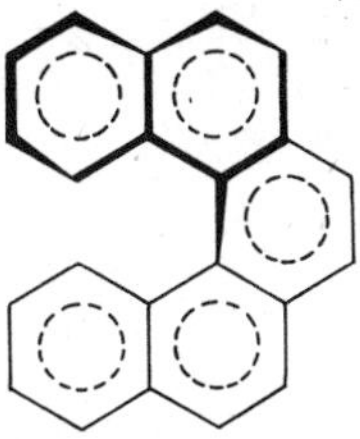   13

the cyclisation of the precursor, 11, by circular irradiation at 365 nm to give a chiral dihydro-intermediate, 12, shown by an enantiomeric excess of the (-)-isomer of the latter with right-circular, or the (+)-isomer with left-circular light (Goedicke and Stegemeyer, 1972). While the major photodiscrimination occurs in the cyclisation of 11 to 12, a further minor photoselection is found in the subsequent photooxidation stage to the product [5]-helicene, 13. The left-circular irradiation of the racemic dihydro-derivative, 12, at 436 nm in the presence of oxygen gives a small enantiomeric excess of (P)-(+)-[5]-helicene, 13 (Goedicke and Stegemeyer, 1972).

10.2.4 Diastereomeric photodiscrimination

The direct photodiscrimination between the enantiomers of a racemic mixture in the fluid phase by circularly-polarized radiation has a basic limitation in the small magnitude of the ratio of the molecular dimensions to the wavelength of the light absorbed by the valence state electronic transitions of the molecule. Where the discriminating agent itself has molecular dimensions, the extent of the selection between the enantiomers of a racemate may become substantial, as in the formation of a *p* and the corresponding *n* diastereomer, followed by a photodiscrimination between the latter with unpolarized radiation.

In a cholesteric liquid crystal, the chirality dimension, the pitch of the macroscopic helix formed by the aligned molecules, has a magnitude of the order of the wavelength of visible radiation. The photocyclisation of 2-styrylbenzo[c] phenanthrene, **14**, dissolved in a right-handed cholesteric mesophase of cholesteryl benzoate or analogous cholesterol derivatives, affords [6]-helicene with an enantiomeric excess of some 1% of the (P)-(+)-isomer, **15** (Nakazaki, Yamamoto and

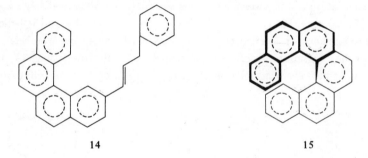

14   15

Fig. 10.3. The photocyclisation of the styryl benzo[c] phenanthrene (*a*) to the [6]-helicene (*b*) in toluene, with iodine, where R = CO_2(-) menthyl. With a chemical yield of 58%, the P-(+)-helicene (*b*) is produced in 96% diastereomeric excess at $-78°C$, but the M-(-)-helicene (*b*) is afforded in 60% diastereomeric excess at $+80°C$ (Vanest and Martin, 1979).

(*a*)   (*b*)

Fujiwara, 1978). Even in a solution of a mechanically-twisted nematic meso-phase, with a helical pitch-length of some 80 μm, the photosynthesis of [6] -helicene from the precursor, **14**, proceeds with an optical yield of 0.04% of the (P)-(+)-isomer, **15**, for a right-handed helical preparation (Nakazaki, Yamamoto, Fujiwara and Maeda, 1979).

Randomly orientated in a fluid chiral solvent, the diarylethylene, **14**, under-goes photocyclisation with unpolarized radiation to [6] -helicene with a 2% enantiomeric excess of the (P)-(+)-isomer, **15**, employing (S)-(+)-ethyl mandelate as the solvent (Laarhoven and Cuppen, 1978). The diastereomeric associations between the precursor, **14**, and a chiral hydrocarbon solvent, such as (+)-α-pinene, which affords an optical yield of 0.2%, are expected to be weak and to produce only a minor selectivity. A larger discrimination is shown if a chiral group is covalently bonded to an annelated *cis*-stilbene precursor, **14**, or an analogue (Cochez, Martin and Jespers, 1976/77). The (-)-menthyl ester derivative of **14** undergoes photocyclisation in toluene solution to give the corresponding [6] -helicene diastereomeric products (fig. 10.3) with a 96% excess of the (+)-dia-stereomer at ‑78 °C or a 60% excess of the (-)-diastereomer at +80 °C (Vanest and Martin, 1979).

10.2.5 Solid-state photodiscrimination

The chirality dimension of an enantiomorphous crystal lattice is of a molecular order, namely, that of the unit cell, as opposed to the helix pitch-length of a cholesteric liquid crystal, or a twisted nematic mesophase. The solid-state photochemical reactions of some achiral molecules, or even of racemic mixtures, in a chiral crystal accordingly provide photoproducts with a substan-tial excess of the enantiomer selected by the particular lattice structure (Green, Lahar and Rabinovich, 1979). The achiral monomer, *trans*-1,3-pentadiene, **16**, forms an inclusion compound with the enantiomers of the *trans, anti*-perhydro-triphenylene, **17**, which crystallise in the chiral space group P6$_3$. The monomer in crystals of the inclusion compound undergoes coordination polymerisation on γ-irradiation to give the isotactic *trans*-1,4-polypentadiene, **18**, with an excess of

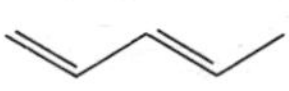

16

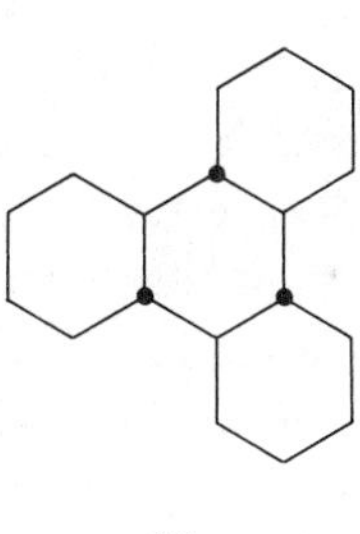

17

18

the (+)-isomer employing the (R)-(−)-host enantiomer, **17** (Farina, Audisio and Natta, 1967).

A chiral host molecule is not essential for a solid-state dissymmetric photo-synthesis. A number of 1,4-diaryl-1,3-butadiene systems crystallise in the chiral space group $P2_12_12_1$ and photodimerise in the solid state to give a substituted cyclobutane photoproduct with mirror symmetry. Two different 1,4-diaryl-1,3-butadiene systems of the set form mixed crystals belonging to the $P2_12_12_1$ space group and give, on irradiation, chiral heterodimers. The irradiation of a poly-crystalline sample of the mixed crystals gives a racemic mixture of the hetero-dimers, but the selection of a large single crystal to ensure a homochiral-lattice packing mode, and the use of a radiation filter allowing the photoexcitation of only one of the two monomer dienes, provides a 70% enantiomeric excess in the heterodimer photoproduct (Elgavi, Green and Schmidt, 1973).

Dimers, trimers, and higher oligomers are formed by the solid-state irradiation of the benzene-1,4-diacrylate, **19**, which crystallises in the space group P1, either

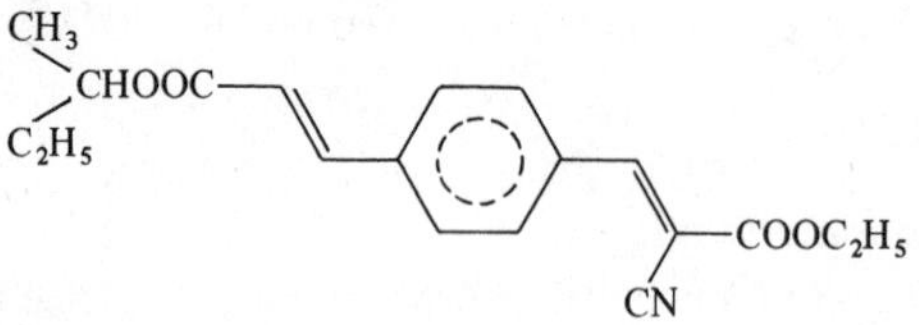

19

as a single enantiomer or as a racemic mixture. A crystal of the racemate has a chiral lattice structure with a disordered distribution of the enantiomeric 2-butyl groups. The irradiation of a large single crystal of the racemate, grown from the melt, gives dimeric photoproducts with a specific rotation, $[\alpha]_D$, up to some 30°, compared to $\sim$ 110° for the photodimers produced by the irradiation of a crystal of a single enantiomer of **19** or of mixed crystals down to an optical purity of 22% (Addadi and Lahav, 1979).

10.3   Optical activation

Each of the chiral photodiscriminations has a thermal counterpart, although the analogues of the photoprocesses based upon circular irradiation have a larger enantioselectivity and were early discovered. The kinetic resolution of a racemate by the selective reaction of one of the enantiomers, Pasteur's (1858) third method of resolving (±)-tartaric acid, using micro-organisms, has formal analogies to the circular photolysis of a racemate. The organic 'asymmetric transformations of the first kind' (Kuhn, 1932), and the 'Pfeiffer effect' in co-ordination chemistry (Pfeiffer and Quehl, 1931), where an optically-labile racemic mixture reequilibrates to an excess of one of the enantiomers in a chiral medium, were earlier parallels of the photostationary reequilibrations produced by circular irradiation. Similarly, the expectation of Le Bel (1874), that an enantiomeric excess from a symmetric substrate is accessible through reactions conducted either in the presence of optically-active substances, or through the agency of circularly-polarized light, was first realised by way of the former route.

10.3.1 *Kinetic resolution*

The isolation of one of the enantiomers of a racemic mixture by the selective reaction of the other isomer follows a kinetic course common to the use of any chiral agency, a micro-organism, an enzyme system or other chiral catalyst or reactant, or circularly-polarized radiation. The course of a kinetic resolution was early investigated in connection with enzymatic enantioselectivity (Bredig and Fajans, 1908), and Kuhn (1936) extended his previous analysis of the circular photolysis of racemates to enzymatic resolution.

Studies of chemical kinetic resolution were initially concerned with the qualitative features of the reaction. Marckwald and McKenzie (1899) found that in the esterification of racemic mandelic acid with a deficiency of (-)-menthol, the *n*-diastereomer, (-)-menthyl (+)-mandelate, **20**, was the more abundant pro-

20

duct and the residual mandelic acid had become optically enriched with the (-)-isomer. Subsequently such enantioselective reactions of a racemate with a chiral reagent were employed to correlate stereochemical configuration in a series of analogous chiral reactants (Eliel, 1962; Morrison and Mosher, 1976).

Most of the reactions employed correlated the configurations of molecules belonging to a common chirality type, based upon the dissymmetric centre of an asymmetric carbon atom, and such reactions generally did not extend to different types, notably, the axial dissymmetry of the hindered biaryl series. A configurational connection between the two chirality types was established by Mislow and coworkers, based upon a chemical kinetic resolution followed to virtual completion, with an analysis of the time-evolution of the optical yield in both the substrate and the product (Newman, Rutkin and Mislow, 1958). More recently the analysis has been extended to the kinetic course of the circular photolysis of a racemate (Kagan *et al.*, 1974).

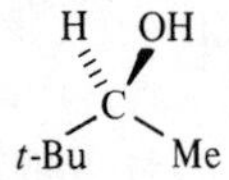

21

In the reaction studied, an alcohol with an asymmetric carbon atom, (S)-(+)-methyl-*t*-butylcarbinol, **21**, was employed to reduce catalytically with aluminium *t*-butoxide the carbonyl group of a racemic-bridged biphenyl ketone, **22**, with axial dissymmetry. The partial reduction of the racemic ketone, **22**, produces an

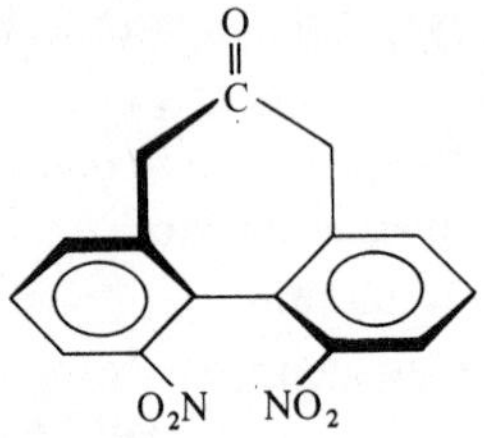

22 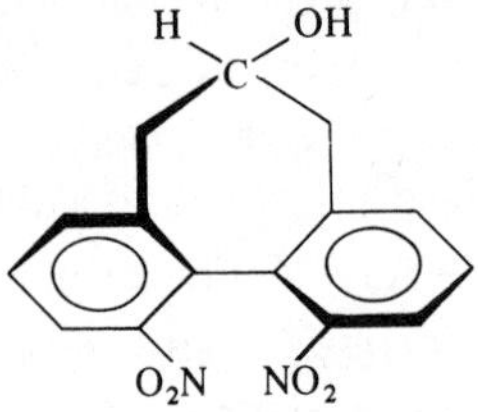   23

excess of the laevorotatory-bridged biphenyl carbinol, **23**, and leaves behind the unreduced ketone with the dextrorotatory isomer in excess. An analysis of the steric interactions in the probable transition state of the reduction (fig. 10.4) indicates that the (R)-(−)-ketone, **22**, is reduced faster than its enantiomer by the (S)-(+)-alcohol, **21**, to give the (R)-(−)-bridged biphenyl carbinol, **23**, as the preferred product (Newman *et al.*, 1958). The chirality of the carbinol, **23**, derives from the axial dissymmetry of the biphenyl system and not from the carbinol centre.

With a large excess of the (S)-(+)-alcohol, **21**, the reduction of the ketone becomes a pseudo-first-order reaction or, more precisely, two parallel first-order reactions, one referring to the (R)-(−)-ketone, **22**, and the other to its (S)-(+)-isomer, with unequal rate constants, k_R and k_S, respectively. Of the two ketone substrate materials, M_R and M_S, the former is consumed the more rapidly, leaving an excess of the latter (fig. 10.5). The absolute excess of the (S)-(+)-ketone, M_S, rises to an optimum at a time, t_{max}, when the two rates, $k_R[M_R]$, and $k_S[M_S]$, are equal, and the optical activity of the kinetically-resolved substrate attains an optimum at that time, falling off subsequently as the absolute excess is reduced (fig. 10.5).

The fractional enantiomeric excess of the substrate material, f_M, tends to unity, however, as the quantity of residual substrate goes to zero at infinite reaction time, according to the hyperbolic tangent relation,

$$f_M = ([M_S] - [M_R])/([M_S] + [M_R]) = \tanh[(k_R - k_S)t/2] \tag{10.2}$$

In contrast, the bridged biphenyl carbinol product, P, **23**, has a maximum fractional enantiomeric excess, f_P, at the initiation of the reaction, and f_P tends to zero as the reaction goes to completion (fig. 10.6).

The fractional enantiomeric excess of the product, f_P, and of the substrate, f_M, are connected by the general relation,

$$f_P = ([P_R] - [P_S])/([P_R] + [P_S]) = f_M([M_R] + [M_S])/([P_R] + [P_S]) \tag{10.3}$$

The particular optimum value of f_P at zero time equals the value of f_M at t_{max}, when the absolute excess of the substrate, $([M_S] - [M_R])$, is a maximum (fig. 10.5). The optimum value of f_P is dependent upon the ratio of the two first-order rate constants, $K = k_R/k_S$, according to the expression,

$$f_P(t_0) = f_M(t_{max}) = (K - 1)/(K + 1) \tag{10.4}$$

For the reduction of the racemic ketone, **22**, by the (S)-(+)-alcohol, **21**, in dioxan solution at 63°C the rate-constant ratio, K, has the value of 2.2, giving

Fig. 10.4. The kinetic resolution of racemic 4′,1″-dinitro-1,2,3,4-dibenz-1,3-cycloheptadiene-6-one, **22**, by (S)-(+)-methyl-*t*-butylcarbinol, **21**, catalysed by aluminium *t*-butoxide, to give the carbinol product **23**. The (R)-(−)-ketone substrate is reduced at a faster rate than the (S)-(+)-ketone on account of the smaller steric hindrance in the transition state of the reduction in reaction (*a*) (Newman, Rutkin and Mislow, 1958).

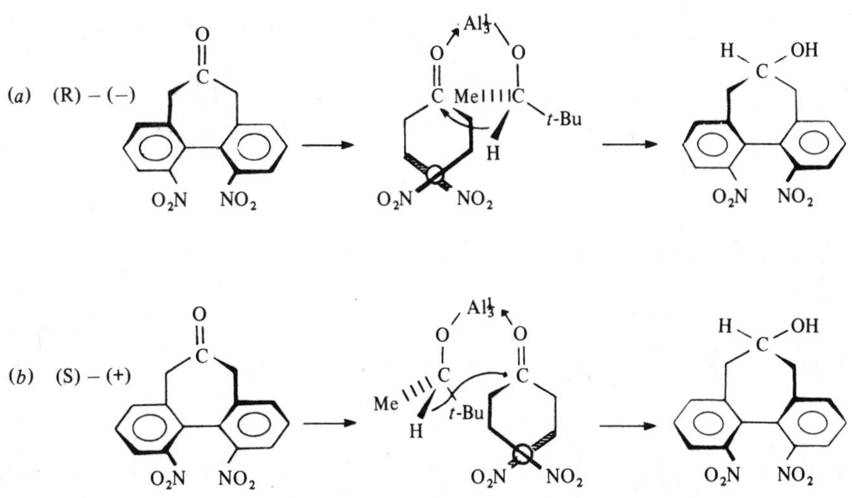

Fig. 10.5. The time-development of the enantiomeric excess ([S] − [R]) and of its optical rotation [α] in a first-order kinetic resolution of a racemate with a rate-constant ratio $(k_R/k_S) = 2$. The concentration [C] refers to the individual enantiomers [R] and [S] and to the enantiomeric excess in the residual substrate.

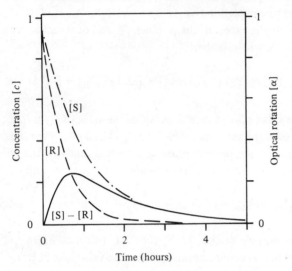

Fig. 10.6. The time-development of the fractional enantiomeric excess of the substrate material f_M, and of the product f_P (eqns 10.2 and 10.3), and of the mole fraction of the residual substrate material x_M, and of the product x_P, in a first-order kinetic resolution of a racemate with the rate constant ratio $(k_R/k_S) = 2.2$ (Newman, Rutkin and Mislow, 1958).

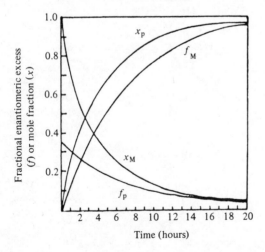

0.365 for the fractional enantiomeric excess, f_P at zero time, and f_M at t_{max} = 3.66 h (Newman *et al.*, 1958).

In a kinetic resolution, whether thermal or photochemical, the optical yield is optimal only when the chemical yield goes to zero, either for the initial product or the residual substrate. The basis for a choice of the quenching point of an incomplete kinetic resolution, in order to obtain satisfactory optical and chemical yields, is provided by the general relations between the fractional enantiomeric excess of the residual substrate, f_M, and the fraction of the product formed or, what is equivalent, the fraction of the unreacted substrate (fig. 10.7). The relationships, given originally for the circular photolysis of a racemate over a range of values for the dissymmetry ratio, $g = [(\epsilon_L - \epsilon_R)/\epsilon]$ (Kagan *et al.*, 1974), apply equally to the thermal kinetic resolution of a racemate through the correspondence,

$$g = 2(K - 1)/(K + 1) \tag{10.5}$$

Fig. 10.7. The relationship between the fractional enantiomeric excess of the substrate material f_M (eqn 10.2), and its residual mole fraction x_M, or the mole fraction of the product x_P, in a first-order kinetic resolution, as a function of the rate constant ratio $K = (k_R/k_S)$, or in a circular photolysis as a function of the dissymmetry ratio $g = (\epsilon_L - \epsilon_R)/\epsilon$. Relation (A) corresponds to $K = 7$, or $g = 1.5$, curve (B) to $K = 2.3$ or $g = 0.8$, and curve (C) to $K = 1.27$ or $g = 0.24$ (Kagan, Balavoine and Moradpour, 1974).

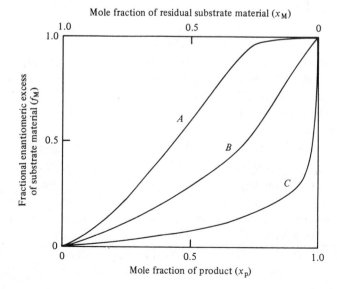

The generally greater efficiency of the thermal kinetic resolution of a racemate, compared to the corresponding circular photolysis, is illustrated by the dissymmetry ratio, $g = 0.02$, for the circular photolysis of the azide, **9**, corresponding to a small thermal kinetic discrimination, $K = 1.02$, whereas the value of $K = 2.2$ for the kinetic resolution of the ketone, **22**, with the alcohol, **21**, corresponds to an exceptionally large dissymmetry ratio, $g = 0.75$, for an isotropic fluid phase. The limiting value of the dissymmetry ratio, $g = 2$, is approximately attained by a cholesteric mesophase for the axial propagation of radiation with a wavelength equal to the helix pitch-length. The corresponding limit of infinity for the thermal discrimination ratio is approximated by many enzyme systems, although some have small K values or do not discriminate between enantiomers (Bentley, 1969).

10.3.2 Asymmetric transformation

In 1846, Dubrunfaut discovered that the optical rotation $[\alpha]_D$ of glucose in aqueous solution changed slowly from an initial value of $+111°$ to an equilibrium value of $+52.5°$. Lowry (1899) coined the term, mutarotation, for the process, and showed that the α- and the β-hemiacetal forms of glucose each give an equilibrium mixture of 38% of the α- and 62% of the β-epimer in aqueous solution.

Subsequent investigations showed that mutarotation in solution is common among the diastereomeric derivatives of enantiomers owing their chirality to restricted rotation around a single bond, or of chiral molecules which readily undergo a dissociation from the asymmetric centre, whether of a hydrogen ion from an organic acid or a polyatomic ligand in a coordination complex (Turner and Harris, 1947; Harris 1958). The mutarotation process involves a discrimination between the forward and the reverse rates of interconversion of the enantiomers in an optically-labile racemate due to a dissymmetric environment, whether a chiral solvent, a chiral counter-ion, or a covalently-bonded chiral group. The discrimination results in an equilibrium between the enantiomers of a labile racemate or, rather, between the corresponding p and n diastereomers, with a constant differing from unity. Kuhn (1932) termed such an interconversion, an asymmetric transformation of the first kind. The corresponding second kind of asymmetric transformation involves the crossing of a phase boundary to give an enantiomeric or diastereomeric separation, either by crystallisation from the solution, or by extraction with an immiscible solvent.

The second kind of transformation may result in the complete conversion of a racemate into one of its enantiomers. Racemic 1,1'-binaphthyl-5,5'-dicarboxylic acid undergoes a 78% conversion to its (+)-isomer by the progressive precipitation of the n-diastereomer with brucine from 2-ethoxyethanol solution at 80°C on gradual dilution with water (Hall, Ridgwell and Turner, 1954). The tris (catechyl) complex of arsenic (V), $[As(cat)_3]^-$, which is optically stable in neutral aqueous solution but labile below pH 3, is converted wholly into its (−)-isomer with cinchonine (Mann and Watson, 1947), or into the (+)-isomer with quinine (Craddock and Jones, 1961).

The first kind of asymmetric transformation in coordination chemistry is frequently termed the 'Pfeiffer effect' from the discovery by Pfeiffer and Quehl (1931) of a substantial optical activity induced in aqueous solutions of the tris (1,10-phenanthroline) complex, $[Zn(phen)_3]^{2+}$, and similar optically-labile complexes, by the addition of an anionic, neutral, or cationic chiral species. If the labile complex is sufficiently stable to allow optical resolution by standard procedures, the effect may be quantified by a comparison of the induced optical activity with that of a resolved enantiomer. The salts of the racemic tris(oxalato) chromium(III) complex, $M_3[Cr(ox)_3]$, in diethyl-(+)-tartrate solution at 30°C reequilibrate to give isomer ratios, $[\Lambda]/[\Delta]$, of the complex anion, $[Cr(ox)_3]^{3-}$, in the range 2 – 5, dependent upon the particular alkali-metal cation (Fig. 10.8). The effective species is a cation-solvent complex, since the addition of the crown-ether, 1,8-crown-6, which forms an achiral complex with the alkali-metal cation,

Fig. 10.8. The relationship between the constant of the equilibrium established between the enantiomers of the complex anion $K = [\Lambda]/[\Delta]$, in the racemic salts, $M_3[Cr(ox)_3]$, dissolved in diethyl-(2R,3R)-(+)-tartrate-water mixtures at 30°C is a function of the mole fraction x of the ester, for $M^+ = Na^+$, K^+ and Rb^+. The free energy of the discrimination between the enantiomers is similarly related to x through $\delta\Delta G = -RTlnK$.

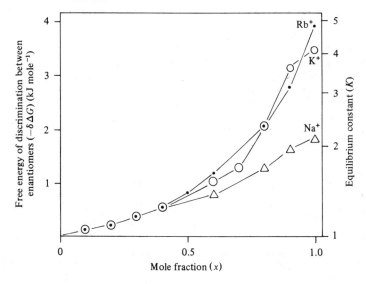

progressively quenches the discrimination between the enantiomers of the complex anion, as does dilution with an achiral solvent (Drake, Levey, Mason and Prosperi, 1982).

Diastereomeric discriminations in the fluid phase extend to optically-stable enantiomers. Dwyer and coworkers found that the solubilities of the (+)- and the (-)-isomer of the ruthenium(II) complex, $[Ru(phen)_3](ClO_4)_2$, while equal in water or aqueous sodium chloride, differ by 3% in a $M/10$ aqueous solution of sodium potassium (+)-tartrate. Moreover, the related tris($2,2'$-bipyridyl) complexes of osmium(II) and osmium(III), each constituting a chemically- and optically-stable species, are connected by a redox potential differing by 2.5 mV for the (+)- and the (-)-isomers in aqueous solutions of ammonium (+)-camphorsulphonate at an ionic strength of 0.1, whereas the corresponding potentials for achiral electrolyte solutions do not differ within the experimental error of 0.2 mV (Gyarfas, 1954).

The diastereomeric discriminations between optically stable enantiomers are more substantial in non-aqueous chiral solvents. The solubility of the complex salt, Λ-(+)-$[Co(en)_3]Cl_3$, in diethyl (+)-tartrate at 25°C is 4.2 times larger than that of the Δ-(-)-isomer, and solutions of the Λ-(+)-enantiomer are more viscous than corresponding solutions of the Δ-(-)-antipode with the same concentration (Yamamoto and Yamamoto, 1975). Successive extractions with cyclohexane of diethyl (+)-tartrate solutions of the racemic tris(pentane-2,4-dionato) complexes, $[M^{(III)}(pd)_3]$, leave a 10% enantiomeric excess of the Λ-isomer in the chiral ester layer at 20°C, the initial cyclohexane extracts being enriched with the Δ-enantiomer of the neutral complex (Mason *et al.*, 1977).

Table 10.1. *The enantiomer equilibrium ratio, $[\Lambda]$ $[\Delta]$, of tris-chelate metal complexes dissolved in diethyl-(2R,3R)-(+)-tartrate, or in mixtures with water containing a mole fraction, x, of the diester at a temperature, T, with the differential enthalpy, $\delta \Delta H$ (kJ mol^{-1}), and entropy, $\delta \Delta S$ (J mol^{-1} K^{-1}), relative to the unit enantiomer ratio of the corresponding racemate in an achiral environment (Drake, Levey, Mason and Prosperi, 1982)*

Complex	x	$T°C$	$[\Lambda]/[\Delta]$	$-\delta\Delta H$	$-\delta\Delta S$
$[Co(en)_3]Cl_3$ [a]	1.0	25	4.2	14	34
$(NH_4)_3[Cr(ox)_3]$	1.0	30	4.8	18	47
$K_3[Cr(ox)_3]$	0.9	30	3.5	12	30
	0.8	30	2.5	10	26
$[Ni(phen)_3]Cl_2$	1.0	25	1.09	0.5	0.9
$[Cr(pd)_3]$	1.0	20	1.2		

(a) From Yamamoto and Yamamoto (1975).

The sign of the diastereomeric discrimination between a given chiral α-hydroxy ester solvent and the enantiomers of a tris-chelate complex is common to a range of cationic, neutral, and anionic complexes (table 10.1). Studied by the enantiomer-reequilibration, the differential-solubility, or the solvent-extraction method, diethyl-(2R,3R)-(+)-tartrate is found to give rise to an enantiomeric excess of the Λ-isomer of the tris-chelate complex, whereas ethyl-(2S)-(−)-lactate produces an enantiomeric excess of the corresponding Δ-isomer (Drake *et al.*, 1982). The differential entropy values obtained indicate that the preferred solute enantiomer orders the chiral solvent molecules more effectively than its antipode, forming the larger solute–solvent intermolecular associations (table 10.1).

10.3.3 Organic asymmetric synthesis

The concept of asymmetric synthesis developed from the work of Fischer on interconversions in the sugar series. The three chiral centres of L-arabinose determined the configuration of the new asymmetric carbon atom introduced by the cyanohydrin synthesis, giving L-mannonic acid as the major product, and L-gluconic acid in small quantity. These and analogous results led Fischer (1894) to the view that, in chemical synthesis, 'there are equal chances for mirror-images only', and, 'once a molecule is asymmetric, its extension also proceeds in an asymmetric sense. This concept completely eliminates the difference between natural and artificial synthesis. The advance of science has removed the last chemical hiding place for the once so highly esteemed *vis vitalis*.'

A more formal definition of asymmetric synthesis was given by Marckwald (1904) as the chemical production of optically-active substances from symmetrically-constituted compounds with the intermediate use of optically-active materials, but with the exclusion of all analytical processes giving a physical optical resolution. The definition was exemplified by the decarboxylation of the mono-brucine salt of ethylmethylmalonic acid at 170°C to give α-methyl-butyric acid containing a 10% enantiomeric excess of the (−)-isomer (Marckwald, 1904). The status of the reaction as an asymmetric synthesis was questioned by, among others, Cohen and Patterson (1904), who maintained that, since the mono-anion of ethylmethylmalonic acid constitutes a racemic mixture, a physical separation of the diastereomeric brucine salts might have occurred on crystallisation, prior to the decarboxylation. Moreover, an incomplete decarboxylation would constitute a kinetic resolution, giving an optically-enriched product, of the type studied earlier by Marckwald and McKenzie (1899).

At the same time, McKenzie (1904) reported the first of a series of studies, extending over more than thirty years, on the asymmetric synthesis of chiral α-hydroxy acids from the corresponding α-keto ester, prepared from a chiral carbinol, by reduction or the addition of achiral Grignard reagents (Ritchie, 1933, 1947). Previous workers had attempted an asymmetric synthesis of mandelic acid by the reduction of the (-)-menthol ester of benzoyl formic acid, but the alkaline hydrolysis of the ester product brought about the racemisation of the mandelic acid enantiomer formed in excess. McKenzie (1904) overcame the racemisation problem by employing the recently-discovered methyl magnesium iodide reagent (Grignard, 1901), to form atrolactic acid with the (-)-isomer in excess, and later, by acetylating the α-hydroxy ester product of the reduction, prior to hydrolysis, when (-)-mandelic acid was obtained as the major product.

The enantiomeric excess in the α-hydroxy acid product, often substantial, was ascribed to an asymmetric induction from the chiral centre of the carbinol group in the α-keto ester substrate. The asymmetric induction appeared to be evident in the substrate, as shown by the anomalous rotatory dispersion associated with the carbonyl group absorption, and by the mutarotation of ethanol solutions of the α-keto ester, due to hemiacetal formation. The sign of the mutarotation, for a variety of α-keto esters, corresponded to the sign of the optical rotation given by the α-hydroxy acid enantiomer produced in excess by a range of Grignard reagents, with the occasional exception of the largest of these, 1-naphthyl magnesium bromide (Ritchie, 1933, 1947). The structural result of the asymmetric induction from the chiral alcohol group was envisaged to be a chiral pyramidal configuration of the carbonyl group and the bonded carbon atoms in the α-keto ester (Ritchie, 1933).

The alternative view of Tiffeneau and coworkers (1935), that the two faces of the plane defined by the carbonyl group and the bonded carbon atoms are differentially attacked by the Grignard reagent, provided the basis for subsequent steric interpretations. On reexamining the numerous studies by McKenzie and coworkers on asymmetric synthesis in the α-keto ester series, Prelog (1953) arrived at the generalisation that the Grignard reagent preferentially attacks the carbonyl group from the particular side of a conventional α-keto-carboxyl plane subject only to the steric constraint of the smallest of the three groups attached to the chiral carbinol carbon atom. The conventional plane is defined by the two carbonyl groups in an antiparallel conformation, the carbinol oxygen and carbon atoms, and the largest of the three groups bonded to the latter atom, with the small and the medium-sized group disposed in staggered positions on either side of the plane (fig. 10.9).

According to the model, the enantiomer of the α-hydroxy acid formed in excess by the addition of a Grignard reagent, a lithium alkyl, or a hydride reducing agent, has the stereochemical configuration represented by an approach of the reagent from the side of smallest carbinol group. If the α-keto ester of the antipodal alcohol is employed in the reaction, the enantiomeric α-hydroxy acid is produced in excess. The reaction, with Prelog's steric analysis, provides a method for correlating configuration. Provided the three groups bonded to the chiral carbinol centre are hydrocarbon residues, or hydrogen for the smallest group, the benzoyl formic ester with methyl magnesium iodide gives an excess of (R)-(−)-atrolactic acid if the alcohol has the (R)-configuration, or of (S)-(+)-atrolactic acid if the carbinol belongs to the enantiomeric (S)-series (Prelog, 1956).

The analogous steric rule of Cram covers the differential formation of diastereomers from the addition reactions of aldehydes or ketones with a chiral centre adjacent to the carbonyl group (Cram and Abd Elhafez, 1952). With a conventional plane defined by the carbonyl group and the largest of the three groups attached to the chiral centre in a *trans*-conformation, an organometallic or metal hydride reagent adds to the carbonyl group preferentially from the particular side flanked by the smaller of the two groups bonded to the asymmetric carbon atom either side of the plane (fig. 10.10). The rule is modified if one of the groups bonded to the chiral centre has a lone-pair of electrons available for coordination to the metal atom of the reagent, notably a hydroxy or amino group. The electron-donating group and the carbonyl group assume a *cis*-conformation, to allow the formation of a five-membered chelate ring with the metal atom of the addition reagent. The mean plane of the chelate ring is flanked on either side by the two other groups bonded to the chiral centre, and the hydride or alkyl group of the reagent adds to the same side of the carbonyl group as that of the smaller flanking group (fig. 10.10). As the two chiral centres in the diastereomeric product are not readily separable, the steric rules of Cram have found a less general application than those of Prelog to the correlation of stereochemical configuration in series of chiral analogues.

Fig. 10.9. The steric model of Prelog for the asymmetric synthesis of atrolactic acid from the benzoyl formic ester of a chiral alcohol by the reaction with a Grignard reagent. A conventional plane is defined by the two carbonyl groups in an antiparallel conformation, the C—O—C ester bonds, and the largest, R_L, of the three groups bonded to the carbinol centre. The Grignard reagent preferentially attacks from the side of the plane flanked by the smallest group, R_S, bonded to the carbinol centre (Prelog, 1953).

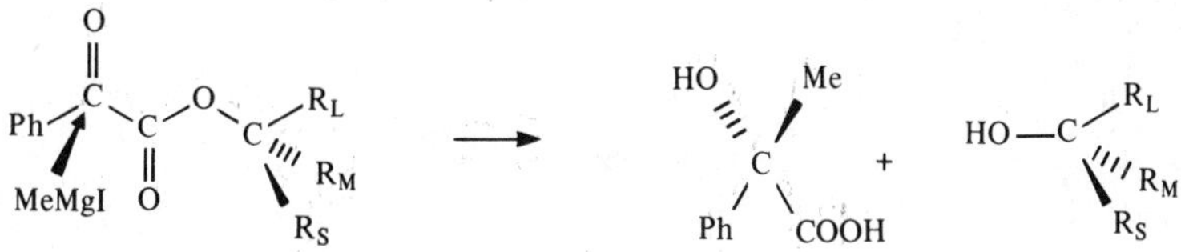

The degree of stereoselectivity in a McKenzie–Prelog or a Cram asymmetric synthesis is measured by the mole ratio of the diastereomeric products, $[P]/[N]$, where P refers to the like combination of configuration at the two chiral centres, (R,R) or (S,S), and N to the corresponding antipodal combination, (R,S) or (S,R). Alternatively, but not equivalently, the diastereomers may be described as *p* or *n* according to the respective like or opposed sign of the optical rotation given by the two separate enantiomeric components of the diastereomer. In a kinetically-controlled asymmetric synthesis, the diastereomeric product ratio, $Q = [P]/[N]$, represents the free energy difference between the corresponding precursor transition states through the relation,

$$\delta\Delta G^{\neq} = -RTlnQ \tag{10.6}$$

In a series of related asymmetric synthesis, 10.6 provides the basis for a linear free energy relationship, developed by Ruch and Ugi (1969).

In the asymmetric atrolactic ester synthesis of Prelog and coworkers (1955, 1956), the newly-formed chiral centre is constant in the series, and the relative

Fig. 10.10. The steric model of Cram for the differential formation of diastereomers in the carbonyl addition reactions of chiral aldehydes and ketones. For the open-chain model, (*a*), where the large, R_L, small, R_S, and medium-sized, R_M, substituents at the chiral centre are alkyl groups, a conventional plane is defined by the carbonyl group and the group, R_L, in a *trans*-conformation. The reagent adds to the carbonyl group preferentially from the side of the plane flanked by the group, R_S, bonded to the adjacent chiral centre. If one of the substituents bonded to the asymmetric centre carries lone-pair electrons, (*b*), the conventional plane is defined by the chelate ring formed by the coordination of the addition reagent to the carbonyl group and to the substituent with non-bonding electrons. Again the reagent preferentially adds to the side of the plane flanked by the group, R_S (Cram and Abd Elhafez, 1952).

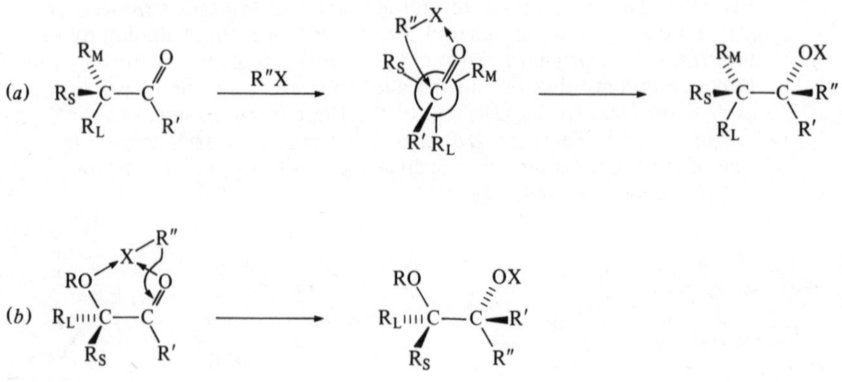

stereoselectivity displayed in the reaction of methyl magnesium iodide with the benzoyl formic ester is dependent upon a combination of steric ligand properties, λ_i, one for each of the three groups bonded to the carbinol centre of the ester, unchanged during the course of the reaction from substrate to product through the transition state (fig. 10.11). The particular combination of single ligand properties proposed by Ruch and Ugi (1969) as the determinant of the relative stereoselectivity in a series of analogous asymmetric syntheses has the form,

$$lnQ = \delta\rho(\lambda_1 - \lambda_2)(\lambda_2 - \lambda_3)(\lambda_3 - \lambda_1) \qquad (10.7)$$

Table 10.2. *The Ruch and Ugi (1969) single-ligand con-straints, λ_L (eqn 10.7) relative to the hydrogen ligand constant as zero ($\lambda_H = 0$), and that of the methyl group as unity ($\lambda_{Me} = 1$), for the carbinol ligands of the ben-zoyl formic esters employed in the Prelog atrolactic acid synthesis (fig. 10.9), together with the observed and calculated mole percentage of the major product. The reaction constant ρ (eqn 10.7) has the value 0.313*

				Product mole %	
L_1	L_2	L_3	λ_1	calc.	obs.
Ph	Me	H	1.24	55	52
1-naphthyl	Me	H	1.28	56	56
Me_3C	Me	H	1.45	62	62
Ph_3C	Me	H	1.70	71	75
(+)-Bornyl	Me	H	1.83	75	77

Fig. 10.11. The general model of Ruch and Ugi (1969) for the differen-tial formation of (R,R') and (R,S') diastereomers from a (R)-enantiomer substrate. The overall discrimination is a function of the single-ligand constants of the groups substituted at the newly-formed chiral centre $(L_1', L_2', L_3',)$, as well as those of the groups bonded to the asymmetric centre of the substrate (L_1, L_2, L_3), but the former are constant for a given reaction type, such as the atrolactic ester asymmetric synthesis, and are assimilated into the reaction constant ρ (eqn 10.7).

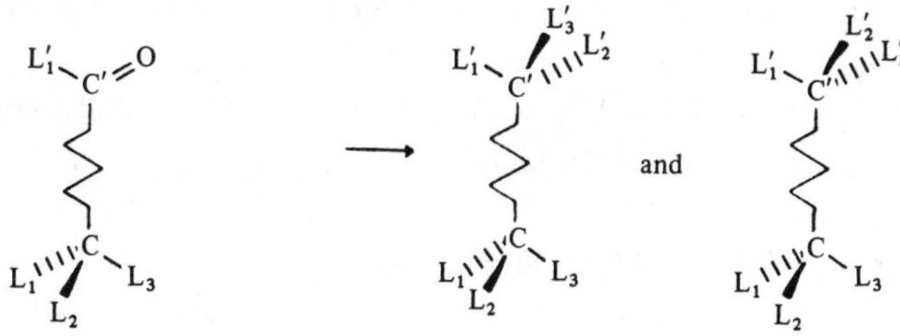

where the carbinol ligands are numbered serially in a clockwise sense for an observer viewing the ester from the carbon to oxygen bond-direction of the carbinol group. The factor, δ, of eqn 10.7 has the value of +1 for the (R)-configuration of the carbinol or -1 for the (S)-configuration, and the factor, ρ, is a constant for a given α-keto acid and particular Grignard reagent. Relative values of the single-ligand constants, λ_L, may be estimated from the free energy difference between the axial and the equatorial conformations of cyclohexane substituted with the ligand, L, or from the steric coefficient of the ligand, defined by Taft (1951). Values of the single-ligand constants, λ_L, relative to hydrogen as zero, $\lambda_H = 0$, and the calculated (eqn 10.7) and the observed diastereomeric excess in the series of atrolactic ester syntheses, are listed in table 10.2.

The scope of the Ruch and Ugi (1969) treatment of asymmetric syntheses has been critically discussed by Horeau and coworkers (Vigneron, Dhaenens and Horeau, 1977). The treatment is found to be eminently satisfactory for the atrolactic ester asymmetric synthesis, but to lack validity for the kinetic resolution of α-phenylbutyric acid by chiral secondary alcohols. The converse kinetic resolution of racemic secondary alcohols by (+)-α-phenylbutyric anhydride, introduced by Horeau (1962) for the correlation of configuration in the carbinol series, gives large stereoselectivity, the Q ratio having the value of 5.7 for methyl-t-butylcarbinol, **21** (Vigneron *et al.*, 1977).

Subsequent and more physically-based theoretical treatments of organic asymmetric synthesis are assessed by Kagan and Fiaud (1978), who review new synthetic developments in the field. New approaches to asymmetric synthesis are reviewed additionally by Valentine and Scott (1978), ApSimon and Seguin (1979), Brown and coworkers (1980), and Bosnich and Fryzuk (1981).

10.3.4 Stereoselection in metal coordination

Initially, the formation of a tris-chelate transition-metal complex from a chiral bidentate ligand was believed to be completely stereospecific, giving a single diastereomeric product. Smirnoff (1920) reported that the reaction of the hexachloroplatinum(IV) ion with (-)-1,2-propylenediamine yielded only the complex, (+)-[Pt(-)-pn$_3$]$^{4+}$, and that, in the preparation of the corresponding cobalt(III) complex, the sole product formed is the (-)-[Co(-)-pn$_3$]$^{3+}$ complex ion. Similar results were reported by Jaeger (1930), who employed the enantiomers of *trans*-1,2-cyclopentanediamine as the chiral ligand, although Lifschitz (1925) had obtained earlier both (+)- and (-)-isomers of the neutral complex, [Co(L-ala)$_3$], using L-alanine as the chiral chelating ligand.

Dwyer and coworkers (1959), in a reexamination of the work of Smirnoff (1920), discovered that two principal isomers are formed in each of the cases previously studied, the reactions being stereoselective, rather than stereospecific. Prepared in ethanol solution, where thermodynamic equilibrium between the product isomers is established, the major (+)- and minor (−)-isomer of $[Pt(−)-pn_3]^{4+}$ were obtained in the ratio of 5.7. The corresponding ratio of the major (−)- to the minor (+)-isomer of $[Co(−)-pn_3]^{3+}$ was found to be 15, following equilibration of the complexes in aqueous solution on charcoal at 25°C.

At the same time, Corey and Bailar (1959) presented a conformational analysis of the chelate rings in the complexes formed by transition-metal ions, with 1,2-diamines, based upon the X-ray diffraction structural study of the double salt, $2\{\Lambda-(+)-[Co(en)_3]Cl_3\}.NaCl.6H_2O$ by Saito and coworkers (1954). The analysis indicated that, in a tris(1,2-diamine) complex, the carbon–carbon bond of each chelate ring is parallel to the threefold rotation axis of the complex in the preferred conformation, lel_3, in contrast to an oblique orientation with respect to that axis, which gives ob_3 as the highest energy conformation, and ob_2lel and lel_2ob as intermediate forms.

Owing to the equatorial preference of the methyl group with respect to the mean plane of the chelate ring, (R)-(−)-1,2-propylenediamine has a λ-conformation in a coordination compound, with the C—C bond of the chelate ring and the N · · · N line-of-centres forming a segment of a left-handed helix. The constraint due to the exocyclic methyl group requires that, in the $[Co(−)-pn_3]^{3+}$ complexes, the overall conformation of the chelate rings is lel_3 if the configuration of the rings around the metal ion has the right-handed propellor form, the Δ-configuration, whereas the overall conformation is ob_3 for the Λ-configuration.

The identification of the major isomer isolated by Dwyer and coworkers (1959) as the lel_3 form, $\Delta(\lambda\lambda\lambda)-(−)-[Co(−)-pn_3]^{3+}$, and the minor isomer as the ob_3 form, $\Lambda(\lambda\lambda\lambda)-(+)-[Co(−)-pn_3]^{3+}$, was subsequently verified by X-ray crystal-structure determinations, including absolute configuration (Saito, 1979). Corey and Bailar (1959) estimated that the lel_3 conformation of a tris(1,2-diamine)cobalt(III) complex is lower in energy than the corresponding ob_3 form by some 7.5 kJ mol^{-1}, compared with the value of 6.7 kJ mol^{-1} given by the $[Co(−)-pn_3]^{3+}$ isomer-ratio of 15 at 25°C, determined by Dwyer and coworkers (1959).

Subsequently the conformational analysis of the chelate rings in transition-metal complexes was extended to six- and seven-membered ring systems, to donor atoms other than nitrogen, and to polydentate ligand systems, employing force fields ranging from the largely-empirical to those based upon *ab initio* molecular orbital procedures. The initial analyses were concerned primarily with the energy differences between conformers of a known or an assumed fixed structure, but more recent analyses refer to the molecular conformational energy surface, or to regions of the surface around the potential energy minima. The subject has been reviewed by Hawkins (1971), Buckingham and Sargeson (1971), and by Niketić and Rasmussen (1977).

The connection between the stereochemical configuration of the chelate rings around the metal ion and the particular conformation preferred by the chelate rings confers a chiral stereoselectivity upon the reactions of ligands which are achiral when not coordinated. The coordination of sarcosine and other N-alkyl derivatives of glycine to a metal ion generates a chiral centre at the coordinated nitrogen atom, and $[Co(NH_3)_4(sarc)]^{2+}$ is resolvable into its enantiomers (Buckingham, Mason, Sargeson and Turnbull, 1966). The replacement of the four ammonia ligands by two ethylenediamine chelate rings gives only two of the four possible diastereomers, ΛR and ΔS, where R or S refers to the configuration of the sarcosine nitrogen atom, and Λ or Δ to the configuration of the three chelate rings around the cobalt(III) ion.

The hydrogen atoms of the CH_2 group in a glycinate chelate ring exchange with deuterium in alkaline D_2O solution and, for the corresponding N-alkyl glycinate ring, the exchange becomes stereoselective between the two hydrogen atoms. The particular complex ΛR-(N-benzylglycinato)bis-(ethylenediamine)cobalt(III) chloride, with a structure and absolute configuration established by X-ray crystallographic analysis, is found to exchange hydrogen for deuterium differentially in D_2O at pH 10.5 to give the (2S)-deutero-derivative, 24, in 80%

24

optical yield (Golding, Gainsford, Herlt and Sargeson, 1976). Similarly, the parent glycinato complex, Λ-(+)-$[Co(en)_2(gly)]^{2+}$, reacts stereoselectively with acetaldehyde in aqueous solution at the (2S)- position of the glycinato ligand to give, after liberation of the ligand with hydrogen sulphide, (2S, 3S)-threonine and (2S, 3R)-allothreonine with 16% and 35% optical yield, respectively (fig. 10.12) (Dabrowiak and Cooke, 1975). The stereoselective reactions of achiral ligands in cobalt(III) complexes are reviewed by Sargeson (1980).

A disadvantage of the use of six-coordinate tris-chelate metal complexes for the stereoselective transformation of achiral ligands into chiral products is the consumption of the optically-resolved coordination compound in the reaction, as in the production of chiral amino acids from the glycinato ligand (fig. 10.12). Transition-metal complexes with a stable lower coordination number more readily form adducts which, after the reaction of the activated additional ligand species, regenerate the original complex. The development of chiral catalysts based upon metal complexes with a low coordination number has taken two main directions. Firstly, established achiral metal-complex catalysts have been dissymmetrically modified, notably, the homogeneous hydrogenation catalyst, [RhCl(PPh$_3$)$_3$], discovered by Wilkinson and coworkers (Osborn, Jardine, Young and Wilkinson, 1966). Secondly, simple chiral analogues of the known, or the inferred, active centre in a metal-enzyme system have been evolved for catalytic asymmetric syntheses (Kagan and Fiaud, 1978).

The original Wilkinson catalyst was initially modified by replacing the triphenylphosphine ligand with an enantiomer of the general type (R$_1$ R$_2$ R$_3$ P), but only low optical yields were obtained by the hydrogenation of prochiral olefins using such catalysts. Considerable improvements were effected by the introduction of chiral chelating biphosphine ligands and by the choice of dehydroamino acids (enamides) as the principal type of prochiral olefin substrate. These developments provide the α-amino acids from the corresponding enamide precursor in more than 90% optical yield by hydrogenation in homogeneous solution employing a catalytic complex of rhodium(I) with one of a range of chiral biphosphine chelates (Brown and coworkers, 1980; Bosnich and Fryzuk, 1981).

Fig. 10.12. The preferential replacement by acetaldehyde in aqueous solution at pH 9.5 of the *pro*-S glycinato C—H group in the complex Λ-(+)-(glycinato)-bis(ethylenediamine)cobalt(III) to give, after decomposition of the complex with hydrogen sulphide, (2S, 3S)-threonine in 16% enantiomeric excess and (2S,2R)-allothreonine in 35% enantiomeric excess (Dabrowiak and Cooke, 1975).

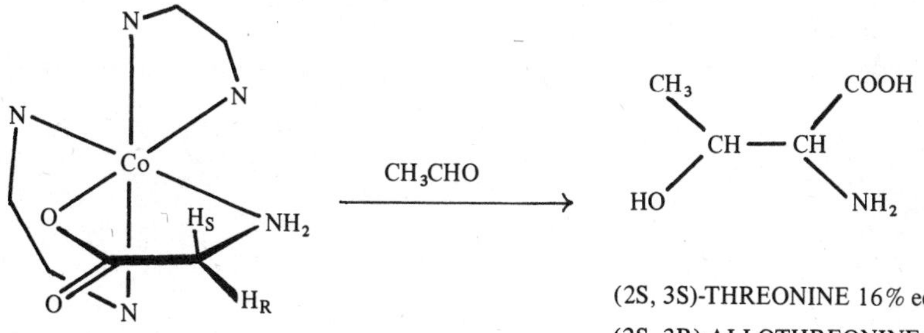

(2S, 3S)-THREONINE 16% ee.

(2S, 3R)-ALLOTHREONINE 35% ee.

10.4 Chiral biodiscrimination

In his Faraday Lecture, Fischer (1907) addressed the London Chemical Society on 'Synthetic chemistry in its relation to biology', restating and illustrating his key-and-lock hypothesis. Examining the action of relatively-crude enzyme preparations on sugar derivatives, Fischer (1894) found that α-methyl-D-glucoside is hydrolysed by maltase but not by emulsin, whereas β-methyl-D-glucoside is cleaved by emulsin but not by maltase, and neither of the corresponding L-glucosides are affected by either of the enzyme preparations. 'This represents a significant extension of Pasteur's observation that micro-organisms alter only one of the two enantiomorphs', Fischer suggested, 'Here, apparently, the geometric structure has such a profound influence upon the action of the chemical affinities that it is permissible to compare the two molecules under reaction with a *key and lock*'.

The conviction of Fischer (1898), that the key-and-lock hypothesis had a particular, physiological significance, supported, and was supported by, Ehrlich's (1897) concept of a template relationship between an antibody and the corresponding antigen. The selective staining of different tissues by the new synthetic organic dyes led Ehrlich to suggest that a complementary configurational relationship obtains between the sidechain groups of the susceptible tissues and those of the dye, or of a chemotherapeutic agent modelled upon it. Ehrlich (1900) proposed, in his Croonian lecture to the Royal Society, that a toxin has two types of sidechain. One, the haptophore, is a binding-group with a key-and-lock configurational relationship to the cell receptor, allowing the other, the toxophore, to produce its toxic action. If the cell survives, the haptophore receptors are replicated in excess and then shed to appear as circulating anti-toxin.

10.4.1 Receptor dissymmetry

The influence of the views of Fischer (1898) on the physiological significance of molecular chirality appears in the textbook on *Stereochemistry* by Stewart (1907), which covers the differential physiological activity of stereoisomers, as well as their chemical and physical properties. Stewart included some of the earlier studies of Cushny (1926) on the differential pharmacological action of enantiomers. Cushny found that (−)-hyoscyamine, **25**, is more than an order of magnitude more potent than the (+)-enantiomer in its effect on motor nerve

25

terminals, and that there is an even larger difference in sympathomimetic effect between natural (−)-adrenaline, **26**, and the relatively ineffective synthetic (+)-isomer. Cushny (1926) called for the examination of each enantiomer of a chiral biologically-active substance, in order to determine the stereochemical 'configur-

HO
HO

H OH

NHMe

26

ation of the tissues'. The pharmacological lock-and-key mechanism for Cushny involved an asymmetric induction similar to that proposed contemporaneously to account for organic asymmetric synthesis, or for the different physico-chemical properties of diastereomers. The receptors themselves are chiral, and the chemically-equivalent enantiomers of a racemic drug become physiologically differentiated in the *p*- and the *n*-diastereomeric drug-receptor complex.

A more direct stereochemical mechanism was proposed in the three-point contact model of Easson and Stedman (1933), who set out to dissociate differ-ential pharmacological activity from asymmetric induction. Taking the asym-metric tetrahedron as their model of a chiral structure, Easson and Stedman suggested that three inequivalent receptor sites formed a complementary match to a triangular tetrahedral face of one drug enantiomer but not to the corres-ponding face of its optical antipode, which failed to bind effectively to the receptor sites (fig. 10.13). If the binding of three different groups of the drug to

Fig. 10.13. A three-point drug receptor (*a*) congruent with a triangular face of an asymmetric tetrahedral active enantiomer (*b*) is not congruent with any tetrahedral face of the corresponding mirror-image isomer (*c*). The achiral analogue (*d*) has one face, like that of (*b*), congruent with the three-point drug receptor (*a*), and another (shaded) which corre-sponds to the incongruent face of the less-active antipode (*c*) (Easson and Stedman, 1933).

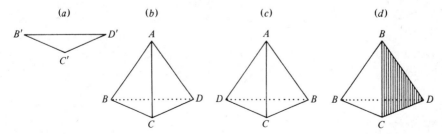

matching receptor sites determines pharmacological activity, an achiral analogue of the more effective drug enantiomer, with the fourth position of the asymmetric tetrahedron replaced by a group identical to one of three forming the binding face of the drug, is expected to be just as effective as the more active enantiomer (fig. 10.13).

In support of their hypothesis, Easson and Stedman (1933) reported a study of the contractile effect of miotic benzylamine derivatives, 27, and 28, on strips of isolated rabbit intestine. Relative to the achiral derivative, 28, R = H, the (+)-

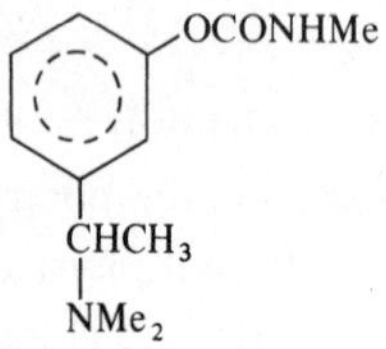

$$\text{27} \qquad\qquad \text{28}$$

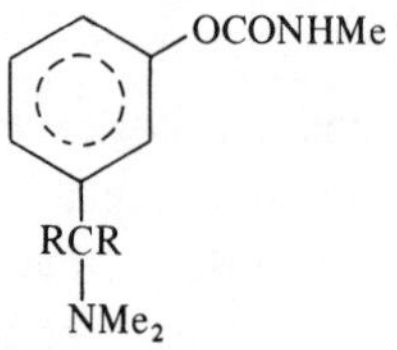

and the (−)-isomer of 27 were found to be 5 and 24 times more active, respectively with racemic 27 displaying an activity some 12 times greater. However, the achiral dimethyl derivative, 28, R = CH_3, proved to be 16 times more active than the analogue, R = H, providing support for the three-point contact model.

The Easson and Stedman model met with little immediate response, receiving no mention at the Faraday Discussion (1943) on the mode of drug action, where the earlier views of Cushny (1926) were commended. Shortly afterwards, however, the three-point contact model was independently proposed by Ogston (1948) in another connection. The enzymatic conversion of L-serine to glycine was presumed to involve an oxidation of the substrate to aminomalonic acid followed by a decarboxylation to give the product. The use of L-serine labelled with ^{13}C in the carboxyl group and ^{15}N in the amino group resulted in a glycine product with an isotope ratio, $^{15}N{:}^{13}C$, identical to that of the L-serine substrate, whereas the ratio is expected to be twice as large in the product if the intermediate aminomalonic acid has equivalent carboxyl groups (Shemin, 1946).

Unaware of the work of Easson and Stedman (1933), Ogston (1948) pointed out that if the enzyme has three complementary receptor sites matching the three functional groups attached to the chiral centre of L-serine, the asymmetric centre is preserved in the isotopically-chiral aminomalonic acid intermediate (fig. 10.14). A distinction between the originally-labelled carboxyl group of L-serine and the carboxyl group derived by oxidation is maintained, and the removal of the latter gives a glycine product with the isotope ratio of the L-serine substrate unchanged.

The Ogston model soon became generally accepted in the biochemical field, with the modification that only one point of attachment is essentially required for the enzymatic transformation of a prochiral substrate or intermediate to a chiral product (Ogston, 1958), as in the analogous but less stereoselective case of the classical organic asymmetric syntheses. In the fields of pharmacology and plant physiology the three-point contact model for substrate–receptor interaction was revived during the 1950s, largely in its original Easson and Stedman (1933) form.

Wain and coworkers (1951, 1952) employed the model to account for the differential auxin activity of the resolved enantiomers of synthetic plant growth-regulating substances, analogous to the naturally-occurring indole-3-acetic acid (Wain, 1977). Draber and Stetter (1979) refer to the strong auxin activity of the (+)-enantiomer of the phenoxypropionic acids, **29a, b** and the anti-auxin activity

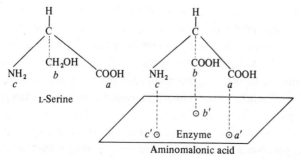

| | (a) | X = 2-Br |
| | (b) | X = 3:4-(CH₃)₂ |

$$\text{29}$$

of the corresponding (−)-isomer, as evidence for the three-point attachment of these substances to the plant receptors.

The Easson and Stedman model was extended to the optical isomers of synthetic analgesics by Beckett and Casy (1954, 1955), who reviewed the general field of stereoisomerism and biological action, covering the plant growth auxins as well as the earlier pharmacological studies of differential enantiomeric activity. Like Cushny (1926), these authors suggest that the use of enantiomeric substances

Fig. 10.14. The three-point contact model of Ogston (1948) for the enzymatic conversion of L-serine, through aminomalonic acid, to glycine with a preservation of the isotopic labelling of the amino and the carboxylate groups.

'enables some of the fine structure of the receptor surfaces to be delineated', a view restated in a survey of a wider range of the differential pharmacological activities of optical isomers (Portoghese, 1970). The view is exemplified by the use of the stereoisomers of synthetic analgesics, such as methadone, **30**, and its analogues, as probes of the opiate receptor topography (Portoghese, 1978). The opiate receptors, located in the brain and possibly in other tissues of vertebrates,

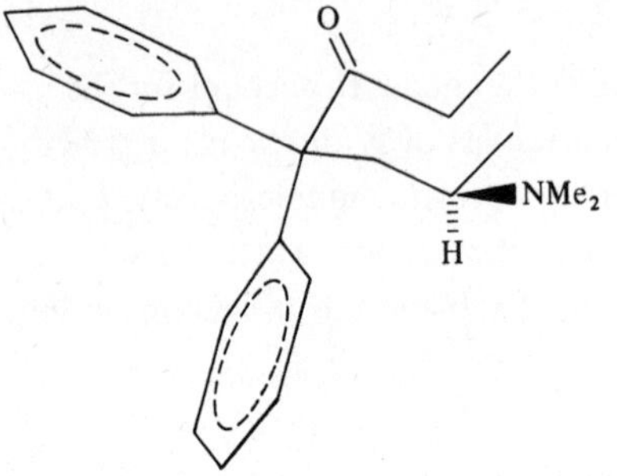

30 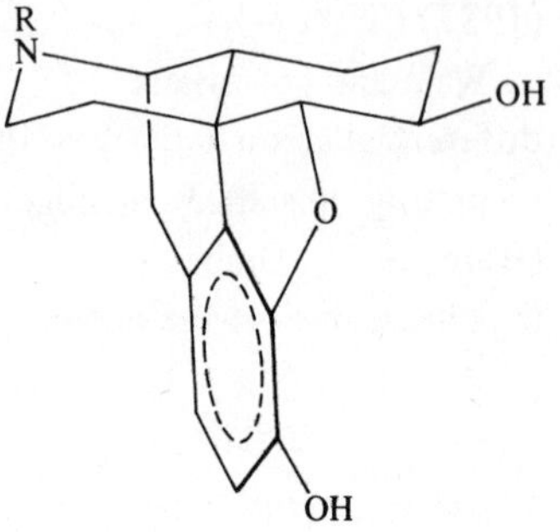   31

bind stereospecifically not only one isomer of a narcotic analgesic but also the corresponding isomer of its antagonist, e.g. the (−)-isomer of morphine, **31**, R = methyl, and the antagonist, nalorphine, **31**, R = allyl (Snyder, 1977).

Where the system of drug and receptor, enzyme and substrate, or antibody and antigen is isolable *in vitro*, photon probes may provide detailed information as to the binding and the structural complementation of the two components. If the system forms a crystalline complex, X-ray diffraction analysis become applicable, but otherwise a range of fluid-phase spectroscopic techniques may be employed. The combination of X-ray crystallography and nuclear magnetic resonance solution studies have proved to be particularly fruitful in the investigation of the binding of substrates and inhibitors to lysozyme and other enzyme systems (Lipscomb, 1972; Porter and Fitzsimons, 1978).

Analogous studies of the antibody combining site in the globular proteins show a constancy in the amino-acid sequences of the framework of the cavity, combined with a hypervariability in the corresponding sequences at the site surface, adapted to the specific hapten combining group of the antigen (Capra and Edmundson, 1977; Bedarkar and Blundell, 1978). The distinction by an antibody between enantiomeric hapten groups has been long established (Landsteiner, 1945; Karush, 1956). These and similar structural studies of pharmacologically-active molecules (Duax, 1978) suggest a mutual stereo-chemical adaptation of receptor and reagent. The distortion of a substrate on binding to an enzyme towards the structure of the transition state for reaction is taken to be a probable major source of the catalytic power of enzymes (Pauling, 1948; Lipscomb, 1972).

10.4.2 Stereoisomers in bioapplications

The more important applications of synthetic substances to biological systems lie in the fields of medical and agricultural or horticultural chemistry. Many of the older synthetic compounds employed in these applications are inorganic, or simple achiral organic substances, such as the sulphonamides. The more recently developed bioactive compounds often contain one or more chiral centres, although the synthetic racemic or diastereomeric mixture is sometimes regarded as a single substance.

The *US Pharmacopeial Dictionary of Drug Names* (Griffith, 1980) gives the graphic structural formulae and systematic chemical description of some 2050 pharmaceutical substances. Of the total, almost one-half (1004) are chiral compounds, of which 518 or some 25% of the total are natural products or semi-synthetics from natural products, such as the β-lactam or steroid derivatives. The remaining chiral pharmaceuticals are made up of 88 synthetic enantiomers and 398 racemic or diastereomeric mixtures. Only a minority of the racemates are so designated by a ($\pm$) or (R,S) prefix.

The enantiomeric mixtures regarded as a single substance include the anti-malarial chloroquine, **32**, which has a (+)-isomer more effective against the malaria parasite and less toxic to the host than the racemate (Blaschke, 1980), and methadone, **30**, where the (−)-isomer is the effective analgesic agent (Beckett and Casy, 1955). Methadone spontaneously resolves on crystallisation and the

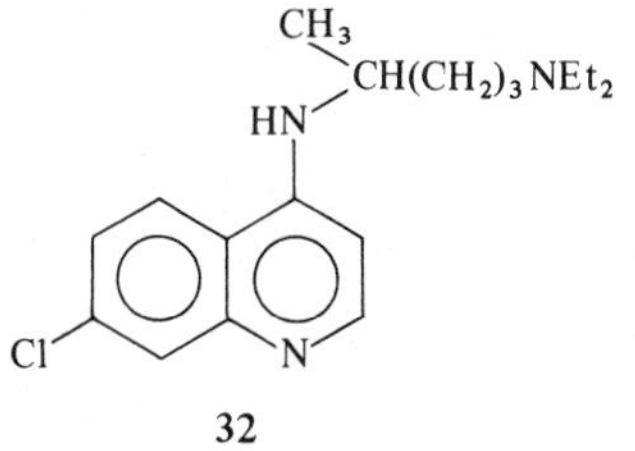

32 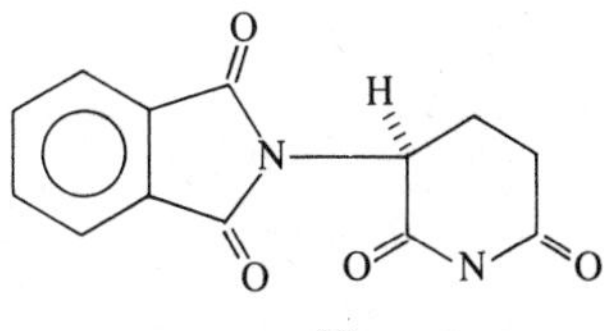   33

enantiomers are readily separated by entrainment (Zaugg, 1955). In the case of the one-time widely-used racemic analgesic thalidomide, **33**, it turns out that only the (S)-(−)-isomer possesses teratogenic properties (Blaschke, 1980).

A survey of the compendium of agricultural and horticultural chemicals, *The Pesticide Manual* (Worthing, 1979), indicates that, of some 550 entries, 103 refer to chiral compounds. Of the latter, 14 are natural products, nicotine, rotenone, gibberellic acid, the natural pyrethrins and the like, The remaining 89 chiral substances are synthetic racemates, and in 10 of these cases the differential ac-

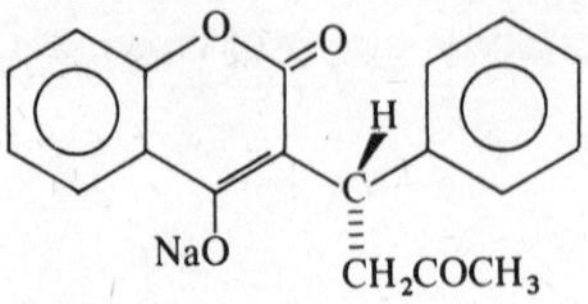

34

tivity of the enantiomers is discussed. The (S)-isomer of warfarin, **34**, for example, has a sevenfold larger rodenticidal activity than the (R)-isomer (Worthing, 1979) although, as a pharmaceutical anticoagulant, the enantiomers of warfarin are not distinguished (Griffith, 1980; British Pharmacopoeia Commission, 1981).

In the development of synthetic pyrethroids as insecticides, it has been found essential, for optimum potency, to preserve the configuration of at least two of

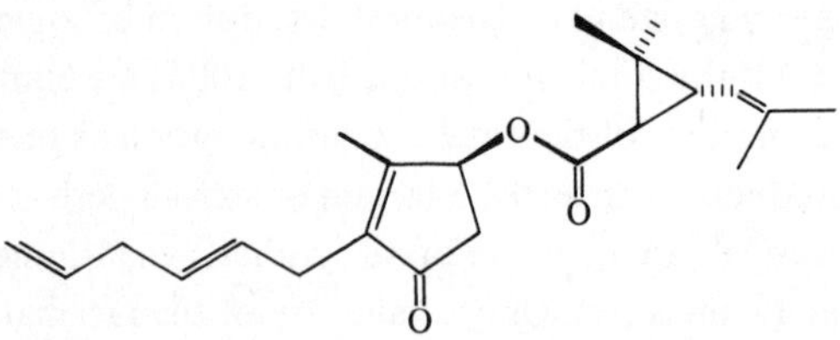

35

the chiral centres found in the most active natural ester, pyrethrin I, **35**, the (1R)-configuration of the crysanthemate ester and the (4S)-configuration of the cyclopentenolone (Elliot and Janes, 1978). The racemic form of the selective herbicide flamprop-isopropyl, **36**, controls wild oats growing among barley but

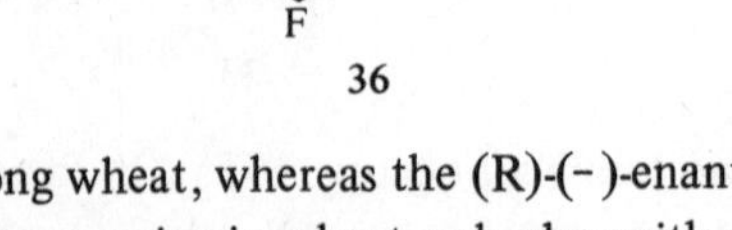

36

not among wheat, whereas the (R)-(−)-enantiomer suppresses wild oats and blackgrass growing in wheat or barley with an activity more than twice that of the corresponding racemic mixture. The competitive inhibition of the activity of an enantiomer by the antipodal isomer in a racemic mixture is not uncommon in pharmacology, one isomer serving as an agonist and the other acting as the antagonist (Ariëns, 1964). The respective auxin and anti-auxin activities of the (+)- and the (−)-isomer of each of the phenoxypropionic acids, **29**, provides an example from plant physiology (Draber and Stetter, 1979).

11 CHIRAL ENERGY DISCRIMINATION

11.1 Magnitudes of the energy differentials

Pasteur (1850) observed that concentrated solutions of (+)- and (−)-tartaric acid on mixing become perceptably warm as the solid racemate precipitates. Berthelot and Jungfleisch (1875) reported that the major part of the heat liberated, some 18 kJ mol^{-1}, involves the separation of the solid racemate, as the mixing of dilute aqueous solutions of the tartaric acid enantiomers, 2.5% by weight, liberates only 500 J mol^{-1}. The latter value is substantially overestimated. Amaya and coworkers (1968) measured, by direct microcalorimetry, an enthalpy change of 1.99 ± 0.04 J mol^{-1} for the mixing of aqueous solutions containing 15.6% by weight of the enantiomers of tartaric acid at 298.6 K. Measurements of the heat of solution in water of the crystalline solids give, for (+)-tartaric acid and the corresponding racemate, 16.2 and 25.6 ± 0.1 kJ mol^{-1}, respectively, at 298.15 K (Matsumoto and Amaya, 1980).

In general, the energy discrimination between a homochiral assembly of enantiomers and the heterochiral assembly of the corresponding racemate is weaker in solution ($\sim$ J mol^{-1}) than in the solid state ($\sim$ kJ mol^{-1}). At 298 K, the heats of sublimation of menthol are 95.8 and 78.7 kJ mol^{-1} for the (−)-isomer and the racemate, respectively, whereas the corresponding values for the (+)-isomer and the racemate of dimethyl tartrate are 97.5 and 151.5 kJ mol^{-1}, respectively (Chickos, Garin, Hitt and Schilling, 1981). In the case of menthol, the enantiomer (melting point 43°C) has a lower lattice enthalpy than the racemate (melting point 28°C) at 298 K, while dimethyl tartrate, with a melting point of 49° and 87°C for the enantiomer and the racemate, respectively, follows the more general trend, where the racemate has the lower-energy lattice structure (§ 9.2).

The energy discriminations between corresponding P and N diastereomers in the fluid phase commonly range up to the kJ mol^{-1} order, particularly for non-aqueous solvents (table 10.1). The thermometric titration of 1-phenyl-ethylamine in dioxan solution with mandelic acid at 298 K gives a heat of neutralisation of 35.20 and 36.64 kJ mol^{-1} for the formation of the N, (R,S) or (S,R), and the P, (R,R) or (S,S), diastereomer, respectively (Arnett and Zingg, 1981). A comparable differential heat of neutralisation is found for the corresponding dimethylsulphoxide solutions, but no significant difference is detected using water as the solvent.

The diastereomers formed by tris-chelate coordination compounds are more directly classified than their organic counterparts as homochiral, (Λ,Λ) or (Δ,Δ), and heterochiral, (Λ,Δ) or (Δ,Λ). As in the case of enantiomeric assemblies, heterochiral diastereomeric associations are the more generally preferred in the fluid and the solid phase, although homochiral associations in both phases have been characterised. An aqueous solution (0.12M) of the 1,10-phenanthroline complex, Δ-(-)-[Ni(phen)$_3$]Cl$_2$, preferentially extracts, from a tetrachloromethane (CCl$_4$) solution of the racemate, the Δ-(-)-isomer of the 2,4-pentandionate complex, [Cr(pd)$_3$] (Iwamoto, Yamamoto and Yamamoto, 1977). The 1.5% enantiomeric excess of Δ-(-)-[Cr(pd)$_3$] in the aqueous solution corresponds to 75 J mol^{-1} in the free energy of the homochiral diastereomeric discrimination at ambient temperature.

In contrast, the tris(catechyl)arsenic(V) enantiomer, Δ-(-)-[As(cat)$_3$]$^-$, preferentially forms a diastereomeric association with Λ-(+)-[Cr(pd)$_3$] in ethanol solution. The variation in the circular dichroism induced in the visible d → d absorption of racemic [Cr(pd)$_3$] by Δ-(-)-K[As(cat)$_3$] with temperature over the 200 – 300 K range gives an enthalpy preference of 10.7 kJ mol^{-1} for the heterochiral diastereomeric association (Drake *et al.*, 1982). Similarly, the tris(oxalato) complex of chromium(III), [Cr(ox)$_3$]$^{3-}$, and the corresponding tris(malonato) complex, [Cr(mal)$_3$]$^{3-}$, are found to antiracemise in dioxan–water solution to an enantiomeric excess of up to 4% of the Δ-isomer with a co-solute of the Λ-enantiomer of one of some twenty different cobalt(III) complexes of the general type, *cis*-[Co(diamine)$_2$XY]$^{n+}$, where X and Y are either monodentate ligands or the coordinating groups of a chelate ligand (Miyoshi, Matsumoto and Yoneda, 1981).

11.2 Intermolecular discrimination energies

In order to accommodate the deviations from the ideal-gas law, van der Waals (1873) proposed a hard-sphere model for the fluid phase in his equation,

$$(P + a/V^2)(V - b) = RT \qquad (11.1)$$

where the term (a/V^2) allows for the mutual attractive forces between the molecules, and b for their finite volume and short-range mutual repulsion. The attractive term implies an intermolecular potential dependent upon R^{-6}, where R is the mean separation between two molecules. Three potentials of the required form were proposed, the first two being confined to molecules with a permanent electric-dipole moment.

Keesom (1921) accounted for the van der Waals attractive forces between dipolar molecules by averaging the Coulombic dipole-dipole potential over all mutual orientations, using the Maxwell-Boltzmann distribution to allow for the preferred low-energy orientations, and obtained the expression,

$$E = -[2\mu_A\mu_B/(3kTR^6)] \tag{11.2}$$

where μ_A and μ_B are the dipole moments of the two interacting molecules. Debye (1920) pointed out that Keesom's dipole orientation energy is not consistent with the observed temperature variation of the a and b terms in the van der Waals equation (eqn 11.1). An additional contribution, temperature-independent, is provided, Debye indicated, by the potential between the dipole moment of one molecule, μ_A, and the dipole induced in the other, proportional to the dipole polarizability, α_B, of the second molecule. On averaging over all mutual orientations, the induction energy is given by the equation,

$$E = -(\alpha_A\mu_B{}^2 + \alpha_B\mu_A{}^2)/R^6 \tag{11.3}$$

Neither Keesom's dipole-dipole nor Debye's dipole-induced dipole potential accounted for the van der Waals attractive interaction between atoms and molecules lacking a permanent dipole moment. Carrying Debye's argument a stage further, London (1930) noted that the proton and the electron of the hydrogen atom constitute a dipole which rapidly reorientates with the electronic motion and averages to zero over a short time interval. The instantaneous nuclear-electronic dipole of one molecule induces a dipole with a low-energy orientation in a neighbouring molecule which, by a reciprocal interaction, induces a dipole with the same phase in the first molecule. The resulting net attractive interaction, with an induced dipole-induced dipole origin, is universal, independent of any permanent molecular moment, dipole or multipolar. London (1930, 1937) expressed the attractive interaction by the approximation,

$$E = -(3\alpha_A\alpha_B/2R^6)[hI_AI_B/(I_A + I_B)] \tag{11.4}$$

where I_A and I_B refer to the ionisation potential of the molecules.

London suggested the term 'dispersion forces' for the universal attractive interaction from the use of the quantum mechanical form of the dispersion equation, due originally to Drude (1900), for the electric-dipole polarizability of a molecule in an applied field with a frequency, ν,

$$\alpha_0(\nu) = (2/3h) \sum_{n \neq 0} |\mu_{0n}|^2 \nu_{0n}/(\nu_{0n}{}^2 - \nu^2) \tag{11.5}$$

where μ_{0n} is the electric-dipole moment and ν_{0n} the frequency of the transition connecting the molecular electronic states, A_0 and A_n. The polarization of the molecule in an electric field involves the mixing-in of contributions from all excited electronic states into the ground state, the perturbation being expressed in terms of virtual electronic transitions, $A_0 \rightarrow A_n$. When the applied electric field derives from the instantaneous nuclear-electronic dipole of a second molecule, B, the perturbation is mutual, giving the intermolecular dispersion energy,

$$E_D = -(2/3) \sum_{n(A)} \sum_{m(B)} D_{0n}^A D_{0m}^B / [R_{AB}^6 (E_{0n}^A + E_{0m}^B)] \tag{11.6}$$

where D_{0n}^A refers to the electric-dipole strength, $|\mu_{0n}|^2$, of the electronic transition, with an energy, E_{0n}^A, of the molecule A, D_{0m}^B and E_{0m}^B having a similar significance for the molecule B, which is located at a separation R_{AB} from the molecule A.

Eqn 11.6 holds generally for isotropic molecules and for the dispersion energy of anisotropic molecules averaged over all mutual orientations. London (1942) considered the dispersion interaction between anisotropic molecules in a locked mutual orientation, suggesting for such cases the calculation of the Coulomb potential, V_E, between the transitional charge density, q_μ^{0n}, on atom μ of molecule A in the electronic promotion $A_0 \rightarrow A_n$, and the corresponding transitional charge density, q_ν^{0m}, on atom ν of molecule B,

$$V_E = \sum_{\mu(A)} \sum_{\nu(B)} q_\mu^{0n} q_\nu^{0m} (1/R_{\mu\nu}) \tag{11.7}$$

where $R_{\mu\nu}$ represents the separation between the atoms μ of A and ν of B. Coulson and Davis (1952) applied the procedure to the dispersion interaction of long-chain polyene molecules over a range of locked mutual orientations and separations, finding that the exponent, n, of the distance-dependency, R_{AB}^{-n}, falls from 6 to 3.5 as the separation between parallel chains is reduced.

11.2.1 *Electromagnetic discrimination*

The London treatment of the dispersion interaction considered only the mutual electric-dipole polarization of two molecules, neglecting the mutual perturbation of the molecules through the magnetic-dipole and higher multipole virtual electronic transitions. Mavroyannis and Stephen (1962) took into account the magnetic-dipole transitions, as well as the electric-dipole promotions in the two molecules, finding electromagnetic cross-terms in their expressions for the dispersion energy. These cross-terms are non-vanishing only for chiral molecules, and correspond to the rotational strengths, R_{on}^A, of the electronic transitions of molecule A, and the corresponding strengths, R_{0m}^B, of molecule B. Since the rotational strengths are signed, the additional electro-magnetic contribution to the dispersion energy discriminates between homo-chiral and heterochiral molecule-pairs. For an average over all orientations of the two molecules, the electromagnetic contribution is given by the relation,

$$\Delta E_{EM} = -(4/3) \sum_{n(A)} \sum_{m(B)} R_{on}^A R_{0m}^B / [(E_{on}^A + E_{0m}^B) R_{AB}^6] \qquad (11.8)$$

In an assembly of two enantiomers, each with a single electronic transition, the homochiral pair has a lower energy than the heterochiral pair, as the two rotational strengths have the same sign in the former case but the opposite sign in the latter.

The electromagnetic treatment of the interactions between molecules is extended to the induction energy and to the dispersion energy for fixed orien-tations of the two molecules, or semi-locked configurations in which the molecules rotate about the intermolecular axis, by Craig and coworkers (1971). For the semi-locked case, where the electric-dipole transition moments of the two enantiomers are parallel along the line of intermolecular centres, the electro-magnetic contribution to the dispersion energy (eqn 11.8) increases by a factor of 6, with the homochiral pair again stabilised relative to the heterochiral pair.

The differential intermolecular energy of a homochiral and a heterochiral pair of enantiomers which have permanent electric-dipole and -multipole moments, or an electric and a magnetic moment, is investigated by Craig and Schipper (1975). For an electrostatic energy discrimination, it is necessary that each molecule has both an even and an odd multipole moment, a 2^n-pole with n odd and n even, except for the charge, $n=0$. The simplest chiral multipole combination consists of an electric-quadrupole moment and an electric dipole which does not lie upon a symmetry axis or plane of the quadrupole moment. If the molecules are freely rotating, the electrostatic discrimination is negligible, but if the molecules have a locked orientation and a small separation, as in a crystal lattice, the leading dipole–quadrupole term, with realistic values for the two moments, gives a differential stabilisation of the homochiral pair of some 300 J mol^{-1}, relative to the heterochiral pair (Craig and Mellor, 1976; Craig, 1979).

A problem in the application of intermolecular chiral discrimination theory to particular cases is the lack of a definitive molecular centre, to which the intermolecular separation can be referred, in the majority of enantiomers. The problem is avoided by considering chiral molecules with dihedral symmetry,

D_p, where the principal p-fold rotation axis, C_p, and the twofold axes perpendicular to it intersect at a point which defines the required molecular centre. A permanent dipole moment is symmetry-forbidden in dihedral molecules, so that the electrostatic and induction intermolecular discriminations are not important for chiral molecules of this symmetry. The universal dispersion interaction, including the electromagnetic discrimination, obtains for dihedral and lower-symmetry chiral molecules equally, and eqn 11.8 is applicable to D_p systems.

The CD spectra of Δ-(-)-[Ni(phen)$_3$]$^{2+}$ and Δ-(-)-[Cr(pd)$_3$] over the near-infrared, visible, and ultraviolet regions (fig. 11.1) provide the rotational strengths and the transition energies of the two species for the evaluation of the expected discrimination energy (eqn 11.8), in connection with the homochiral discrimination of 75 J mol^{-1} observed by Iwamoto and coworkers (1977). The calculated discrimination is heterochiral, with the value of 5×10^{-7} J mol^{-1} if the two species are effectively in contact in solution, the molecular centres being separated by some 10 Å, or two orders of magnitude smaller if the mean separation of the molecular centres corresponds to the species concentration.

Fig. 11.1. The electronic absorption spectra (upper curves) and the circular dichroism spectra (lower curves) of (A) Δ-(-)-[Ni(phen)$_3$]$^{2+}$ (full line), and (B) Δ-(-)-[Cr(pd)$_3$] (broken line).

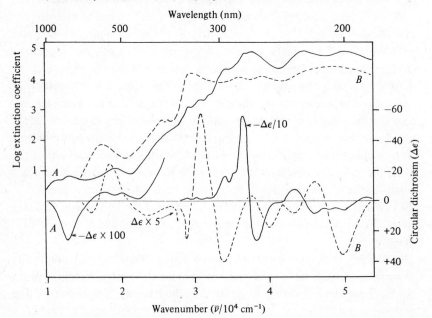

While the electromagnetic discrimination is appreciable for a model assembly of two enantiomers, each with one transition and, thus, a single rotational strength, the general sum-rule for the rotational strengths over the spectrum of a chiral molecule, $\sum_n R_{0n} = 0$, ensures that the sum over states in eqn 11.8 is small, for both the homochiral and the heterochiral molecule-pair. The sum rule is illustrated, over a limited but important spectral manifold, for the two species considered (fig. 11.1). For either the homochiral or the heterochiral molecule-pair, each positive term in the sum of eqn 11.8, due to rotational strengths of like sign, is accompanied by a negative term of comparable magnitude, arising from rotational strengths of opposite sign, and the sum over the successive terms becomes small, although the sum does not vanish, since each term is weighted by the appropriate energy denominator.

The electromagnetic dispersion discrimination (eqn 11.8) remains too small for the case of Δ-(-)-[Ni(phen)$_3$]$^{2+}$ and Δ-(-)-[Cr(pd)$_3$] if it is assumed that the two molecules have a fixed mutual orientation and are in contact, either with the threefold rotational axes of the two species collinear, or with a twofold rotational axis of each directed along the line of molecular centres, owing to the rotational strength sum-rule. Such fixed mutual orientations, or semi-locked analogues, where the molecules rotate about the intermolecular axis, are probable in fluid-phase conditions allowing the detection of chiral energy discrimination. As yet, chiral discrimination in the gas phase, where free rotation of the two species and their mean random mutual orientation is expected, has not been observed. Measurements of chiral energy discriminations in the fluid phase become feasible for chiral solutes in chiral liquids (table 10.1) and for chiral cosolutes, at a small dilution, in an achiral solvent.

11.2.2 *Extended transition monopole discrimination*

Slow-neutron scattering and X-ray diffraction studies of molecular liquids indicate a short-range order, extending over several molecular diameters, in the liquid phase analogous, in some cases, to the long-range order of the corresponding high-temperature crystal phase, such as submicroscopic 'icebergs' in water (Powles, 1973; Blum, Narten, Karnicky and Pings, 1976). The high-temperature crystal of CCl_4 is a face-centred cubic plastic phase in which the molecules align head-to-tail with a common threefold rotation axis. The on-axis chlorine atom of one molecule is located in the cavity formed by the three off-axis chlorine atoms of the nearest neighbouring molecule. The molecules undergo restricted rotation in the plastic crystal, switching cooperatively from one orentation to another. The melting of the plastic phase involves little loss of order, the entropy change being only 7.5 J mol^{-1} K^{-1}, and the short-range order in the liquid phase of carbon tetrachloride averages to a body-centred cubic cluster with two of the twelve nearest neighbour molecules orientationally correlated in the head-to-tail collinear C_3 axis packing (Nishikawa, Tohji and Murata, 1981).

In contrast to the spherical-top tetrahedral structure of carbon tetrachloride, all known chiral molecules have an anisotropic structure, with a symmetric-top dihedral symmetry at the highest. In a liquid-phase assembly of chiral molecules, a local structure at least comparable to that of carbon tetrachloride is expected, or more ordered in view of the reduced molecular symmetry. The major discriminatory interactions then arise from the semi-locked mutual orientations of the anisotropic chiral molecules, requiring the use of the London (1942) Coulomb potential between the transitional charge densities on each pair of atoms, one of the molecule A and the other of molecule B, the two molecules having a fixed mutual orientation (eqn 11.7). On this basis, the origin of the intermolecular energy discrimination between a homochiral and the corresponding heterochiral assembly is primarily stereochemical, the finite extension of the molecule,

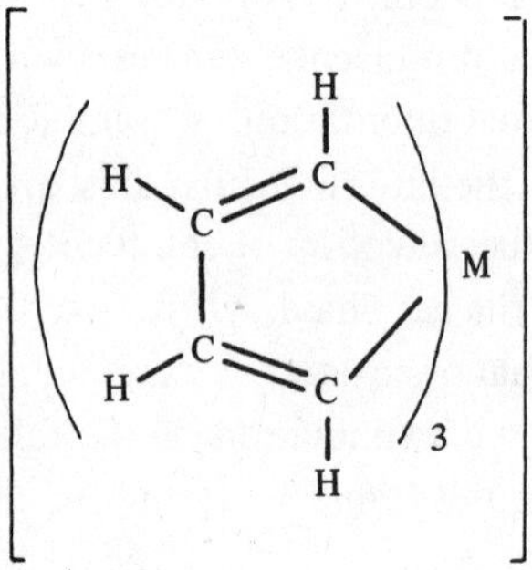

$$\left[\left(\begin{array}{c} \text{H} \\ \text{H} \diagdown \text{C} \diagup \overset{|}{\text{C}} \\ \text{H} \diagdown \underset{\text{C}}{\text{C}} \diagup \overset{\text{C}}{\underset{|}{\text{C}}} \\ \text{H} \end{array} \right)_3 \text{M} \right]$$

1

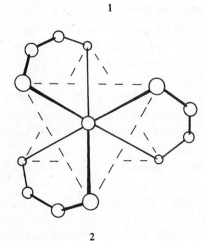

2

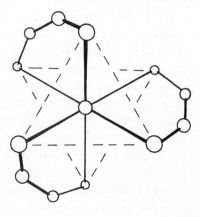

3

relative to the separation between the molecular centres, being taken into account. From the electromagnetic viewpoint (eqn 11.8), on the other hand, the molecular electric and magnetic transition moments forming the rotational strengths are regarded as point dipoles, located at the molecular centres, connected by the interchange of virtual circularly-polarized photons.

The extended transition monopole treatment of London (1942) gives the dispersion energy between two molecules, A and B, through the relation,

$$E_D = -2 \sum_{n(A)} \sum_{m(B)} [(A_0 B_0 | V_E | A_n B_m)]^2 / [E_{0n}^A + E_{0m}^B] \qquad (11.9)$$

where V_E refers to the Coulomb potential between the atom μ of molecule A and the atom ν of molecule B, with the atom–atom separation, $R_{\mu\nu}$ (eqn 11.7). If each of the transitional charge distributions, $(A_0 | A_n)$ and $(B_0 | B_m)$, is approximated as a point transition dipole located at the molecular centre of A and B, respectively, eqn 11.6 is recovered. Otherwise the transitional charge distributions are expanded over the atomic centres, to obtain the transition monopoles, q_μ^{0n} on atom μ in molecule A and q_ν^{0m} on atom ν in molecule B, for the evaluation of the potential, V_E (eqn 11.7).

Table 11.1. *The dispersion energy, E_D (eqns 11.7 and 11.9) of two tris(butadienyl) complexes, 1, for the homochiral pair (Δ,Δ), and the dispersion discrimination energy between a homochiral and a heterochiral pair, $\Delta E_D = E_D(\Delta,\Delta) - E_D(\Lambda,\Delta)$. The values refer to averages over rotations about the intermolecular line of centres, or about an axis perpendicular to that line, for mutual orientations with either the C_3 molecular axes collinear, or with a C_2 axis of each molecule directed along the line of centres*

Orientation	C_3 collinear		C_2 collinear	
Rotation axis	C_3	C_2	C_3	C_2
Contact[a]				
$-E_D / \text{J mol}^{-1}$	3933	1536	1148	1204
$-\Delta E_D / \text{J mol}^{-1}$	+233	+27	+25	−2.3
$R_{AB} = 10 \text{ Å}$				
$-E_D / \text{J mol}^{-1}$	214	228	286	217
$-\Delta E_D / 10^{-1} \text{ J mol}^{-1}$	+58	+57	+185	+9.5
$R_{AB} = 20 \text{ Å}$				
$-E_D / 10^{-2} \text{ J mol}^{-1}$	322	281	258	244
$-\Delta E_D / 10^{-4} \text{ J mol}^{-1}$	+195	+151	+580	+43

[a] The contact separation between the molecular centres, R_{AB}, lies in the range 5.91 to 7.90 Å.

The extended transition monopole model has been applied to the dihedral tris(butadienyl) model complex, **1**, in order to evaluate the dispersion energy between a homochiral molecule-pair, both right-handed with the Δ-configuration, **2**, and between a heterochiral molecule-pair, one right-handed and the other left-handed with the Λ-configuration, **3** (Kuroda *et al.*, 1978, 1981). The model complex **1** consists of three mutually perpendicular butadienyl chelate rings bonded to the central atom with octahedral coordination. The bond lengths and bond angles of the complex have standard values, and each atom is assigned the conventional hard-sphere van der Waals radius in order to define inter-molecular contact. Following the procedure adopted by Coulson and Davis (1952) for the calculation of the dispersion energy of the linear polyenes, the use of eqns 11.7 and 11.9 is confined to the transitions of the π-electron manifold of the butadienyl chelate rings in the model complex, **1**.

For a given separation between the molecular centres, R_{AB}, the dispersion energy of a homochiral molecule pair and a heterochiral molecule pair is estimated through eqns 11.7 and 11.9 for a particular mutual orientation and

Fig. 11.2. The variation of the distance R between the molecular centres of two D_3 enantiomers (**1**) (bottom curves), the dispersion energy, E_D (middle curves), and the chiral discrimination energy, $\Delta E_D = E_D(\Delta\Delta)$ $- E_D(\Delta\Lambda)$ (top curves), with the angle of rotation about the line of centres for (A) collinear C_3 molecular rotation axes and (B) collinear C_2 molecular rotation axes, at a van der Waals contact separation for the homochiral molecular pair ($\Delta\Delta$, full line) and the heterochiral molecular pair ($\Delta\Lambda$, broken line). The dispersion-energy discrimination is homochiral for positive values of ($- \Delta E_D$), and heterochiral for negative values.

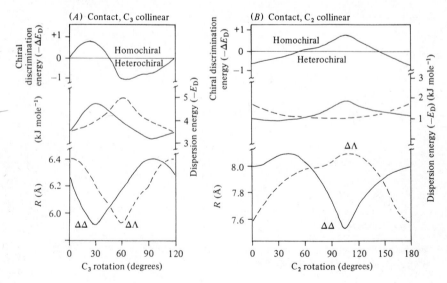

for successive angular increments of rotation of one molecule around the inter-molecular line of centres, or of both molecules about an axis perpendicular to that line. The variation with the rotation of the dispersion energy, and the dispersion discrimination energy between the homochiral and the heterochiral molecule pair, are illustrated for two principal mutual molecular orientations with a contact separation between the molecular centres (fig. 11.2) and with a larger separation of 10 Å (fig. 11.3). The principal mutual molecular orientations considered are, firstly, the case where the threefold rotation axes of the two molecules are collinear and, secondly, an orientation where a twofold rotation axis of each molecule is directed along the line of centres. Average values of the dispersion energy, and of the dispersion discrimination energy, over the molecular rotations in each of these principal mutual molecular orientations are recorded in table 11.1.

At a separation between the molecular centres corresponding to van der Waals contact of the dihedral tris-chelate complexes, 1, the direction of the dispersion discrimination to a homochiral or a heterochiral preference is markedly dependent upon the particular fixed mutual orientation of the two molecules (fig. 11.2)

Fig. 11.3. The variation of the dispersion energy, E_D (lower curves), and the chiral discrimination energy, $\Delta E_D = E_D(\Delta\Delta) - E_D(\Delta\Lambda)$ (top curves), with the angle of rotation about the line of centres of a homo-chiral pair ($\Delta\Delta$) and a heterochiral ($\Delta\Lambda$) pair of D_3 enantiomers (1) at a separation of 10 Å between the molecular centres for (A) collinear molecular C_3 rotation axes, and (B) collinear molecular C_2 rotation axes (Kuroda, Mason, Rodger and Seal, 1981).

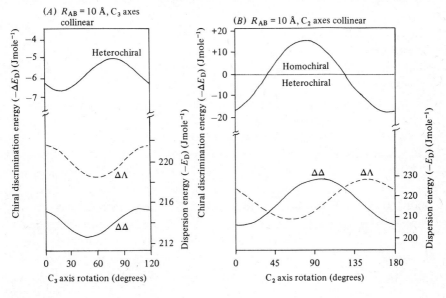

although a strong bias towards the heterochiral preference is already evident in the average values over molecular rotations (table 11.1). At larger separations the dispersion discrimination favouring heterochiral molecule pairs becomes general and ordered, decreasing as the separation increases in proportion to R_{AB}^{-8}. The relationship simulates a dispersion discrimination arising from a transition dipole-transition quadrupole combination, although the dispersion energy estimates are based upon the extended-transition monopole model, and the discrimination has primarily a stereochemical origin. While not directly comparable to free energy or enthalpy differences measured at or near to ambient temperature, the estimated dispersion chiral discrimination energies are of the correct order of magnitude, given that encounters between chiral molecules in solution result in extended contacts with a preferred mutual orientation.

11.3   Origins of molecular chirality

A basis for the transition from racemic chemistry to chiral biochemistry remains a major problem of chemical evolution (Kasha and Pullman, 1962; Calvin, 1969). Plausible mechanisms have been demonstrated for the synthesis under primitive terrestrial conditions of the racemic monomers required for the assembly of the chiral biopolymers, and for the production of diastereomeric mixtures of macromolecules. But a testable mechanism for the natural selection of specifically the L-amino acids and the D-sugars remains to be demonstrated.

Miller and Urey (1959) showed that a simulated primitive atmosphere, composed of methane, ammonia, water and hydrogen, subjected to an electric discharge gives four of the amino acids in small yield, with a substantial tarry residue. The initial substances formed, as also in the vacuum ultraviolet photo-lysis of the gas mixture, are cyanogen, hydrogen cyanide, urea, formaldehyde, and other carbonyl compounds. Subsequently the amino acids are generated by the Strecker synthesis from hydrogen cyanide and the aldehydes, while the formose condensation provides a mixture of carbohydrates from formaldehyde. Adenine, the most readily formed of the nucleic-acid bases, emerges as a simple pentamer in the polymerisation of hydrogen cyanide (Oró and Kimball, 1961).

A wider range of the natural amino acids are formed by circulating the primitive atmosphere gas mixture over silica or alumina at an elevated temperature 950–1050°C, and the amino acids spontaneously polymerise at temperatures no higher than 130°C to give 'thermal proteinoids'. The proteinoids, containing some 200 amino-acid units, are produced at 60°C in the presence of poly-phosphates. When heated with water at 130–180°C, the proteinoids form micro-spheres, $1-2\,\mu$m in diameter, which grow at the expense of the dissolved proteinoid and subdivide into daughter microspheres (Fox, 1964).

In the presence of polyphosphate esters, sugars undergo dehydration polymerization to polysaccharides, and ribose condenses with adenine, although the β-1' bonding of adenosine, as yet, is not established in appreciable yield (Schramm, Grotsch and Pollmann, 1962). The multiply-bonded products from the electric-discharge experiments, cyanogen, cyanoacetylene, and cyanamide, serve as additional coupling agents (Calvin, 1969).

Observations on the synthesis and the hydrolysis of polypeptides indicate that the adoption of the α-helical secondary structure promotes homochirality and stereoselection in reactivity. The polymerisation of γ-benzyl-L-glutamate-N-carboxyanhydride proceeds at a rate faster by a factor of 20 than that of the corresponding racemate, and the addition of a small fraction of the D-anhydride substantially retards the former polymerisation (Blout and Idelson, 1956). The partial polymerisation of non-racemic mixtures of D- and L-leucine-N-carboxy-anhydride gives a polyleucine product in which the original enantiomeric excess is enhanced while that of the residual monomer is depleted. When the polymer product is partially hydrolysed, the optical purity of the residual polymer is further increased (Blair, Dirbas and Bonner, 1981).

The faster rate of polymerisation of the optically-pure anhydrides indicates the advantages, in a natural-selection sense, of homochiral chemical processes. Further, the enantioselection in polypeptide formation and hydrolysis provides a model for a progressive transition from racemic to homochiral biochemistry, given an initial enantiomeric excess, however small. The enantioselection processes are required to be effectively competitive with the corresponding spontaneous racemisation rates. The racemisation half-life of L-aspartic acid is of the order of 10^4 years, affording a means of dating fossil bones and possible centenarians. For Kuhn (1955), the process of aging is due, in no small measure, to biochemical racemisation.

A number of classical mechanisms have been proposed for the production of an initial enantiomeric excess in the transition from racemic to homochiral biochemistry, but none of these account, in principle, for the particular choice of L-amino acids and D-sugars. Adsorption or surface catalysis by enantiomorph-ous mineral crystals, such as quartz, the differential circular photolysis of race-mates by reflected solar radiation, or the Coriolis forces due to the rotation of the earth, which differ in the two hemispheres, might equally generate an initial enantiomeric excess of D-amino acids or L-sugars. The particular outcome is viewed generally as 'a matter of chance' (Dickerson, 1978).

11.3.1 Non-classical interactions

The classical interactions of gravitation and electromagnetism are indifferent to chirality in a general sense. Neither impose a necessary or preferred relation between the linear and the angular momentum of a particle. The left-circular photon, with the spin component, $M_J = +1$, is accompanied by the right-circular photon, with the angular momentum component, $M_J = -1$. The two types of photon are interconverted by specular reflectance at normal incidence, or by transmission through a half-wave plate, and they have an equivalence in the electromagnetic interaction (fig. 11.4).

The strong nuclear interaction is equally indifferent to particle chirality. The parity-violating weak nuclear interaction displays discrimination, however, preferring left-handed particles, with antiparallel linear and angular momenta, and right-handed antiparticles, with parallel linear and angular momenta, provided that the rest-mass of the particle or the antiparticle is non-vanishing. The weak nuclear interaction connects through the weak neutral current with the electro-magnetic interaction, so that the electronic binding energy of an enantiomer and its antipode no longer have an exact equivalence. In principle at least, the parity-violating weak nuclear interaction provides a basis for a determinate transition from racemic chemistry to homochiral biochemistry in chemical evolution, although the particular choice of D-sugars and L-amino acids for the world of matter, and the implied adoption of L-sugars and D-amino acids in an antiworld of antimatter, has not been established as yet.

11.3.2 Parity non-conservation in the weak interaction

Apart from a minor chemical tradition with Pasteur's predilection for cosmic dissymmetry, it was generally held until the 1950s that the laws of physics and chemistry would remain unchanged in a mirror-reflected world of antipodal chirality. Initially the concept of parity was developed in quantum

Fig. 11.4. The right-handed, or left-handed, chirality of a particle, dependent upon the respective parallel, or antiparallel, relation between the linear momentum vector and the angular momentum axial vector.

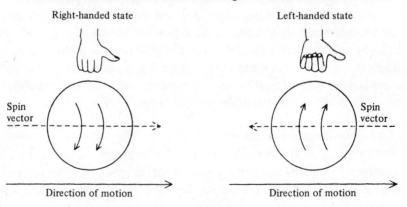

| Right-handed state | Left-handed state |

theory for the spatial-symmetry classification of electronic states, as in the rule of Laporte (1924), which allows electric-dipole transitions in an atom between states of opposite parity, one symmetric and the other antisymmetric with respect to inversion through the nucleus, and forbids such transitions connecting states of like parity.

The parity concept became generalised in the quantum theory of observation and measurement, notably by Wigner (1952), who proposed that quantum-mechanical operators and their observable expectation values conserve parity. Wigner accepted the difficulty of justifying theoretically the conservation of parity, and conceded the possibility that parity (P) and charge conservation (C), the operation changing a particle into the corresponding antiparticle, e.g. an electron into a positron, are coupled, so that the combined operation, CP, is the more exact symmetry conservation law (Wick, Wightman and Wigner, 1952). Subsequently the CP coupling proved to lack an adequate generality, requiring the addition of time-reversal, T, or the reversal of velocity, and the combination CPT remains the most general conservation principle as yet unbroken (Wilczek, 1980; Georgi, 1981).

From a survey of the evidence for parity conservation, Lee and Yang (1956) concluded that parity violations are probable in the weak nuclear interactions, and they designed experiments involving radioactive β-decay to investigate the question. Wu and her coworkers (1957) carried out such an experiment on cobalt-60, which decays to nickel-60, with the emission of an electron (e^-) and a neutrino (ν),

$$^{60}\text{Co} \rightarrow {}^{60}\text{Ni} + e^- + \nu \qquad (11.10)$$

The rotation axes of the cobalt-60 nuclei were aligned by cooling to less than 0.1 K in a magnetic field, in order to define a parity reference axis. If parity is conserved the numbers of electrons emitted along the two directions of the reference axis are expected to be identical. In the event, the number of electrons emitted in the particular direction requiring the linear and the angular momentum of the electron to align antiparallel was found to be the larger by some 40%.

An analogous experiment was carried out with cobalt-58, which decays to iron-58 with the emission of a positron (e^+) and an antineutrino ($\bar{\nu}$),

$$^{58}\text{Co} \rightarrow {}^{58}\text{Fe} + e^+ + \bar{\nu} \qquad (11.11)$$

In this case, the positrons were preferentially emitted in the opposite direction, corresponding to the right-handed parallel alignment of the linear and the angular momentum (Rodberg and Weisskopf, 1957). All subsequent experiments on β-decay and on other weak nuclear interactions have shown a bias towards left-chirality in the electron, and right-chirality in the positron, approximately proportional to the ratio of the velocity of the particle, or the antiparticle, to the velocity of light.

11.3.3 Parity non-conserving circular radiolysis

The implications of parity non-conservation in the weak interaction for the origin of molecular chirality were investigated initially by Ulbricht (1959) and Vester and coworkers (1959) who suggested that the circularly-polarized γ-ray and X-ray Bremsstrahlung, produced by the progressive retardation of the longitudinally-polarized electrons in radioactive β-decay, should give rise to an enantio-differentiating photolysis or radiolysis. The radiolysis products of ten different reactions, carried out with a variety of β-emitters, proved to have a zero optical rotation, however, within experimental error (Ulbricht and Vester, 1962). Goldhaber and coworkers (1957) found that, while high-energy γ-ray Bremsstrahlung from a strontium-90 β-source are almost completely left-circularly polarized, the degree of circular polarization falls off progressively with increasing wavelength to the 0.01% level in the X-ray region, becoming negligible for the ultraviolet and visible-wavelength ranges. The chiral photo-discriminations are expected to attain an optimum in the visible and ultraviolet-wavelength region, where electromagnetic radiation interacts with the molecular valency electrons, and the optical activity of an enantiomer is at a maximum. The optical rotation of quartz, for example, is only $3.5 \pm 1.3°$ cm^{-1} at the X-ray wavelength of 1.540 Å, smaller than the rotation of $5.25°$ cm^{-1} at 3.2 μm in the infrared region, compared to $3709°$ cm^{-1} at 185 nm, near to the vacuum ultraviolet valence-shell electronic absorption of silica (Hart and Rodrigues, 1981).

Following the initial investigation of Ulbricht and Vester (1962), studies of the differential β-radiolysis of enantiomers have centred upon the amino acids. While the β-emitting radioisotopes produce only longitudinally-polarized electrons with antiparallel linear and angular momentum, the use of a linear accelerator provides the choice of either parallel- or antiparallel-polarized electrons and a choice of electron energy. The irradiation of racemic leucine with 120 keV longitudinally-polarized electrons from a linear accelerator produces a slight preferential degradation of the D-isomer with antiparallel-polarized electrons, the intrinsic polarization in β-decay, while the parallel-polarized electrons give essentially an equivalent opposite radiolysis, selectively degrading the L-enantiomer (Bonner *et al.*, 1976). It appears improbable that the differential radiolysis of the enantiomers in racemic amino acids is due to the Bremsstrahlung mechanism, since γ-irradiation racemises the enantiomers (Bonner, Blair and Lemmon, 1979).

11.3.4 Universal optical activity

The strong, the weak, and the electromagnetic interaction are now unified, leaving only gravitation as a singular interaction (Georgi, 1981; Hawking, 1981). The particular connection between the weak nuclear interaction and electromagnetism through the weak neutral current, proposed by Weinberg (1967) and Salam (1968), has the consequence that weak optical activity is universal, although very small. Spherical atoms are expected to rotate the plane of polarization, not only of electromagnetic radiation, but also of matter-waves. Moreover, the weak neutral-current coupling of the electromagnetic and the weak interaction is expected to break the exact energy degeneracy of an enantiomeric pair of molecules, based upon a pure electromagnetic hamiltonian.

In an atom, the $s_{1/2}$ and the relativistic $p_{1/2}$ electrons have a finite density probability at the atomic nucleus, where they interact with the nucleons through the short-range weak nuclear current. The electronic states of opposite parity in an atom become mixed, and Laporte's (1924) parity rule, mutually excluding a simultaneous electric- and magnetic-dipole moment in an electronic transition of an atom, ceases to be exact. The parity mixing is the more readily detected as optical activity in atomic electronic transitions which are magnetic-dipole allowed by Laporte's rule, just as molecular optical activity is the more readily accessible in the magnetic-dipole transitions of enantiomers, on account of the large dissymmetry ratio.

The parity non-conserving potential of the weak nuclear interaction (V_{pnc}) is proportional to Z^3, where Z is the atomic number and, accordingly, the magnetic-dipole electronic transitions of the heavier atoms are of particular interest, especially those transitions in the visible and adjacent regions covered by tune-able lasers. Rotations on the order of 10^{-7} radians for a unit pathlength are expected close to the nuclear hyperfine components of the $^4S_{3/2} \rightarrow {}^2D_{5/2}$ electronic transition of bismuth-209 near 648 nm or of the $^4S_{3/2} \rightarrow {}^2D_{3/2}$ transition near 876 nm. Small optical rotations of the correct sign, corresponding to a negative circular birefringence, and of the correct magnitude, have been detected in samples of bismuth vapour located between crossed polarizers from the relative intensity of the laser radiation transmitted (Barkov and Zolotorev, 1979; Lewis et al., 1979).

More recently the rotation of the spin axis of neutrons propagated through crystalline tin has been detected (Stodolsky, 1981). The effect, due to the neutral weak-current interaction between the neutron and the nucleus of the tin atom, is of the order of 5×10^{-6} radians cm^{-1} for native tin and some 7 times larger for the isotope, tin-117. Although detected with cold neutrons, with millivolt energies, the rotation is expected to be independent of the particular energy of the neutron beam, unlike natural optical activity, which is sensitive to the photon energy, as the variation in the optical rotation of quartz with wavelength illustrates. The observed rotation of the neutron spin axis is larger than expected theoretically (10^{-8} radians cm^{-1}), and so too is the differential rate of transmission of left- and right-handed neutrons through crystalline tin, namely, neutrons with the spin axis antiparallel and parallel, respectively, to the direction of the neutron beam.

11.3.5 Enantiomeric energy inequivalence

In the early years of quantum mechanics, Hund (1927) and Rosenfeld (1928) commented on the paradoxical status of optically-stable enantiomers. Optical isomers are interconverted by space inversion, and they have the same electromagnetic energy, but the wavefunction of an enantiomer cannot represent a stationary energy state of the molecule, as it lacks the full symmetry of the electromagnetic hamiltonian, which is parity-conserving. The true stationary states are the parity-symmetric and the parity-antisymmetric combinations of the wavefunctions of the left- and the right-handed isomers,

$$|\psi \pm> = [|\psi_L> \pm |\psi_R>]/\sqrt{2} \tag{11.12}$$

An initially-resolved enantiomer, as an isolated molecule, would be expected to undergo spontaneous inversion into its antipode by tunnelling through the energy barrier separating the minima of the double-bottomed potential well. The two potential minima represent the non-stationary enantiomer states, and the optical activity of the isolated molecule would be expected to vary sinusoidally from a positive to a negative value of equal magnitude in time. The tunnelling racemisation of an isolated molecule is distinct from the thermal racemisation of an assembly of enantiomers, due to activated collisions, where the energy-barrier between the two isomers is surmounted. The paradox was early accommodated by the assumption that the tunnelling time is very long for enantiomers such as the L-amino acids and D-sugars, so that they have an effective optical stability.

The electromagnetic energy degeneracy of enantiomers is lifted by the weak neutral-current coupling of the electromagnetic to the weak nuclear interaction. The two energy minima of the double-bottomed potential well for optical isomers are no longer isoenergetic, being separated by a parity non-conserving energy difference,

$$2\Delta E_{pnc} = <\psi_L |V_{pnc}| \psi_L> - <\psi_R |V_{pnc}| \psi_R> \qquad (11.13)$$

$$= 2<\psi_L |V_{pnc}| \psi_L>$$

If the energy difference, $2\Delta E_{pnc}$, is greater than the parity-conserving electro-magnetic coupling energy between the optical isomers, $<\psi_L |H_{em}| \psi_R>$, the tunnelling-time between the enantiomers becomes very large. Further, the wave-functions, $|\psi_L>$ and $|\psi_R>$, now correspond to stationary states of the augmented hamiltonian, $[H_{em} + V_{pnc}]$ (Harris and Stodolsky, 1978, 1981). According to this view, the true isoenergetic antipode of an enantiomer com-posed of electrons and other particles is the corresponding mirror-image structure of positrons and other antiparticles in an antiworld (Barron, 1981). Thus, all racemates and achiral molecules are optically active, due to the circular bire-fringence produced by the coupling of the weak neutral current in the individual atoms, and those structures composed of particles have enantiomeric counter-parts made up of antiparticles.

The magnitude of the parity non-conserving energy difference between optical isomers was estimated initially by Zel'dovich and coworkers (1977) who showed that, in eqn 11.13, the main contribution derives from the coupling of the electron spin with the charge of the atomic nucleus, involving the spin–orbit interaction, V_{so}. In a closed-shell molecule, the energy difference has the form,

$$\Delta E_{pnc} \sim <S|V_{pnc}|T><T|V_{so}|S>/\Delta E_{ST} \qquad (11.14)$$

where $|S>$ and $|T>$ refer to singlet and triplet states, respectively, separated by the energy, ΔE_{ST}. As the spin–orbit interaction is proportional to Z^2, and V_{pnc} to Z^3, where Z is the atomic number, the energy difference is expected to be appreciable in enantiomers containing heavy atoms. An approximate, order of magnitude, estimate is provided by the relation,

$$\Delta E_{pnc} \sim (\eta Z^5) 10^{-8} \text{ (eV)} \qquad (11.15)$$

where η is a chirality factor dependent upon the parity mixing of the state functions and the electronic charge density near to the atomic nucleus. Zel'dovich and coworkers (1977) adopted a crystal-field model for the parity mixing of the atomic orbitals of a heavy atom in a chiral molecule, and suggested, from dimensional considerations, that the chirality factor, η, might be of the order of 10^{-2}, giving $\Delta E_{pnc} \sim 10^{-12}$ eV for heavy atom molecules.

The more detailed estimates of ΔE_{pnc} by Rein and coworkers (1979, 1980) are considerably smaller. These estimates are based upon the molecular

4   5

orbitals of a chiral ethylene, **4**, twisted through $+10°$ to D_2 symmetry (Bouman and Hansen, 1977), and of the sulphide chromophore, $C—S—C$, in A-nor-2-thiacholestane, **5** (Rosenfeld and Moscowitz, 1972). Rein and coworkers (1979, 1980) found that the molecular orbitals of the chiral ethylene, **4**, which make the largest contribution to the enantiomeric energy difference, ΔE_{pnc}, were just the molecular orbitals that contribute the most substantially to the optical acitivity. The connection is not wholly coincidental since V_{pnc} is pseudo-scalar, and the development of eqn 11.14 gives an expression analogous to that for the rotational strength or the optical rotation (eqns 2.2 and 2.3).

The expansion of the molecular orbitals over the atomic orbitals results in one set of terms referring to a single atomic centre and another set covering two atomic centres. For a basis restricted to s and relativistic p orbitals, which are important in affording a non-vanishing electron density at the nucleus, the one-centre terms are found to sum to zero. A non-vanishing contribution to ΔE_{pnc} from the one-centre terms requires either 3d or higher nl atomic orbitals at the centre, or a relaxation of the generally-adopted assumption that all three 2p orbitals of the centre have the same radial function. The two-centre terms give for the (R)-isomer of the chiral ethylene, **4**, a chirality factor, η (eqn 11.15) of $+3 \times 10^{-4}$, and a parity non-conserving energy higher by 10^{-18} eV than the (S)-enantiomer (eqn 11.13).

For A-nor-2-thiacholestane, **5**, the chirality factor is even smaller, with $\eta = -2 \times 10^{-8}$, and the parity non-conserving energy is now lower than that of the enantiomer, but only by -6×10^{-21} eV (Rein *et al.*, 1979, 1980). The chirality factor and ΔE_{pnc} are considerably smaller for **5** than **4** on two main grounds. Firstly, only the one-centre terms relating to the sulphur atom of A-nor-2-thiacholestane, **5**, were considered, the non-vanishing sum of those terms resulting from the contribution of the sulphur 3d orbitals to the molecular orbitals of the $C—S—C$ chromophore. Secondly, the C_{2v} symmetry of the $C—S—C$ chromophore requires a perturbation from the other groups of the A-nor-2-thiacholestane molecule to give a non-vanishing chirality factor, η. The third-order perturbation treatment required is expected to give a smaller η

factor and ΔE_{pnc} energy for **5** than the corresponding second-order treatment of the two-centre terms for the chiral ethylene **4**.

Yamagata (1966) and others, notably Letokhov (1975), invoked a small parity non-conserving energy difference between enantiomers to account for the transition from racemic chemistry to chiral biochemistry in chemical evolution. The cumulative effect of a large number of repeated syntheses and polymerisations, each with a small energy bias towards a particular enantiomer or homochiral polymer, results in a convergent trend to optical purity. Taking the relative energy difference between enantiomers, $(\Delta E_{pnc}/E)$, to be of the order of 10^{-16}, Letokhov (1975) estimated a selection time for the attainment of biochemical optical purity on the order of a hundred to a million years. The relative energy difference assumed is evidently larger than warranted by the more recent and detailed estimates of ΔE_{pnc}, while the competing effect of the spontaneous racemisation of biomolecules is expected to extend substantially the selection period for biochemical optical purity.

SUGGESTIONS FOR FURTHER READING

The following books and reviews, selected from the bibliography and reference list, extend the discussion of the section or chapter indicated.

Chapters 1, 2 and 3

For the subjects of § 1.1, 1.2, 1.3 and 2.1, Lowry (1935) remains the most comprehensive source. Partington (1953) covers some early wave theories of optical rotation not mentioned by Lowry, and Partington (1964) contains more factual detail on the development of nineteenth century structure theory. More recent assessment of the foundations of classical stereochemistry are given in Mason (1976) and in several contributions to the *van't Hoff–Le Bel Centennial* (Ramsey, 1975).

The original articles on the configurational notation conventions (§ 1.5) are recommended for details (Cahn *et al.*, 1966; IUPAC Convention, 1970). The organic convention and the earlier Fischer convention are covered by many organic and biochemical texts, e.g. Bentley (1969), who gives a clear account of the extension of the organic convention to pro-chirality. The IUPAC configurational convention for coordination compounds is described by Saito (1979).

The book by Caldwell and Eyring (1971) is one of the more detailed of the many surveys available of the quantum theories of optical activity and the general mechanisms (§ 2.2; chapter 3). The review by Schellman (1975) covers band shapes and the evaluation of transition strengths (§ 2.3). The principles and the practice of photoelectric CD spectrometry (§ 2.4) are discussed in detail by Velluz *et al.*, (1965), together with the stereochemical applications of the technique, notably to chiral carbonyl compounds (§ 4.2 and 4.3). Vacuum ultraviolet CD spectroscopy is reviewed by Johnson (1978) and by Schnepp (1979). The luminescence techniques (§ 2.4) are surveyed by Turner (1978), on fluorescence-detected CD and by Steinberg (1978), on circular intensity differentials in the luminescence of chiral molecules.

Schellman (1968) gives a clear and concise survey of the general mechanisms for optical activity (chapter 3). The sector rules based upon the one-electron mechanism (§ 3.2) are due to Schellman (1966), which is recommended for

further reading. Weigang and Höhn (1966, 1968) and Weigang (1979) are recommended for further details of the dynamic polarization mechanisms applied
to chiral organic molecules (§ 3.3 and 3.4).

Chapters 4, 5, 6, 7 and 8

Comprehensive surveys of the applications of CD to chiral organic
molecules, biopolymers, and coordination compounds are provided by the multi-
authored proceedings of international symposia held from time to time on
optical activity and allied topics. The proceedings give a representative cross-
section of studies in the field at the time of the symposium. The editors and dates
of the publication of the more recent symposia are Snatzke (1967); Nyholm
(1967); Ciardelli and Salvadori (1973); and Mason (1979). The proceedings of
the most recent symposium cover vibrational optical activity and the luminescence techniques, as well as the electronic CD of chiral organic molecules,
biopolymers, and coordination compounds, together with contributions on chiral
discrimination.

Chapters 9, 10 and 11

The book by Jacques, Collet and Wilen (1981) is recommended for
chiral phase equilibria (§ 9.2 and 9.3) and for optical resolution procedures (§
10.1). Kitaigorodsky (1973) is suggested for enantiomer and racemate crystal-
packing modes (§ 9.4) and Saito (1979) for absolute stereochemical configuration by diffraction methods (§ 9.5).

Izumi and Tai (1977) cover both the chiral photodiscriminations (§ 10.2)
and the corresponding thermal discriminations (§ 10.3). Morrison and Mosher
(1976) provide more detail on asymmetric organic reactions (§ 10.3).
Parascandola (1975) is suggested for the pharmacological topics in § 10.4 .

The topics of § 11.1 and 11.2 are discussed by Craig and others in Mason
(1979). Two recent symposia cover the subjects of § 11.3 and 9.1, Walker (1979),
and Thiemann (1981).

BIBLIOGRAPHY AND REFERENCES

Addadi, L. and Lahav, M. (1979) *J. Am. Chem. Soc.*, **101**, 2152.

Albert, A. (1973) *Selective Toxicity*, 5th edn., Chapman, London.

Allen, F.H., Neidle, S. and Rodgers, D. (1970) *J. Chem. Soc. C*, 2340.

Allen, F.H. and Rodgers, D. (1966) *J. Chem. Soc. Chem. Comm.*, 838.

Alworth, W.L. (1972) *Stereochemistry and its Application in Biochemistry*, Wiley-Interscience, New York.

Amaya, K., Takagi, S. and Fujishiro, R. (1968) *J. Chem. Soc. Chem. Comm.*, 480.

Amiard, G. (1956) *Bull. Soc. Chim. France*, 447.

ApSimon, J.W. and Seguin, R.P. (1979) *Tetrahedron*, **35**, 2797.

Arago, D.F.J. (1811) *Mém. Inst. France*, **Pt. I**, 93.

Ariëns, E.J. (ed.) (1964) *Molecular Pharmacology. The Mode of Action of Biologically Active Compounds*, Vol. 1, p. 240, Academic Press, New York.

Arnett, E.M. and Zingg, S.P. (1981) *J. Am. Chem. Soc.*, **103**, 1221.

Atkins, P.W. (1980) *Chem. Phys. Letters*, **74**, 358.

Audebert, R. (1979) *J. Liquid Chromatogr.*, **2**, 1063.

Baessler, H. and Labes, M.M. (1970) *Molec. Cryst. Liq. Cryst.*, **6**, 419.

Bailar, J.C., Jr. and Auten, R.W. (1934) *J. Am. Chem. Soc.*, **56**, 774.

Ballard, R.E., McCaffery, A.J. and Mason, S.F. (1965) *J. Chem. Soc.*, 883.

Ballhausen, C.J. (1979) *Molecular Electronic Structures of Transition Metal Complexes*, McGraw-Hill, New York.

Barkov, L.M. and Zolotorev, M.S. (1979) *Phys. Lett.*, **85B**, 308.

Barnett, C.J., Drake, A.F. and Mason, S.F. (1979) *Bull. Soc. Chim. Belg.*, **88**, 853.

Barnett, C.J., Drake, A.F. and Mason, S.F. (1980) *J. Chem. Soc. Chem. Comm.*, 43.

Barrett, G.C. (1972) in *Techniques of Chemistry*, Vol. IV, Part 1, ed. A. Weissberger, Ch. 8. Wiley, New York.

Barron, L.D. (1977) *J. Chem. Soc. Perkin Trans.* **II**, 1074, 1790.

Barron, L.D. (1978) *Adv. Infrared and Raman Spectrosc.*, **4**, 271.

Barron, L.D. (1979) in *Optical Activity and Chiral Discrimination*, ed. S.F. Mason, Ch. 9, p. 219, Reidel, Dordrecht.

Barron, L.D. (1980) *Acc. Chem. Res.*, **13**, 90.

Barron, L.D. (1981) *Chem. Phys. Letters*, **79**, 392.

Barron, L.D. (1981) *Molec. Phys.*, **43**, 1395.

Barron, L.D. (1982) *Molecular Light Scattering and Optical Activity*, Cambridge University Press.

Barron, L.D., Bogaard, M.P. and Buckingham, A.D. (1973a) *J. Am. Chem. Soc.*, **95**, 603.

Barron, L.D., Bogaard, M.P. and Buckingham, A.D. (1973b) *Nature*, **241**, 113.

Barron, L.D. and Buckingham, A.D. (1971) *Molec. Phys.*, **20**, 1111.

Barron, L.D. and Buckingham, A.D. (1974) *J. Am. Chem. Soc.*, **96**, 4769.

Barron, L.D. and Buckingham, A.D. (1975) *Ann. Rev. Phys. Chem.*, **26**, 381.

Barron, L.D. and Buckingham, A.D. (1979) *J. Am. Chem. Soc.,* **101,** 1979.
Barth, G. and Djerassi, C. (1981) *Tetrahedron,* **37,** 4123.
Beckett, A.H. and Casy, A.F. (1954) *J. Pharm. Pharmacol.,* **6,** 986.
Beckett, A.H. and Casy, A.F. (1955) *J. Pharm. Pharmacol.,* **7,** 433.
Bedarkar, S. and Blundell, T.L. (1978) *Molecular Structure by Diffraction Methods,* Vol. **6,** p. 241, Specialist periodical report, Chem. Soc., London.
Beecham, A.F., Hurley, A.C., Mathieson, A. McL. and Lamberton, J.A. (1973) *Nature, Phys.,* **244,** 30.
Beevers, C.A. and Lipson, H. (1932) *Z. Krist.,* **83,** 123.
Belsky, V.K. (1974) *J. Struct. Chem.,* **15,** 631.
Belsky, V.K. and Zorkii, P.M. (1977) *Acta Cryst.,* **A33,** 1004.
Benfey, O.T. (ed.) (1963) *Classics in the Theory of Chemical Combination,* Dover, New York.
Bentley, R. (1969) *Molecular Asymmetry in Biology,* vol. 1, Academic Press, New York.
Bentley, R. (1970) *Molecular Asymmetry in Biology,* vol. 2, Academic Press, New York.
Bernal, I. and Palmer, R.A. (1981) *Inorg. Chem.,* **20,** 295.
Bernstein, W.J., Calvin, M. and Buchardt, O. (1972) *J. Am. Chem. Soc.,* **94,** 494.
Bernstein, W.J., Calvin, M. and Buchardt, O. (1973) *J. Am. Chem. Soc.,* **95,** 527.
Berthelot, M. and Jungfleisch, E.C. (1875) *Ann. Chim. Phys.,* [5], **4,** 147.
Bertsch, K. and Jochims, J.C. (1977) *Tetrahedron Lett.,* 4379.
Bethe, H.A. (1929) *Ann. Physik,* [5] **3,** 133.
Bijvoet, J.M. (1949) *Proc. Koninkl. Ned. Akad. Wetenschap,* **B52,** 313.
Bijvoet, J.M., Peerdeman, A.F. and van Bommel, A.J. (1951) *Nature,* **168,** 271.
Billardon, M. and Badoz, J. (1966) *Compt. Rend. Paris,* **262,** 1672.
Billardon, M., Rivoal, J.C. and Badoz, J. (1969) *Rev. Phys. Appl.,* **4,** 353.
Biot, J.B. (1812) *Mém. Inst. France,* **1,** 1.
Biot, J.B. (1817) *Mém. Acad. Sci. France,* **2,** 41.
Biot, J.B. (1838) *Mém. Acad. Sci. France,* **15,** 93.
Biot, J.B. (1842) *Compt. Rend. Paris,* **15,** 619.
Blair, N.E., Dirbas, F.M. and Bonner, W.A. (1981) *Tetrahedron,* **37,** 27.
Blaschke, G. (1980) *Angew. Chem. (Int. Edn.),* **19,** 13.
Blout, E.R. and Idelson, M. (1956) *J. Am. Chem. Soc.,* **78,** 3857.
Blout, E.R. and Idelson, M. (1957) *J. Am. Chem. Soc.,* **79,** 3948.
Blout, E.R. and Idelson, M. (1958) *J. Am. Chem. Soc.,* **80,** 2387.
Blum, L., Narten, A.H., Karnicky, J.F. and Pings, J.C. (1976) *Adv. Chem. Phys.,* **34,** 157 and 203.
Boltzmann, L. (1874) *Pogg. Ann. Phys. Chem. Jubelband,* 128.
Bonner, W.A. (1979) in *Origins of Optical Activity in Nature,* ed. D.C. Walker, p. 5, 47, Elsevier, Amsterdam.
Bonner, W.A., Blair, N.E. and Lemmon, R.M. (1979) *J. Am. Chem. Soc.,* **101,** 1049.
Bonner, W.A., Van Dort, M.A. and Yearian, M.R. (1976) *Nature,* **264,** 198.
Born, M. (1918) *Ann. Physik,* [4] **55,** 177.
Bosnich, B. (1969) *Acc. Chem. Res.,* **2,** 266.
Bosnich, B. and Fryzuk, M.D. (1981) *Topics in Stereochem.,* **12,** 119.
Bouman, T.D. and Hansen, A.E. (1977) *J. Chem. Phys.,* **66,** 3460.
Bouman, T.D. and Moscowitz, A. (1968) *J. Chem. Phys.,* **48,** 3115.
Boyle, H.P. (1971) *Quart. Rev. Chem. Soc.,* **25,** 323.
Boys, S.F. (1934) *Proc. Roy. Soc. London,* **A144,** 655, 675.
Brahms, J. and Brahms, S. (1970) in *Fine Structure of Proteins and Nucleic Acids,* ed. G.D. Fasman and S.N. Timasheff, Ch. 3, p. 191, Dekker, New York.
Brahms, S. and Brahms, J. (1980) *J. Mol. Biol.,* **138,** 149.
Bredig, G. and Fajans, K. (1908) *Chem. Ber.,* **41,** 752.

Brewster, J.H. (1967) *Topics in Stereochem.*, **2**, 1.

Brickell, W.S., Brown, A., Kemp, C.M. and Mason, S.F. (1971*a*) *J. Chem. Soc. A*, 751, 756.

Brickell, W.S., Brown, A., Kemp, C.M. and Mason, S.F. (1971*b*) *Molec. Phys.*, **20**, 787.

Bridge, N.J. and Buckingham, A.D. (1966) *Proc. Roy. Soc. London*, **295**A, 334.

British Pharmacopoeia Commission (1981) *Approved Names*, Her Majesty's Stationery Office, London.

Brocki, T., Moskovits, M. and Bosnich, B. (1980) *J. Am. Chem. Soc.*, **102**, 495.

Brongersma, H.H. and Mul, P.M. (1973) *Chem. Phys. Letters*, **19**, 217.

Brown, J.M., Chaloner, P.A., Murrer, B.A. and Parker, D. (1980) in *Stereochemistry of Optically Active Transition Metal Compounds*, ed. B.E. Douglas and Y. Saito, p. 169, Am. Chem. Soc., Washington, D.C.

Buchardt, O. (1974) *Angew. Chem. Int. Edn.*, **13**, 179.

Buckingham, A.D. (1967) *Adv. Chem. Phys.*, **12**, 107.

Buckingham, A.D. and Joslin, C.G. (1981) *Chem. Phys. Letters*, **80**, 615.

Buckingham, A.D. and Stiles, P.J. (1974) *Acc. Chem. Res.*, **7**, 258.

Buckingham, D.A., Mason, S.F., Sargeson, A.M. and Turnbull, K.R. (1966) *Inorg. Chem.*, **5**, 1649.

Buckingham, D.A. and Sargeson, A.M. (1971) *Topics in Stereochem.*, **6**, 219.

Burke, J.G. (1966) *Origins of the Science of Crystals*, University of California, Berkeley.

Cahn, R.S., Ingold, C. and Prelog. V. (1966) 'Specification of Molecular Chirality', *Angew. Chem. Internat. Edn.*, **5**, 385.

Caldwell, D.J. and Eyring, H. (1971) *The Theory of Optical Activity*, Wiley-Interscience, New York.

Callomon, J.H. and Innes, K.K. (1963) *J. Molec. Spectrosc.* **10**, 166.

Calvin, M. (1969) *Chemical Evolution*, Oxford University Press, London.

Capra, J.D. and Edmundson, A.B. (1977) *Scientific American*, **236**, No. 1, 50.

Cesario, M., Guilhem, J., Pascard, C., Collet, A. and Jacques, J. (1978) *Nouv. J. Chim.*, **2**, 343.

Chabay, I. (1972) *Chem. Phys. Lett.*, **17**, 283.

Charney, E. (1979) *The Molecular Basis of Optical Activity, Optical Rotatory Dispersion and Circular Dichroism*, Wiley, New York.

Cheng, J.C., Nafie, L.A., Allen, S.D. and Braunstein, A.I. (1976) *Appl. Opt.*, **15**, 1960.

Chickos, J.S., Garin, D.L., Hitt, M. and Schilling, G. (1981) *Tetrahedron*, **37**, 2259.

Chion, B., Lajzerowitz, J., Bordeaux, D., Collet, A. and Jacques, J. (1978) *J. Phys. Chem.*, **82**, 2682.

Chung, S.Y. and Holzwarth, G. (1975) *J. Mol. Biol.*, **92**, 449.

Ciardelli, F. and Salvadori, P. ed. (1973) *Fundamental Aspects and Recent Developments in Optical Rotatory Dispersion and Circular Dichroism*, Heyden, London.

Cochez, Y., Martin, R.H. and Jespers, J. (1976/77) *Israel J. Chem.*, **15**, 29.

Cohan, N.V. and Hameka, H.F. (1966) *J. Am. Chem. Soc.*, **88**, 2136.

Cohen, J.B. and Patterson, T.S. (1904) *Chem. Ber.*, **37**, 1012.

Collet, A., Brienne, M-J. and Jacques, J. (1980) *Chem. Rev.*, **80**, 215.

Collman, J.P., Blair, R.P., Slade, A.L. and Marshall, R.L. (1963) *Inorg. Chem.*, **2**, 576.

Condon, E.U. (1937) *Rev. Mod. Phys.*, **9**, 432.

Condon, E.U., Altar, W. and Eyring, H. (1937) *J. Chem. Phys.*, **5**, 753.

Conti, R., Bucksbaum, P., Chu, S., Commins, E. and Hunter, L. (1979) *Phys. Rev. Lett.*, **42**, 343.

Cookson, R.C. (1954) *J. Chem. Soc.*, 282.

Corey, E.J. and Bailar, J.C. (1959) *J. Am. Chem. Soc.*, **81**, 2620.

Coster, D., Knol, K.S. and Prins, J.A. (1930) *Z. Phys.*, **63**, 345.

Cotton, A. (1895) *Compt. Rend. Paris*, **120**, 989, 1044.

Cotton, A. (1896) *Ann. Chim. Phys.*, [7] **8**, 347.
Cotton, A. (1909) *J. Chim. Physique*, **7**, 81.
Cotton, A. (1930) *Trans. Faraday Soc.*, **26**, 377.
Coulson, C.A. and Davis, P.L. (1952) *Trans. Faraday Soc.*, **48**, 777.
Coulson, C.A. and Streitwieser, A. Jr. (1965) *A Dictionary of π-electron Calculations*, Pergamon, Oxford.
Cowman, M.K. and Fasman, G.D. (1978) *Proc. Nat. Acad. Sci. USA*, **75**, 4759.
Crabbé, P. (1965) *Optical Rotatory Dispersion and Circular Dichroism in Organic Chemistry*, Holden-Day, London.
Craddock, J.H. and Jones, M.M. (1961) *J. Am. Chem. Soc.*, **83**, 2839.
Craig, D.P., (1979), in *Optical Activity and Chiral Discrimination*, ed. S.F. Mason, Ch. 12, p. 293, Reidel, Dordrecht.
Craig, D.P. and Mellor, D.P. (1976) *Topics in Current Chemistry*, **63**, 1.
Craig, D.P., Power, E.A. and Thirunamachandran, T. (1971) *Proc. Roy. Soc. London*, **A322**, 165.
Craig, D.P. and Schipper, P.E. (1975) *Proc. Roy. Soc. London*, **A342**, 19.
Craig, D.P. and Walmsley, S.H. (1968) *Excitons in Molecular Crystals*, Benjamin, New York.
Cram, D.J. and Abd Elhafez, F.A. (1952) *J. Am. Chem. Soc.*, **74**, 5828, 5851.
Cram, D.J. and Cram, J.M. (1978) *Acc. Chem. Res.*, **11**, 8.
Crum Brown, A. (1890) *Proc. Roy. Soc. Edin.*, **17**, 181.
Cushny, A.R. (1926) *Biological Relations of Optically Isomeric Substances*, Williams and Wilkins, Baltimore.
Cvitas, T., Hollas, J.M. and Kirby, G.H. (1970) *Molec. Phys.*, **19**, 305.

Dabrowiak, J.C. and Cooke, D.W. (1975) *Inorg. Chem.*, **14**, 1305.
Davankov, V.A. (1980) *Adv. Chromatogr.*, **18**, 139.
Debye, P. (1920) *Physik. Z.*, **21**, 178.
de Gennes, P.G. (1974) *The Physics of Liquid Crystals*, Clarendon Press, Oxford.
Dekkers, H.P.J.M. and Closs, L.E. (1976) *J. Am. Chem. Soc.*, **98**, 2210.
Delépine, M. (1921) *Bull. Soc. Chim. France*, [4], **29**, 656.
Delépine, M. (1934) *Bull. Soc. Chim. France*, [5], **1**, 1256.
Delépine, M. and Charonnat, R. (1930) *Bull. Soc. France Mineral.*, **53**, 73.
Denning, R.G. (1967) *J. Chem. Soc. Chem. Comm.*, 120.
de Vries, Hl. (1951) *Acta Cryst.*, **4**, 219.
Dickerson, R.E. (1978) *Scientific American*, **239**, No. 3, 62.
Diem, M., Photos, E., Khouri, H. and Nafie, L.A. (1979) *J. Am. Chem. Soc.*, **101**, 6829.
Dingle, R. and Ballhausen, C.J. (1967) *Matt. Fys. Medd. Dan. Vid. Selsk.*, **35**, No. 12.
Djerassi, C. (1960) *Optical Rotatory Dispersion: Applications to Organic Chemistry*, McGraw-Hill, New York.
Djerassi, C. and Klyne, W. (1957) *J. Am. Chem. Soc.*, **79**, 1506.
Dörr, F. (1971) in *Creation and Detection of the Excited State*, ed. A.A. Lamola, Vol. 1, p. 53, Dekker, New York.
Dougherty, R.C. (1980) *J. Am. Chem. Soc.*, **102**, 380.
Dougherty, R.C., Edwards, D. and Cooper, K. (1980) *J. Am. Chem. Soc.*, **102**, 381.
Douglas, B.E. and Saito, Y. ed. (1980) *Stereochemistry of Optically Active Transition Metal Compounds*, ACS Symposium 119, Am. Chem. Soc., Washington, D.C.
Draber, W. and Stetter, J. (1979) in *Chemistry and Agriculture*, Spec. Publ. No. 36, Chem. Soc., London.
Drake, A.F., Gould, J.M. and Mason, S.F. (1980) *J. Chromatogr.*, **202**, 239.
Drake, A.F., Hirst, S.J., Kuroda, R. and Mason, S.F. (1982) *Inorg. Chem.*, **21**, 533.
Drake, A.F., Kuroda, R., Mason, S.F., Peacock, R.D. and Stewart, B. (1981) *J. Chem. Soc. Dalton*, 976.

Drake, A.F., Levey, J.R., Mason, S.F. and Prosperi, T. (1982) *Inorg. Chim. Acta,* 57, 151.
Drake, A.F. and Mason, S.F. (1977) *Tetrahedron,* 33, 937.
Drake, A.F. and Mason, S.F. (1978) *J. Physique,* 38, C4-212.
Drude, P. (1893) *Ann. Physik,* 50, 595.
Drude, P. (1900) *The Theory of Optics,* trans. C.R. Mann and R.A. Millikan, Dover Reprint, New York, (1959).
Duax, W.L. (1978) *Molecular Structure by Diffraction Methods,* Vol. 6, Specialist periodical report, Chem. Soc., London, p. 261.
Dudley, R.J., Mason, S.F. and Peacock, R.D. (1972) *J. Chem. Soc. Chem. Comm.,* 1084.
Dudley, R.J., Mason, S.F. and Peacock, R.D. (1975) *J. Chem. Soc. Faraday Trans.* II, 71, 997.
Duesler, E.N. and Raymond, K.N. (1971) *Inorg. Chem.,* 10, 1486.
Duggan, M., Ray, N., Hathaway, B., Tomlinson, G., Brint, P. and Pelin, K. (1980) *J. Chem. Soc. Dalton Trans.,* 1342.
Duschinsky, R. (1934) *Chem. Ind.,* 53, 10.
Dwyer, F.P. and Garvan, F.L. (1959) *J. Am. Chem. Soc.,* 81, 1043.
Dwyer, F.P., Garvan, F.L. and Shulman, A. (1959) *J. Am. Chem. Soc.,* 81, 290.
Dwyer, F.P. and Mellor, D.P. ed. (1964) *Chelating Agents and Metal Chelates,* Academic Press, New York.
Dwyer, F.P. and Sargeson, A.M. (1959) *J. Am. Chem. Soc.,* 81, 5269, 5272.

Easson, L. and Stedman, E. (1933) *Biochem. J.,* 27, 1257.
Ehrlich, P. (1900) *Proc. Roy. Soc. London,* 66, 424.
Elgavi, A., Green, B.S. and Schmidt, G.M.J. (1973) *J. Am. Chem. Soc.,* 95, 2058.
Eliel, E.L. (1962) *Stereochemistry of Carbon Compounds,* McGraw-Hill, New York.
Elliot, M. and Janes, N.F. (1978) *Chem. Soc. Rev.,* 7, 473.
Elsbernd, H. and Beattie, J.K. (1969) *Inorg. Chem.,* 8, 893.
Endo, I., Horikoshi, S. and Utsuno, S. (1981) *J. Chem. Soc. Chem. Comm.,* 296.

Fajans, K. (1923) *Naturwissenschaften,* 11, 165.
Faraday Discussion (1914) 'Optical Rotatory Power', *Trans. Faraday Soc.,* 10, 14–138.
Faraday Discussion (1930) 'Optical Rotatory Power', *Trans. Faraday Soc.,* 26, 265–461.
Faraday Discussion (1943) 'Modes of Drug Action', *Trans. Faraday Soc.,* 39, 319–444.
Faraday M. (1846) *Phil. Mag.,* 28, 294.
Farina, M., Audisio, G. and Natta, G. (1967) *J. Am. Chem. Soc.,* 89, 5071.
Faulkner, T.R., Marcott, C., Moscowitz, A. and Overend, J. (1977) *J. Am. Chem. Soc.,* 99, 8160, 8169.
Faulkner, T.R., Moscowitz, A., Holzwarth, G., Hsu, E.C. and Mosher, H.S. (1974) *J. Am. Chem. Soc.,* 96, 252.
Fay, R.C., Girgis, A.Y. and Klabunde, U. (1970) *J. Am. Chem. Soc.,* 92, 7056.
Feofilov, P.P. (1961) *The Physical Basis of Polarized Emission,* Consultants Bureau, New York.
Findlay, A. (1951) *The Phase Rule,* 9th edn, ed. A.N. Campbell and N.O. Smith, Dover, New York.
Finkelstein, R. and Van Vleck, J.H. (1940) *J. Chem. Phys.,* 8, 790.
Fischer, E. (1891) *Chem. Ber.,* 24, 2683.
Fischer, E. (1894) *Chem. Ber.,* 27, 2985, 3231.
Fischer, E. (1898) *Z. Physiol. Chem.,* 26, 60.
Fischer, E. (1907) *J. Chem. Soc.,* 91, 1749.
Flammang-Barbieux, M., Nasellski, J. and Martin, R.M. (1967) *Tetrahedron Lett.,* 743.
Flook, R.J., Freeman, H.C. and Scudder, M.L. (1977) *Acta Cryst.,* B33, 801.
Fox, S.W. ed. (1964) *Origins of Prebiological Systems and of their Molecular Matrices,* Academic Press, New York.

Fredga, A. (1944) *The Svedberg, 1884–1944*, p. 261, Almquist and Wiksells, Uppsala.
Fredga, A. (1960) *Tetrahedron*, 8, 126.
Fresnel, A. (1816) *Ann. Chim.*, 1, 239.
Fresnel, A. (1819) *Ann. Chim.*, 10, 288.
Fresnel, A. (1821) *Ann. Chim.*, 17, 80, 102, 167.
Fresnel, A. (1824) *Bull. Soc. Philomat.*, 9, 150.
Freudenberg, K. ed. (1933) *Stereochemie*, Deuticke, Leipzig.
Friedel, G. (1913) *Compt. Rend. Paris*, 157, 1533.
Friedel, G. (1922) *Ann. Physique*, 18, 273.
Furberg, S. and Hassel, O. (1950) *Acta Chem. Scand.*, 4, 1020.
Furlani, C. (1968) *Coord. Chem. Rev.*, 3, 141.

Gardner, M. (1979) *The Ambidextrous Universe*, 2nd edn, Scribners, New York.
Gause, G.F. (1941) *Optical Activity and Living Matter*, Biodynamica, Normandy, Missouri.
Gawroński, J. and Gawroński, K. (1980) *J. Chem. Soc. Chem. Comm.*, 346.
Georgi, H. (1981) *Scientific American*, 244, No. 4, 40.
Gerlach, H. (1968) *Helv. Chim. Acta*, 51, 1587.
Gernez, D. (1866) *Compt. Rend. Paris*, 63, 843.
Gil-Av, E. and Nurok, D. (1974) *Adv. Chromatogr.*, 10, 99.
Goedicke, Ch. and Stegemeyer, H. (1970) *Tetrahedron Lett.*, 937.
Goedicke, Ch. and Stegemeyer, H. (1972) *Chem. Phys. Lett.*, 17, 492.
Goldhaber, M., Grodzins, L., Sunyar, A.W. and McVoy, K.W. (1957) *Phys. Rev.*, 106, 826 and 828.
Golding, B.T., Gainsford, G.J., Herlt, A.J. and Sargeson, A.M. (1976) *Tetrahedron*, 32, 389.
Gorin, E., Walter, J. and Eyring, H. (1938) *J. Chem. Phys.*, 6, 824.
Gottarelli, G. and Samori, B. (1972) *J. Chem. Soc. Perkin II*, 1998.
Graham, T. (1842) *Elements of Chemistry*, p. 157. Bailliere, London.
Green, B.S., Lahav, M. and Rabinovich, D. (1979) *Acc. Chem. Res.*, 12, 191.
Griffith, J.S. (1961) *The Theory of Transition Metal Ions*, Cambridge University Press.
Griffith, M.C. ed. (1980) *USAN 1981: USAN and USP Dictionary of Drug Names, 1961–1980, Cumulative List*, US Pharmacopeial Convention Inc., Rockville, Maryland.
Grignard, V. (1901) *Ann. Chim.*, [7] 24, 433.
Grinter, R., Harding, M.J. and Mason, S.F. (1970) *J. Chem. Soc. A*, 667.
Groen, M.B., Schadenberg, H. and Wynberg, H. (1971) *J. Org. Chem.*, 36, 2797.
Grosjean, M. and Legrand, M. (1960) *Compt. Rend. Paris*, 251, 2150.
Guye, P.A. (1890) *Compt. Rend. Paris*, 110, 714.
Gyarfas, E.C. (1954) *Rev. Pure Appl. Chem.*, 4, 73.

Hall, D.M., Ridgwell, S. and Turner, E.E., (1954) *J. Chem. Soc.*, 2498.
Harada, N., Chen, S.L. and Nakanishi, K. (1975) *J. Am. Chem. Soc.*, 97, 5345.
Harada, N. and Nakanishi, K. (1972) *Acc. Chem. Res.*, 5, 257.
Harada, N. and Nakanishi, K. (1982) *Circular Dichroic Spectroscopy: Exciton Coupling in Organic and Bioorganic Stereochemistry*, University Science Books, Mill Valley, California.
Harada, N., Takuma, Y. and Uda, H. (1977) *Bull. Chem. Soc. Japan*, 50, 2033.
Harada, N., Takuma, Y. and Uda, H. (1978) *J. Am. Chem. Soc.*, 100, 4029.
Harada, N., Tamai, Y., Takuma, Y. and Uda, H. (1980) *J. Am. Chem. Soc.*, 102, 501 and 506.
Harding, M.J., Mason, S.F. and Peart, B.J. (1973) *J. Chem. Soc. Dalton Trans.*, 955.
Harris, M.M. (1958) *Prog. Stereochem.*, 2, 157.
Harris, R.A. and Stodolsky, L. (1978) *Phys. Lett.*, 78B, 313.
Harris, R.A. and Stodolsky, L. (1981) *J. Chem. Phys.*, 74, 2145.
Hart, M. and Rodrigues, A.R.D. (1981) *Phil. Mag.*, B43, 321.

Hawking, S.W. (1981) *Phys. Bull.*, **32**, 15.

Hawkins, C.J. (1971) *Absolute Configuration of Metal Complexes*, Wiley-Interscience, New York.

Hellwinkel, D. and Mason, S.F. (1970) *J. Chem. Soc. B*, 640.

Herpin, P. (1958) *Bull. Soc. France Mineral. Cryst.*, 81, 201.

Herschel, J.F.W. (1822) *Trans. Cambr. Phil. Soc.*, 1, 43.

Hess, H., Burger, G. and Musso, H. (1978) *Angew. Chem. Int. Edn.*, 17, 612.

Hesse, G. and Hagel, R. (1973) *Chromatographia*, 6, 277.

Hesse, G. and Hagel, R. (1976) *Ann.*, 996

Hesse, G. and Hagel, R. (1976) *Chromatographia*, 9, 62.

Heyn, M.P. (1975) *J. Phys. Chem.*, 79, 2424.

Hilmes, G. and Richardson, F.S. (1976) *Inorg. Chem.*, 15, 2582.

Hoffmann, R. and Woodward, R.B. (1968) *Acc. Chem. Res.*, 1, 17.

Hoffmann, R. and Woodward, R.B. (1969) *Angew. Chem. Int. Edn.*, 8, 781.

Holzwarth, G., Chabay, I. and Holzwarth, N.A.W. (1973) *J. Chem. Phys.*, 58, 4816.

Holzwarth, G. and Doty, P. (1965) *J. Am. Chem. Soc.*, 87, 218.

Holzwarth, G. and Holzwarth, N.A.W. (1973) *J. Opt. Soc. Am.*, 63, 324.

Holzwarth, G., Hsu, E.C., Mosher, H.S., Faulkner, T.R. and Moscowitz, A. (1974) *J. Am. Chem. Soc.*, 96, 251.

Horeau, A. (1962) *Tetrahedron Lett.*, 506, 965.

Hsu, E.C. and Holzwarth, G. (1973) *J. Chem. Phys.*, 59, 4678.

Hudec, J. and Kirk, D.N. (1976) *Tetrahedron*, 32, 2475.

Hug, W., Kamatari, A., Srinivasan, K., Hansen, H.J. and Sliwka, H.R. (1980) *Chem. Phys. Lett.*, 76, 469.

Hug, W., Kint, S., Bailey, G.F. and Schener, J.R. (1975) *J. Am. Chem. Soc.*, 97, 5589.

Hug, W. and Surbeck, H. (1979) *Chem. Phys. Lett.*, 60, 186.

Hund, F. (1927) *Z. Physik*, 43, 805.

Huygens, C. (1690), *Traité de la Lumière*, Leiden, trans. S.P. Thompson, (1912), *Treatise on Light*, Macmillan, London. Dover reprint, New York (1962).

Ihde, A.J. (1964) *The Development of Modern Chemistry*, Harper, New York.

Inskeep, W.H., Miles, D.W. and Eyring, H. (1970) *J. Am. Chem. Soc.*, 92, 3866.

Ito, T., Kobayashi, A., Marumo, F. and Saito, Y., (1971) *Inorg. Nucl. Chem. Lett.*, 7, 1097.

Ito, T. and Shibata, M. (1977) *Inorg. Chem.*, 16, 108.

IUPAC Convention (1970) 'Nomenclature for the absolute configuration of six-coordinate complexes based on the octahedron', *Inorg. Chem.*, 9, 1.

Iwamoto, E., Yamamoto, M. and Yamamoto, Y. (1977) *Inorg. Nuc. Chem. Lett.*, 13, 339.

Iwasaki, H. (1974) *Acta Cryst.*, A30, 173.

Izumi, Y. and Tai, A. (1977) *Stereo-differentiating Reactions*, Kodansha, Tokyo.

Jacques, J., Collet, A. and Wilen, S.H. (1981) *Enantiomers, Racemates, and Resolutions*, Wiley, New York.

Jacques, J., Gros, C. and Boursier, S. (1977) in *Stereochemistry*. Vol. 4., ed. H. Kagan, Thieme, Stuttgart.

Jacques, J., Leclerq, M. and Brienne, M-J. (1981) *Tetrahedron*, 37, 1727.

Jaeger, F.M. (1919) *Rec. Trav. Chim.*, 38, 196.

Jaeger, F.M. (1930) *Optical Activity and High Temperature Measurement*, McGraw-Hill, New York.

Jaeger, F.M. (1937) *Bull. Soc. Chim. France*, 4, 1201.

Jain, P.C. and Lingafelter (1967) *J. Am. Chem. Soc.*, 89, 6131.

Japp, F.R. (1898) *Nature*, 58, 482.

Jasperson, S.N. and Schnatterly, S.E. (1969) *Rev. Sci. Instrum.*, 40, 761.

Jensen, H.P. and Galsbøl, F. (1977) *Inorg. Chem.*, 16, 1294.

Jirgensons, B. (1973) *Optical Activity of Proteins and other Macromolecules*, 2nd edn, Springer, New York.

Johnson, W.C. Jr. (1971) *Rev. Sci. Instrum.*, **42**, 1283.

Johnson, W.C. Jr. (1978) *Ann. Rev. Phys. Chem.*, **29**, 93.

Johnson, W.C. Jr. and Tinoco, I. Jr. (1972) *J. Am. Chem. Soc.*, **94**, 4389.

Judkins, R.R. and Royer, D.J. (1970) *Inorg. Nuc. Chem. Lett.*, **6**, 305.

Judkins, R.R. and Royer, D.J. (1974) *Inorg. Chem.*, **13**, 945.

Jungfleisch, M.E., (1882), *J. Pharm. Chim.*, [5] **5**, 346.

Kagan, H. and Fiaud, J.C. (1978) *Topics in Stereochem.*, **10**, 175.

Kagan, H., Moradpour, A., Nicoud, J.F., Balavoine, G., Martin, R.H. and Cosyn, J.P. (1971) *Tetrahedron Lett.*, 2479.

Kagan, H., Moradpour, A., Nicoud, J.F., Balavoine, G. and Tsoucaris, G. (1971), *J. Am. Chem. Soc.*, **93**, 2353.

Kagan, H.B., Balavoine, G. and Moradpour, A. (1974) *J. Mol. Evol.*, **4**, 41.

Kagan, H.B., Balavoine, G. and Moradpour, A. (1974) *J. Am. Chem. Soc.*, **96**, 5152.

Karagounis, G. and Coumoulos, G. (1938) *Nature*, **142**, 162.

Karush, F. (1956) *J. Am. Chem. Soc.*, **78**, 5519.

Kasha, M. and Pullman, B. ed. (1962) *Horizons in Biochemistry*, Albert Szent-Györgyi dedicatory volume, Academic Press, New York.

Kauzmann, W.J., Walter, J.E. and Eyring, H. (1940) *Chem. Rev.*, **26**, 339.

Keesom, W.H. (1921) *Physik Z.*, **22**, 129.

Keiderling, T.A. (1981) *Appl. Spectrosc. Rev.*, **17**, 189.

Keiderling, T.A. and Stephens, P.J. (1977) *J. Am. Chem. Soc.*, **99**, 8061.

Keiderling, T.A. and Stephens, P.J. (1979) *J. Am. Chem. Soc.*, **101**, 1396.

Kelvin, Ld, (1904) *Baltimore Lectures (1884) on Molecular Dynamics and the Wave Theory of Light*, p. 449, Clay and Sons, London.

Kemp, J.C. (1969) *J. Opt. Soc. Am.*, **59**, 950.

Kipping, F.S. and Pope, W.J. (1898) *J. Chem. Soc.*, **73**, 606.

Kirk, D.N., Klyne, W. and Mose, W.P. (1972) *Tetrahedron Letters*, 1315.

Kirkwood, J.G. (1937) *J. Chem. Phys.* **5**, 479.

Kirkwood, J.G., Wood, W.W. and Fickett, W. (1952) *J. Chem. Phys.*, **20**, 561.

Kitaigorodsky, A.I (1973) *Molecular Crystals and Molecules*, Academic Press, London and New York.

Klemm, L.H. and Read, D. (1960) *J. Chromatogr.*, **3**, 364.

Klyne, W. and Buckingham, J. (1978) *Atlas of Stereochemistry: Absolute Configurations of Organic Molecules*, 2nd edn, 2 vols, Chapman and Hall, London.

Korte, E.-H. and Schrader, B. (1981) *Adv. Infrared and Raman Spectrosc.*, **8**, 226.

Krebs, H. and Rasche, R. (1954) *Z. Anorg. Chem.*, **276**, 236.

Kress, R.B., Duesler, E.N., Etter, M.C., Paul, I.C. and Curtin, D.Y. (1980) *J. Am. Chem. Soc.*, **102**, 7709.

Krull, I.S. (1978) *Adv. Chromatogr.*, **16**, 175.

Kuhn, R. (1932) *Chem. Ber.*, **65**, 49.

Kuhn, W. (1925) *Z. Physik*, **33**, 408.

Kuhn, W. (1930) *Trans. Faraday Soc.*, **26**, 293.

Kuhn, W. (1935) *Z. Phys. Chem.*, **B31**, 18.

Kuhn, W. (1936) *Angew. Chem.*, **49**, 215.

Kuhn, W. (1955) *Experientia*, **11**, 429.

Kuhn, W. (1958) *Ann. Rev. Phys. Chem.*, **9**, 417.

Kuhn, W. and Bein, K. (1934) *Z. Physik Chem.*, **B24**, 335.

Kuhn, W. and Braun, F. (1929) *Naturwissenschaften*, **17**, 227.

Kuhn, W. and Knopf, E. (1930) *Z. Phys. Chem.*, **B7**, 292.

Kuramoto, M., Kushi, Y. and Yoneda, H. (1980) *Bull. Chem. Soc. Japan*, **53**, 125.

Kuroda, R. and Mason, S.F. (1979) *J. Chem. Soc. Dalton,* 273.
Kuroda, R. and Mason, S.F. (1981*a*) *J. Chem. Soc. Perkin II,* 167.
Kuroda, R. and Mason, S.F. (1981*b*) *J. Chem. Soc. Perkin II,* 870.
Kuroda, R. and Mason, S.F. (1981*c*) *Tetrahedron,* 37, 1995.
Kuroda, R. and Mason, S.F. (1981*d*) *J. Chem. Soc. Dalton,* 1268.
Kuroda, R., Mason, S.F., Prosperi, T., Savage, S. and Tranter, G.E. (1981) *J. Chem. Soc. Dalton,* 2565.
Kuroda, R., Mason, S.F., Rodger, C.D. and Seal, R.H. (1978) *Chem. Phys. Lett.,* 57, 1.
Kuroda, R., Mason, S.F., Rodger, C.D. and Seal, R.H. (1981) *Molec. Phys.* 42, 33.
Kuroda, R. and Saito, Y. (1976) *Bull. Chem. Soc. Japan,* 49, 433.

Laarhoven, W.H. and Cuppen, J.H.M. (1978) *J. Chem. Soc. Perkin II,* 315.
Landolt, H. (1899) *Optical Activity and Chemical Composition,* trans. J. McCrae, Whittaker, London.
Landsteiner, K. (1945) *The Specificity of Serological Reactions,* 2nd edn, Harvard University Press, Cambridge, Massachusetts.
Laporte, O. (1924) *Z. Physik,* 51, 512.
Larsen, E., Mason, S.F. and Searle, G.H. (1966) *Acta Chem. Scand.,* 20, 191.
Laurent, A. (1855) *Chemical Method,* trans. W. Odling, Cavendish Society, London.
Le Bel, J.A. (1874) *Bull. Soc. Chim. France,* 22, 337.
Le Bel, J.A. (1892) *Bull. Soc. Chim. France,* [3] 7, 613.
Le Bel, J.A. (1894) *Bull. Soc. Chim. France,* [3] 11, 292.
Leclerq, M., Collet, A. and Jacques, J. (1976) *Tetrahedron,* 32, 821.
Leclerq, M. and Jacques, J. (1979) *Nouv. J. Chem.,* 3, 629.
Lee, S.-F., Barth, G. and Djerassi, C. (1981) *J. Am. Chem. Soc.,* 103, 295.
Lee, S.-F., Edgar, M., Pak, C.S., Barth, G. and Djerassi, C. (1980) *J. Am. Chem. Soc.,* 102, 4784.
Lee, T.D. and Yang, C.N. (1956) *Phys. Rev.,* 88, 101.
Le Fèvre, R.J.W. (1965) *Adv. Phys. Org. Chem.,* 3, 1.
Letokhov, V.S. (1975) *Phys. Lett.,* 53A, 275.
Letokhov, V.S. (1977) *Lett. Nuovo Cimento,* 20, 107.
Lewis, L.L., Apperson, G.R., Emmons, T.P., Hollister, J.H., Vold, T.G. and Fortson, E.N. (1979) in *Origins of Optical Activity in Nature,* ed. D.C. Walker, p. 35, Elsevier, Amsterdam
Lifschitz, I. (1925) *Z. Physik. Chem.,* 114, 485.
Lightner, D.A. and Chang, T.C. (1974) *J. Am. Chem., Soc.,* 96, 3015.
Lightner, D.A., Bouman, T.D., Gawroński, J.K., Gawroński, K., Chappuis, J.L., Crist, B.V. and Hansen, A.E. (1981) *J. Am. Chem. Soc.,* 103, 5314.
Lightner, D.A. and Jackman, D.E. (1974) *J. Chem. Soc. Chem. Comm.,* 344.
Linder, K.R. and Mannschreck, A. (1980) *J. Chromatogr.,* 193, 308.
Lindman, K.F. (1925) *Ann. Physik,* [4] 77, 337.
Lipscomb, W.N. (1972) *Chem. Soc. Rev.,* 3, 319.
London, F. (1930) *Z. Physik. Chem.,* B11, 222.
London, F. (1937) *Trans. Faraday Soc.,* 33, 8.
London, F. (1942) *J. Phys. Chem.,* 46, 305.
Lorentz, H.A. (1880) *Ann. Physik,* 9, 641.
Lowry, T.M. (1899) *J. Chem. Soc.,* 75, 211.
Lowry, T.M. (1935) *Optical Rotatory Power,* Longmans, Green and Co., London. Dover Reprint, New York, (1964).

McCaffery, A.J. and Mason, S.F. (1963) *Molec. Phys.,* 6, 359.
McCaffery, A.J. and Mason, S.F. (1964) *Nature,* 204, 468.
McKenzie, A. (1904) *J. Chem. Soc.,* 85, 1249.

Malus, E.T. (1809) *Mém. Soc. Arceuil*, **2**, 143 and 254.
Mandel, R. and Holzwarth, G. (1972) *J. Chem. Phys.*, **57**, 3469.
Mann, F.G. and Watson, J. (1947) *J. Chem. Soc.*, 505.
Marckwald, W. (1896) *Chem. Ber.*, **29**, 42.
Marckwald, W. (1904) *Chem. Ber.*, **37**, 349, 1368.
Marckwald, W. and McKenzie, A. (1899) *Chem. Ber.*, **32**, 2130.
Marcott, C., Blackburn, C.C., Faulkner, T.R., Moscowitz, A. and Overend, J. (1978) *J. Am. Chem. Soc.*, **100**, 5262.
Martin, R.H. (1974) *Angew. Chem. Int. Edn.*, **13**, 649.
Mason, S.F. (1953) *A History of the Sciences: Main Currents of Scientific Thought*, Routledge and Kegan Paul, London. Revised edns., Abelard-Schuman, New York (1956); Collier Macmillan, New York (1962).
Mason, S.F. (1962) *Molec. Phys.* **5**, 343.
Mason, S.F. (1963) *Quart. Rev. Chem. Soc.*, **17**, 20.
Mason, S.F. (1967) *Newer Physical Methods in Structural Chemistry*, ed. R. Bonnett and J.G. Davis, p. 149, United Trade Press, London.
Mason, S.F. (1968) *Inorg. Chim. Acta Rev.*, **2**, 89.
Mason, S.F. (1973*a*) *J. Chem. Soc. Chem. Comm.*, 239.
Mason, S.F. (1973*b*) in *Fundamental Aspects and Recent Developments in Optical Rotatory Dispersion and Circular Dichroism*, ed. F. Ciardelli and P. Salvadori, Ch. 3.6, p. 196, Heyden, London.
Mason, S.F. (1976) *Topics in Stereochem.*, **9**, 1.
Mason, S.F. ed. (1979) *Optical Activity and Chiral Discrimination*, NATO ASI Series C, Reidel, Dordrecht.
Mason, S.F. (1980) in *(IUPAC) Coordination Chemistry – 20*, ed. D. Banerjea, p.235, Pergamon, Oxford.
Mason, S.F. (1980) *Struct. Bonding*, **39**, 43.
Mason, S.F. (1981) *Adv. Infrared and Raman Spectrosc.*, **8**, 283.
Mason, S.F. Peacock, R.D. and Prosperi, T. (1977) *J. Chem. Soc. Dalton*, 702.
Mason, S.F. and Peart, B.J. (1973) *J. Chem. Soc. Dalton*, 949.
Mason, S.F. and Seal, R.H. (1976) *Molec. Phys.*, **31**, 755.
Mason, S.F., Seal, R.H. and Roberts, D.R. (1974) *Tetrahedron*, **30**, 1671.
Mathieu, J.-P. (1936) *J. Chim. physique*, **33**, 78;
Mathieu, J.-P. (1936) *Bull. Soc. Chim. France*, **3**, 476.
Mathieu, J.-P. (1939) *Bull. Soc. France Min. Cryst.*, **62**, 174.
Mathieu, J.-P. (1944) *Ann. Phys.*, **19**, 335.
Mathieu, J.-P. (1946) *Les Théories Moleculaires du Pouvoir Rotatoire Naturel*, Gauthier-Villars, Paris.
Mathieu, J.-P. (1957) 'Activité optique naturelle', in *Encyclopedia of Physics*. Vol. XXVIII, Spectroscopy II, ed. S. Flugge, p. 333, Springer, Berlin.
Matsumoto, M. and Amaya, K. (1980) *Bull. Chem. Soc. Japan*, **53**, 3510.
Mavroyannis, C. and Stephen, M.J. (1962) *Molec. Phys.*, **5**, 629.
Maxwell, J.C. (1865) *Phil. Trans. Roy. Soc. London*, **155**, 459.
Maxwell, J.C. (1873) *Treatise on Electricity and Magnetism*, Vol. 2, p. 383, Clarendon Press, Oxford.
Mead, C.A. and Moscowitz, A. (1980) *J. Am. Chem. Soc.*, **102**, 7301.
Meier, G., Sackmann, E. and Grabmaier, J.G. (1975) *Applications of Liquid Crystals*, Springer, Berlin.
Meyerhoffer, W. (1904) *Chem. Ber.*, **37**, 2604.
Meyers, A.I. (1978) *Acc. Chem. Res.*, **11**, 375.
Mikes, F. and Boshart, G. (1978) *J. Chromatogr.*, **149**, 455.
Mikes, F., Boshart, G. and Gil-Av, E. (1976) *J. Chromatogr.*, **122**, 205.

Miller, S.L. and Urey, H.C. (1959) *Science,* 130, 245.
Mitchell, S. (1933) *The Cotton Effect,* Bell and Sons, London.
Mitscherlich, E. (1844) *Compt. Rend. Paris,* 19, 719.
Miyoshi, K., Matsumoto, Y. and Yoneda, H. (1981) *Inorg. Chem.,* 20, 1057.
Mjojo, C.C. (1979) *J. Chem. Soc. Faraday II,* 75, 692.
Moeller, T. and Gulyas, E. (1958) *J. Inorg. Nucl. Chem.,* 5, 245.
Moffitt, W. (1956a) *Proc. Nat. Acad. Sci. U.S.A.,* 42, 596.
Moffitt, W. (1956b) *J. Chem. Phys.,* 25, 467.
Moffitt, W. (1956c) *J. Chem. Phys.,* 25, 1189.
Moffitt, W. and Moscowitz, A. (1959) *J. Chem. Phys.,* 30, 648.
Moffitt, W., Woodward, R.B., Moscowitz, A., Klyne, W. and Djerassi, C. (1961) *J. Am. Chem. Soc.,* 83, 4013.
Mondine, F.A. (1979) *Rev. Sci. Instrum.,* 50, 386.
Moradpour, A., Nicoud, J.F., Balavoine, G., Kagan, H. and Tsoucaris, G. (1971) *J. Am. Chem. Soc.,* 93, 2353.
Moriarty, R.M., Paaren, H.E., Weiss, U. and Whalley, W.B. (1979) *J. Am. Chem. Soc.,* 101, 6804.
Morrison, J.D. and Mosher, H.S. (1976) *Asymmetric Organic Reactions,* Am. Chem. Soc., Washington, D.C.
Moscowitz, A. (1961) *Tetrahedron,* 13, 48.
Moscowitz, A. (1962) *Adv. Chem. Phys.* 4, 67.
Moscowitz, A. (1965) in *Modern Quantum Chemistry,* ed. O. Sinanoglu, Part III, Ch. 2, p. 31, Academic Press, New York.

Nafie, L.A. and Diem, M. (1979) *Acc. Chem. Res.,* 12, 296.
Nafie, L.A., Diem, M. and Vidrine, D.W. (1979) *J. Am. Chem. Soc.,* 101, 496.
Nafie, L.A., Keiderling, T.A. and Stephens, P.J. (1976) *J. Am. Chem. Soc.,* 98, 2715.
Nakazaki, M., Yamamoto, K. and Fujiwara, K. (1978) *Chem. Lett.,* 863.
Nakazaki, M., Yamamoto, K., Fujiwara, K. and Maeda, M. (1979) *J. Chem. Soc. Chem. Comm.,* 1086.
Neubert, L.A. and Carmack, M. (1974) *J. Am. Chem. Soc.,* 96, 943.
Newman, M.S., Darlak, R.S. and Tsai, L. (1967) *J. Am. Chem. Soc.,* 89, 6191.
Newman, M.S. and Lednicer, D. (1956) *J. Am. Chem. Soc.,* 78, 4765.
Newman, P. (1978) *Optical Resolution Procedures for Chemical Compounds,* Optical Resolution Information Center, Manhattan College, New York.
Newman, P., Rutkin, P. and Mislow, K. (1958) *J. Am. Chem. Soc.,* 80, 6036.
Niketić, S.R. and Rasmussen, Kj. (1977) *The Consistent Force Field,* Lecture Notes in Chemistry 3, Springer, Heidelberg.
Nishikawa, K., Tohji, K. and Murata, Y. (1980) *Tec. Rep. Inst. Solid State Phys. Tokyo,* Ser. A, No. 1085.
Nishikawa, K., Tohji, K. and Murata, Y. (1981) *J. Chem. Phys.,* 74, 5817.
Nishikawa, S. and Matsukawa, K. (1928) *Proc. Imp. Acad. Japan,* 4, 96.
Nordén, B. and Jonás, I. (1976) *Inorg. Nuc. Chem. Lett.,* 12, 34, 43.
Numan, H. and Wynberg, H. (1978) *J. Org. Chem.,* 43, 2232.
Nyholm, R.S. ed. (1967) 'Discussion on Circular Dichroism; Electronic and Structural Principles', *Proc. Roy. Soc. London,* A297, 1–172.

O'Conner, B.H. and Dale, D.H. (1966) *Acta Cryst.,* 21, 705.
Ogston, A.G. (1948) *Nature,* 162, 963.
Ogston, A.G. (1958) *Nature,* 181, 1462.
Oparin, A.I. ed. (1959) *Origin of Life on the Earth,* Pergamon, London.
Oparin, A.I. (1961) *Life, its Nature, Origin and Development,* Oliver and Boyd, London.

Oró, J. and Kimball, A.P. (1961) *Arch. Biochem. Biophys.*, **94**, 217.
Oró, J. and Kimball, A.P. (1962) *Arch. Biochem. Biophys.*, **96**, 293.
Osborn, J.A., Jardine, F.H., Young, J.F. and Wilkinson, G. (1966) *J. Chem. Soc. A*, 1711.

Parascandola, J. (1975) in *van't Hoff – Le Bel Centennial*, ed. O.B. Ramsey, ACS Symposium 12, p. 143, Am. Chem. Soc., Washington, D.C.
Partington, J.R. (1953) *An Advanced Treatise on Physical Chemistry. Vol. 4. Physicochemical Optics*, p. 334 Longmans Green, London.
Partington, J.R. (1964) *A History of Chemistry, Vol. 4. The Nineteenth Century*, Macmillan, London.
Pasteur, L. (1848) *Compt. Rend. Paris*, **26**, 535.
Pasteur, L. (1850) *Ann. Chim. Phys.*, [3], **28**, 56.
Pasteur, L. (1853) *Ann. Chim. Phys.*, [3], **38**, 437.
Pasteur, L. (1858) *Compt. Rend. Paris*, **46**, 615.
Pasteur, L. (1860) *Researches on molecular asymmetry*. Alembic Club Reprint, No. 14, Edinburgh (1948).
Pasteur, L. (1884) *Bull. Soc. Chim. France*, **41**, 219.
Pasteur Vallery-Radot ed. (1922) *Oeuvres de Pasteur. Tome Premier. Dissymétrie Moléculaire*, Masson, Paris.
Pauling, L. (1948) *Nature*, **161**, 707.
Peart, B.J. (1970) Ph.D. Thesis, University of East Anglia.
Perkin, W.H. and Duppa, B.F. (1860) *Chem. News*, **244**.
Perkin, W.H. and Duppa, B.F. (1861) *J. Chem. Soc.*, **13**, 102.
Peterson, D.L. and Simpson, W.T. (1955) *J. Am. Chem. Soc.*, **77**, 3929.
Peterson, D.L. and Simpson, W.T. (1957) *J. Am. Chem. Soc.*, **79**, 2375.
Pfeiffer, P. and Quehl, K. (1931) *Chem. Ber.*, **64**, 2667.
Pickard, R.H. and Kenyon, J. (1911) *J. Chem. Soc.*, **99**, 45.
Piper, T.S. (1961) *J. Am. Chem. Soc.*, **83**, 3908.
Pirkle, W.H., House, D.W. and Finn, J.M. (1980) *J. Chromatogr.*, **192**, 143.
Polavarapu, P.L. and Nafie, L.A. (1980) *J. Chem. Phys.*, **73**, 1567.
Pope, W.J. and Peachey, S.J. (1899) *Trans. Chem. Soc.*, **75**, 1066.
Porter, R. and Fitzsimons, D.W. ed. (1978) *Molecular Interactions and Activity in Enzymes*, Ciba Foundation Symposium 60 (new series), Excerpta Medica, Amsterdam.
Portoghese, P.S. (1970) *Ann. Rev. Pharmacol.*, **10**, 51.
Portoghese, P.S. (1978) *Acc. Chem. Res.*, **11**, 21.
Powles, J.G. (1973) *Adv. Phys.*, **22**, 1.
Prasad, P.L., Nafie, L.A. and Burow, D.F. (1979) *J. Raman Spectrosc.*, **8**, 255.
Prelog, V. (1953) *Helv. Chim. Acta*, **36**, 308.
Prelog, V. (1956) *Bull. Soc. Chim. France*, 987.
Prelog, V., Ceder, O. and Wilhelm, M. (1955) *Helv. Chim. Acta*, **38**, 303.
Prelog, V., Philbin, E., Watanabe, E. and Wilhelm, M. (1956) *Helv. Chim. Acta*, **39**, 1086.
Prelog, V. and Wieland, P. (1944), *Helv. Chim. Acta*, **44**, 1127.
Prigogine, I. and Defay, R. (1967) *Chemical Thermodynamics*, 4th edn., Longmans, London.

Ramsey, O.B. ed. (1975) *van't Hoff – Le Bel Centennial*, ACS Symposium 12, Am. Chem. Soc., Washington, D.C.
Rein, D.W., Hegstrom, R.A. and Sandars, P.G.H. (1979) *Phys. Lett.*, **71A**, 499.
Rein, D.W., Hegstrom, R.A. and Sandars, P.G.H. (1979) in *Origins of Optical Activity in Nature*, ed. D.C. Walker, p. 21, Elsevier, Amsterdam.
Rein, D.W., Hegstrom, R.A. and Sandars, P.G.H. (1980) *J. Chem. Phys.*, **73**, 2329.
Richardson, F.S. (1979) in *Optical Activity and Chiral Discrimination*, ed. S.F. Mason, Ch. 6, p. 107, and Ch. 8, p. 189, Reidel, Dordrecht.

Richardson, F.S. and Riehl, J.P. (1977) *Chem. Rev.,* 77, 773.

Ritchie, P.D. (1933) *Asymmetric Synthesis and Asymmetric Induction,* Oxford University Press, London.

Ritchie, P.D. (1947) *Adv. Enzymol.,* 7, 65.

Robin, M.B. (1975) *Higher Excited States of Polyatomic Molecules,* 2 vols, Academic Press, New York.

Rodberg, L.S. and Weisskopf, V.F. (1957) *Science,* 125, 627.

Roozeboom, H.W.B. (1899) *Z. Physikal. Chem.,* 28, 494.

Rosanoff, M.A. (1906) *J. Am. Chem. Soc.,* 28, 114.

Rosenfeld, J.S. and Moscowitz, A. (1972) *J. Am. Chem. Soc.,* 94, 4797.

Rosenfeld, L. (1928) *Z. Physik,* 52, 161.

Ruch, E. (1972) *Acc. Chem. Res.,* 5, 49.

Ruch, E. and Ugi, I. (1969) *Topics in Stereochem.,* 4, 99.

Russell, C.A. (1971) *The History of Valency,* Leicester University Press.

Russell, M.F., Billardon, M. and Badoz, J. (1972) *Appl. Opt.,* 11, 2375.

Saito, Y. ed. (1977) *Absolute Configuration of Metal Complexes determined by X-ray Analysis',* Research Group, 'Sogo-Kenkyu B', Osaka.

Saito, Y. (1979) *Inorganic Molecular Dissymmetry,* Springer, Berlin.

Saito, Y., Nakatsu, K., Shiro, M. and Kuroya, H. (1954) *Acta Cryst.,* 7, 636.

Saito, Y., Nakatsu, K., Shiro, M. and Kuroya, H. (1955) *Acta Cryst.,* 8, 729.

Salam, A. (1968) in *Proc. 8th Nobel Symposium; Elementary particle theory,* ed. N. Svartholm, Almquist and Wiksell, Stockholm.

Sargeson, A.M. (1980) in *Stereochemistry of Optically Active Transition Metal Compounds,* ed. B.E. Douglas and Y. Saito, p. 115, Am. Chem. Soc., Washington, D.C.

Sarneski, J.E. and Urbach, F.L. (1971) *J. Am. Chem. Soc.,* 93, 884.

Scacchi, A. (1865) *Atti Accad. Sci. Napoli,* 4, 250.

Schellman, J.A. (1966) *J. Chem. Phys.,* 44, 55.

Schellman, J.A. (1968) *Acc. Chem. Res.,* 1, 144.

Schellman, J.A. (1975) *Chem. Rev.,* 75, 323.

Schellman, J.A. and Oriel, P. (1962) *J. Chem. Phys.,* 37, 2114.

Schnepp, O. (1979) in *Optical Activity and Chiral Discrimination,* ed. S.F. Mason, Ch. 5, p. 87, Reidel, Dordrecht

Schnepp, O., Allen, S. and Pearson, E.F. (1970) *Rev. Sci. Instrum.,* 41, 1136.

Schrader, B. and Korte, E.H. (1972) *Angew. Chem. Int. Edn.,* 11, 226.

Schramm, G., Grotsch, H. and Pollmann, W. (1962) *Angew. Chem. Int. Edn.,* 1, 1.

Schröder, I. (1893) *Z. Physikal. Chem.,* 11, 449.

Scott, A.I. and Wrixon, A.D. (1970) *Tetrahedron,* 26, 3695.

Shemin, D. (1946) *J. Biol. Chem.,* 162, 297.

Shinn, T. (1979) 'The French Science Faculty System, 1808–1914', *Hist. Stud. Phys. Sci.,* 10, 271.

Sing, L.Y., Lindley, M., Sundararaman, P., Barth, G. and Djerassi, C. (1981) *Tetrahedron,* 37, Supp. No. 1, 181.

Sing, Y.L., Numan, H., Wynberg, H. and Djerassi, C. (1979) *J. Am. Chem. Soc.,* 101, 5155.

Singh, B.K. and Perti, O.N. (1963) *Optical Activity and Chemical Constitution,* Asia Publishing House, Bombay.

Singh, R.D. and Keiderling, T.A. (1981) *J. Chem. Phys.,* 74, 5347.

Smirnoff, A.P. (1920) *Helv. Chim. Acta,* 3, 177.

Smith, M.S. and Wain, R.L. (1951) *Proc. Roy. Soc. London,* 139B, 118.

Snatzke, G. ed. (1967) *Optical Rotatory Dispersion and Circular Dichroism in Organic Chemistry,* Heyden, London.

Snatzke, G. (1979) in *Optical Activity and Chiral Discrimination*, ed. S.F. Mason, Reidel, Dordrecht.
Snatzke, G. and Eckhardt, G. (1968) *Tetrahedron,* 24, 4543.
Snatzke, G., Ehrig, B. and Klein, H. (1969) *Tetrahedron,* 25, 5601
Snyder, P.A. and Johnson, W.C. Jr. (1973) *J. Chem. Phys.,* 59, 2618.
Snyder, S.H. (1977) *Scientific American,* 236, No. 3, 44.
Spinat, P., Brouty, C. and Whuler, A. (1980) *Acta Cryst.,* B36, 544.
Steinberg, I.Z. (1978) *Methods in Enzymology,* 49, 179.
Steinberg, I.Z. and Gafni, A. (1972) *Rev. Sci. Instrum.,* 43, 409.
Steinstrasser, R. and Pohl, L. (1973) *Angew. Chem. Int. Edn.,* 12, 617.
Stephens, P.J. and Clark, R. (1979) in *Optical Activity and Chiral Discrimination,* ed. S.F. Mason, Ch. 10, p. 263, Reidel, Dordrecht.
Stevens, E.S. (1978) *Methods in Enzymology,* 49, 214.
Stevenson, K.L. (1972) *J. Am. Chem. Soc.,* 94, 6652.
Stevenson, K.L. and Verdieck, J.F. (1969) *Mol. Photochem.,* 1, 271.
Stewart, A.W. (1907) *Stereochemistry,* p. 544 Longmans, Green and Co., London.
Stodolsky, L. (1981) *Nature,* 290, 735.
Su, C.N., Heintz, V.J. and Keiderling, T.A. (1980) *Chem. Phys. Lett.,* 73, 157.
Su, C.N. and Keiderling, T.A. (1980) *J. Am. Chem. Soc.,* 102, 511.
Su, C.N. and Keiderling, T.A. (1981) *Chem. Phys. Lett.,* 77, 494.
Sudmeier, J.L., Blackmer, G.L., Bradley, C.H. and Anet, F.A.L. (1972) *J. Am. Chem. Soc.,* 94, 757.
Sugano, S. (1960) *J. Chem. Phys.,* 33, 1883.

Taft, R.W. (1951) *J. Am. Chem. Soc.,* 74, 3120.
Tanaka, J. (1972) *Acta Cryst.,* A28, 229.
Tanaka, J., Katayama, C., Ogura, F., Tatemitsu, H. and Nakagawa, M. (1973) *J. Chem. Soc. Chem. Comm.,* 21.
Tanaka, J., Ogura, F., Kuritani, M. and Nakagawa, M. (1972) *Chimia,* 26, 271.
Theilacker, W. and Winkler, H.G. (1954) *Chem. Ber.,* 87, 690.
Thiemann, W. ed. (1981) 'Generation and amplification of chirality in chemical systems', *Origins of Life,* Vol. 11, Nos. 1/2, Reidel, Dordrecht.
Thomas, W. (1925) *Naturwiss.,* 13, 627.
Thulstrup, E.W. and Michl, J. (1980) *J. Phys. Chem.,* 84, 82.
Thulstrup, E.W. Michl, J. and Eggers, J.H. (1970) *J. Phys. Chem.,* 74, 3868, 3878.
Tiffeneau, M., Lévy, J. and Ditz, E. (1935) *Bull. Soc. Chim. France,* 5, 1848.
Timmermans, J. (1929) *Rec. Trav. Chim.,* 48, 890.
Tinoco, I. Jr. (1962) *Adv. Chem. Phys.,* 4, 113.
Tinoco, I. Jr. (1973) in *Fundamental Aspects and Recent Developments in Optical Rotatory Dispersion and Circular Dichroism,* ed. F. Ciardelli and P. Salvadori, Ch. 2.4, p. 66, Heyden, London.
Tinoco, I. Jr. (1979) in *Optical Activity and Chiral Discrimination,* ed. S.F. Mason, Ch. 4, p. 57, Reidel, Dordrecht.
Tinoco, I. Jr., Ehrenberg, B. and Steinberg, I.Z. (1977) *J. Chem. Phys.,* 66, 916.
Tinoco, I. Jr. and Freeman, M.P. (1957) *J. Phys. Chem.,* 61, 1196.
Toftlund, H. and Pedersen, E. (1972) *Acta Chem. Scand.,* 26, 4019.
Tsuchida, R., Kobayashi, M. and Nakamura, A. (1936) *Bull. Chem. Soc. Japan,* 11, 38.
Turner, D.H. (1978) *Methods in Enzymology,* 49, 199.
Turner, E.E. and Harris, M.M. (1947) *Quart. Rev. Chem. Soc.,* 1, 299.
Turner, F.E. and Lonsdale, K. (1950) *J. Chem. Phys.,* 18, 156.

Ulbricht, T.L.V. (1959) *Quart. Rev. Chem. Soc.,* 13, 48.

Ulbricht, T.L.V. (1981) *Origins of Life*, 11, 55.
Ulbricht, T.L.V. and Vester, F. (1962) *Tetrahedron*, 18, 629.

Vaira, M. Di. and Orioli, P.L. (1967) *Inorg. Chem.*, 6, 955.
Vaira, M. Di. and Orioli, P.L. (1968) *Acta Cryst.*, B24, 595 and 1269.
Valentine, D. and Scott, J.W. (1978) *Synthesis*, 329.
Vallery-Radot, R. (1885, 1901) *The Life of Pasteur*, 1885 edn, trans. Lady Claud Hamilton, Longmans, Green and Co., London; 1901 edn, trans, Mrs R.L. Devonshire, Dover, New York, 1960.
van der Waals, J.D. (1873) Doctoral dissertation, Leiden.
Vanest, J.-M. and Martin, R.H. (1979) *Rec. Trav. Chim.*, 98, 113.
van Laar, J.J. (1903) *Arch. Neerl.* II, 8, 264.
van Laar, J.J. (1908) *Z. Physikal. Chem.*, 63, 216.
van Laar, J.J. (1908) *Z. Physikal. Chem.*, 64, 257.
van Laar, J.J. (1909) *Z. Physikal. Chem.*, 66, 197.
van't Hoff, J.H. (1874) *Arch. Neéland Sci. Exact. Nat.*, 9, 445. English translation given in Benfey (1963).
van't Hoff, J.H. (1891) *Chemistry in Space*. trans. J.E. Marsh, Clarendon Press, Oxford.
van't Hoff, J.H. (1896) *Studies in Chemical Dynamics*, trans. T. Ewan, Williams and Northgate, London.
van't Hoff, J.H. (1899) *Lectures on Theoretical and Physical Chemistry*, trans. R.A. Lehfeldt, Arnold, London.
van't Hoff, J.H. and van Deventer, C.M. (1887) *Z. Physikal. Chem.*, 1, 165.
Van Vleck, J.H. (1937) *J. Phys. Chem.*, 41, 67.
Velluz, L., Legrand, M. and Grosjean, M. (1965) *Optical Circular Dichroism Principles, Measurements, and Applications*, Verlag Chemie, and Academic Press, New York.
Vester, F., Ulbricht, T.L.V. and Krauch, H. (1959) *Naturwiss.*, 46, 48.
Vigneron, J.P., Dhaenens, M. and Horeau, A. (1977) *Tetrahedron*, 33, 497.

Wagnière, G. (1966) *J. Am. Chem. Soc.*, 88, 3937.
Wagnière, G. (1971) in *Aromaticity, Pseudo-aromaticity and Anti-aromaticity*, Jerusalem Symposium III, Israel Acad. Sci., Jerusalem.
Wain, R.L. (1977) *Chem. Soc. Rev.*, 6, 261.
Walden, P. (1895) *Z. Physik. Chem.*, 17, 245.
Walden, P. (1896a) *Chem. Ber.*, 29, 133.
Walden, P. (1896b) *Chem. Ber.*, 29, 1692.
Walden, P. (1897) *Chem. Ber.*, 30, 3146.
Walden, P. (1908) *Z. Elektrochem.*, 14, 713.
Walden, P. (1941) *Geschichte der Organischen Chemie seit 1880*, Springer, Berlin. Continuation of Graebe, C. (1920), *Geschichte der Organischen Chemie*, Springer, Berlin. Reprinted 1972.
Walker, D.C. ed. (1979) *Origins of Optical Activity in Nature*, Elsevier, Amsterdam.
Wallach, O. (1895) *Annalen*, 286, 140.
Waser, J. (1949) *J. Chem. Phys.*, 17, 498.
Watanabe, T. and Eyring, H. (1964) *J. Chem. Phys.*, 40, 3411.
Weber, G. (1953) *Adv. Protein Chem.*, 8, 415.
Weber, G. (1966) in *Fluorescence and Phosphorescence Analysis*, ed. D.M. Hercules, Ch. 8, p. 217, Interscience, New York.
Weigang, O.E. Jr. (1979) *J. Am. Chem. Soc.*, 101, 1965.
Weigang, O.E. Jr. and Höhn, E.G. (1966) *J. Am. Chem. Soc.*, 88, 3673.

Weigang, O.E. Jr. and Höhn, E.G. (1968) *J. Chem. Phys.*, **48**, 1127.
Weinberg, S. (1967) *Phys. Rev. Lett.*, **19**, 1264.
Weiner, C. ed. (1977) *History of Twentieth Century Physics*, Proc. Int. School Physics, 'Enrico Fermi', Course 57, Academic Press, New York.
Weinstein, S., Feibush, B. and Gil-Av, E. (1976) *J. Chromatogr.*, **126**, 97.
Weinstein, S. and Leiserowitz, L. (1980) *Acta Cryst.*, **B36**, 1406.
Weinstein, S., Leiserowitz, L. and Gil-Av, E. (1980) *J. Am. Chem. Soc.*, **102**, 2768.
Werner, A. (1911) *Chem. Ber.*, **44**, 1887; selected papers of Werner, A., *Classics in Coordination Chemistry*, Kauffman, G.B., trans. and ed., Dover, New York, (1968).
Werner, A. (1912*a*) *Annalen*, **386**, 1.
Werner, A. (1912*b*) *Bull. Soc. Chim. France*, **11**, 1.
Werner, A. (1912*c*) *Chem. Ber.*, **45**, 3061.
Werner, A. (1914) *Chem. Ber.*, **47**, 2171.
Wick, G.C., Wightman, A.S. and Wigner, E.P. (1952) *Phys. Rev.*, **88**, 101.
Wigner, E.P. (1952) *Z. Physik.*, **133**, 101.
Wilczek, F. (1980) *Scientific American*, **243**, No. 6, 60.
Wilen, S.H. (1971) *Topics in Stereochem.*, **6**, 107.
Wilen, S.H. (1972) *Tables of Resolving Agents and Optical Resolutions*, University of Notre Dame Press, Notre Dame, Indiana.
Wilen, S.H., Collet, A. and Jacques, J. (1977) *Tetrahedron*, **33**, 2725.
Winther, C. (1895) *Chem. Ber.*, **28**, 3000.
Wislicenus, J. (1869) *Chem. Ber.*, **2**, 550, 619.
Wood, J.S. (1972) *Prog. Inorg. Chem.*, **16**, 227.
Woodward, L.A. (1967) in *Raman Spectroscopy*, ed. H.A. Szymanski, p. 4, Plenum, New York.
Woody, R.W. (1973) *Tetrahedron*, **29**, 1273.
Woody, R.W. (1977) *J. Polymer Sci. Macromol. Rev.*, **12**, 181.
Worthing, C.R. ed. (1979) *The Pesticide Manual*, 6th edn, British Crop Protection Society, Croydon.
Wu, C.S., Ambler, E., Hayward, R.W., Hoppes, D.D. and Hudson, R.P. (1957) *Phys. Rev.*, **105**, 1413.
Wurtz, A. (1869) *History of Chemical Theory from the Age of Lavoisier to the Present Time*, trans. and ed. H. Watts, p. 1, Macmillan, London.

Yamagata, Y. (1966) *J. Theoret. Biol.*, **11**, 495.
Yamamoto, M. and Yamamoto, Y. (1975) *Inorg. Nuc. Chem. Lett.*, **11**, 833.
Yamanari, K., Hidaka, J. and Shimura, Y. (1973) *Bull. Chem. Soc. Japan*, **46**, 3724.
Yoneda, H. (1979) *J. Liquid Chromatogr.*, **2**, 1157.
Yoshikawa, Y. and Yamasaki, K. (1979) *Coord. Chem. Rev.*, **28**, 205.
Young, T. (1802) *Phil. Trans.*, **92**, 12.
Yuki, H., Okamoto, Y. and Okamoto, I. (1980) *J. Am. Chem. Soc.*, **102**, 6356.

Zaugg, H.E. (1955) *J. Am. Chem. Soc.*, **77**, 2910.
Zel'dovich, B. Ya., Saakyan, D.B. and Sobel'man, I.I. (1977) *JETP Lett.*, **25**, 94.
Zorkii, P.M., Razumaeva, A.E. and Belsky, V.K. (1977) *Acta Cryst.*, **A33**, 1001.

AUTHOR INDEX

Addadi, L., 202, 250
Allen, F. H., 178, 250
Amaya, K., 227, 250, 259
Amiard, G., 187, 250
ApSimon, J. W., 216, 250
Arago, D. F. J., 1ff, 250
Ariëns, E. J., 226, 250
Arnett, E. M., 228, 250
Atkins, P. W., 152, 250

Badoz, J., 27, 251, 262
Baessler, H., 101, 250
Bailar, J. C., Jr., 108, 163, 217, 250, 252
Ballard, R. E., 106, 250
Ballhausen, C. J., 107, 250, 253
Barkov, L. M., 243, 250
Barnett, C. J., 49, 81, 141, 250
Barron, L. D., 143ff, 245, 250, 251
Barth, G., 56, 251
Beattie, J. K., 163, 254
Beckett, A. H., 223ff, 251
Bedarkar, S., 224, 251
Beecham, A. F., 183, 251
Beevers, C. A., 97, 251
Belsky, V. K., 165ff, 251, 265
Bentley, R., 208, 248, 251
Bernal, I., 106, 251
Bernstein, W. J., 198ff, 251
Berthelot, M., 227, 251
Bertsch, K., 192, 251
Bethe, H. A., 116, 251
Bijvoet, J. M., 10, 68, 177ff, 251
Billardon, M., 27, 251, 262
Biot, J. B., 1ff, 251
Blair, N. E., 239, 251
Blaschke, G., 192, 225, 251
Blout, E. R., 239, 251
Blum, L., 233, 251
Blundell, T. L., 224, 251
Boltzmann, L., 18, 251
Bonner, W. A., 242ff, 251
Born, M., 20ff, 251
Bosnich, B., 86, 145, 216, 219, 251, 252
Bouman, T. D., 54, 246, 251

Boyle, H. P., 189, 251
Boys, S. F., 16, 175, 251
Brahms, J., 89ff, 251
Brahms, S., 89ff, 251
Bredig, G., 203, 251
Brewster, J. H., 16, 252
Brickell, W. S., 70, 252
Bridge, N. J., 56, 142, 252
Brienne, M-J., 159, 164, 187, 252
Brocki, T., 145, 252
Brongersma, H. H., 183, 252
Brouty, C., 173, 263
Brown, J. M., 216, 219, 252
Buchardt, O., 198, 251, 252
Buckingham, A. D., 56, 142ff, 152, 250, 251, 252
Buckingham, D. A., 218, 252

Cahn, R. S., 11, 177, 248, 252
Caldwell, D. J., 248, 252
Callomon, J. H., 52, 252
Calvin, M., 198, 237ff, 251, 252
Capra, J. D., 224, 252
Carmack, M., 64, 260
Cesario, M., 166, 171, 252
Chabay, I., 101, 252
Charney, E., 65, 252
Cheng, J. C., 27, 252
Chickos, J. S., 227, 252
Chion, B., 164, 252
Chung, S. Y., 94, 252
Ciardelli, F., 249, 252
Cochez, Y., 200, 252
Cohan, N. V., 138, 252
Cohen, J. B., 211, 252
Collet, A., 157ff, 164ff, 186ff, 249, 252, 258
Collman, J. P., 191, 252
Condon, E. U., 25, 34, 252
Cookson, R. C., 52, 252
Corey, E. J., 108, 217, 252
Coster, D., 179, 252
Cotton, A., 1, 19ff, 103, 151, 252, 253
Coulson, C. A., 69, 230, 235, 253

Cowman, M. K., 97, 253
Craddock, J. H., 209, 253
Craig, D. P., 73, 231ff, 253
Cram, D. J., 193, 213ff, 253
Crum Brown, A., 16, 253
Cushny, A. R., 220ff, 253
Cvitas, T., 82, 253

Dabrowiak, J. C., 218, 253
Debye, P., 229, 253
Dekkers, H. P. J. M., 30, 253
Delépine, M., 163, 253
Denning, R. G., 106, 253
de Vries, Hl., 102, 253
Dickerson, R. E., 239, 253
Dingle, R., 107, 253
Djerassi, C., 53, 56, 177, 251, 253, 260
Dörr, F., 82, 253
Dougherty, R. C., 152, 253
Draber, W., 223, 226, 253
Drake, A. F., 49, 55, 60, 141, 193ff, 210,
    228, 253, 254
Drude, P., 18ff, 254
Duax, W. L., 224, 254
Dudley, R. J., 24, 101, 254
Duesler, E. N., 110, 254, 257
Duggan, M., 133, 254
Duschinsky, R., 187, 254
Dwyer, F. P., 210, 216ff, 254

Easson, L., 221ff, 254
Ehrlich, P., 220, 254
Elgavi, A., 202, 254
Eliel, E. L., 189, 203, 254
Elliot, M., 226, 254
Elsbernd, H., 163, 254
Endo, I., 134, 254
Eyring, H., 35, 54, 68, 175, 248, 252, 264

Fajans, K., 39, 203, 251, 254
Faraday, M., 149, 254
Farina, M., 202, 254
Fasman, G. D., 97, 251, 253
Faulkner, T. R., 139, 254
Fay, R. C., 191, 254
Feofilov, P. P., 31, 254
Findlay, A., 154, 254
Finkelstein, R., 116, 254
Fischer, E., 9ff, 162, 177, 211, 220ff, 254
Flook, R. J., 185, 254
Fredga, A., 163, 255
Fresnel, A., 1ff, 255
Friedel, G., 98, 178, 255
Furlani, C., 134, 255

Gawroński, J., 66, 255, 258
Georgi, H., 241, 243, 255

Gerlach, H., 191, 255
Gernez, D., 186, 255
Gil-Av, E., 195ff, 255, 265
Goedicke, Ch., 188, 198ff, 255
Goldhaber, M., 242, 255
Golding, B. T., 218, 255
Gorin, E., 175, 182, 255
Gottarelli, G., 164, 255
Graham, T., 6, 255
Green, B. S., 201, 255
Griffith, M. C., 225, 255
Grignard, V., 212, 255
Grinter, R., 97, 255
Groen, M. B., 188, 255
Grosjean, M., 1, 26, 177, 255
Guye, P. A., 16, 255
Gyarfas, E. C., 210, 255

Hall, D. M., 209, 255
Hameka, H. F., 138, 252
Harada, N., 79, 83, 255
Harding, M. J., 97, 106, 255
Harris, M. M., 208, 255, 263
Harris, R. A., 245, 255
Hart, M., 242, 255
Hawking, S. W., 243, 255
Hawkins, C. J., 218, 255
Hellwinkel, D., 88, 256
Herpin, P., 173, 256
Herschel, J. F. W., 2, 6, 256
Hess, H., 192, 256
Hesse, G., 192, 256
Heyn, M. P., 50, 256
Hilmes, G., 120, 256
Hoffman, R., 199, 256
Höhn, E. G., 39, 57, 264, 265
Hollas, J. M., 83, 253
Holzwarth, G., 92ff, 97, 101, 139, 252,
    254, 256, 259
Horeau, A., 215ff, 256, 264
Hsu, E. C., 97, 256
Hudec, J., 63, 256
Hug, W., 143ff, 256
Hund, F., 244, 256
Huygens, C., 3, 256

Ingold, C. K., 11, 177, 252
Inskeep, W. H., 68, 256
Ito, T., 88, 125, 256
Iwamoto, E., 228, 256
Iwasaki, H., 183, 256
Izumi, Y., 151, 249, 256

Jacques, J., 157ff, 164ff, 177, 187ff, 249,
    252, 256, 258
Jaeger, F. M., 163, 216, 256
Jain, P. C., 133, 256

Japp, F. R., 150, 256
Jasperson, S. N., 27, 256
Jensen, H. P., 106, 256
Johnson, W. C. Jr., 27, 89, 175, 248, 257, 263
Judkins, R. R., 106, 257
Jungfleisch, M. E., 186, 227, 251, 257

Kagan, H., 197ff, 203ff, 216, 219, 257
Karagounis, G., 191, 257
Karush, F., 224, 257
Kasha, M., 237, 257
Keesom, W. H., 229, 257
Keiderling, T. A., 49, 140ff, 147, 257, 262, 263
Kelvin, Ld., 7, 257
Kemp, J. C., 27, 257
Kipping, F. S., 188, 257
Kirk, D. N., 54, 63, 256, 257
Kirkwood, J. G., 42ff, 175, 257
Kitaigorodsky, A. I., 166ff, 249, 257
Klemm, L. H., 193, 257
Klyne, W., 53ff, 178, 253, 257, 260
Korte, E.-H., 100ff, 257, 262
Krebs, H., 191, 257
Kress, R. B., 186, 257
Kuhn, R., 202, 208, 257
Kuhn, W., 20ff, 39ff, 103, 151, 175, 197, 203, 239, 257
Kuramoto, M., 173, 257
Kuroda, R., 49, 106, 137, 171ff, 234, 257, 258

Laarhoven, W. H., 200, 258
Labes, M. M., 101, 250
Lahav, M., 201ff, 250, 255
Landsteiner, K., 224, 258
Laporte, O., 22, 35, 241, 243, 258
Larsen, E., 88, 258
Laurent, A., 7ff, 258
Le Bel, J. A., 1, 6ff, 149ff, 258
Leclerq, M., 157ff, 187ff, 258
Le Fèvre, R. J. W., 56, 258
Legrand, M., 1, 26, 177, 255
Leiserowitz, L., 171, 196, 265
Letokhov, V. S., 247, 258
Lewis, L. L., 243, 258
Lifschitz, I., 216, 258
Lightner, D. A., 54, 66, 258
Linder, K. R., 192, 258
Lindman, K. F., 98, 258
Lipscomb, W. N., 224ff, 258
London, F., 229ff, 258
Lonsdale, K., 177, 263
Lorentz, H. A., 24ff, 258
Lowry, T. M., 5, 26, 103, 208, 248, 258

McCaffery, A. J., 94, 106, 250, 258
McKenzie, A., 203, 211ff, 258, 259
Malus, E. T., 1, 259
Mandel, R., 92, 259
Mann, F. G., 209, 259
Marckwald, W., 189, 203, 211, 259
Marcott, C., 139ff, 254, 259
Martin, R. H., 198, 200, 254, 257, 259, 264
Mason, S. F., 49, 85ff, 108, 132, 172, 177, 210, 258, 259
Mathieu, J.-P., 98ff, 103ff, 259
Matsumoto, M., 227, 259
Mavroyannis, C., 230ff, 259
Maxwell, J. C., 18, 259
Mead, C. A., 152, 259
Meier, G., 73, 259
Meyerhoffer, W., 158, 259
Mikes, F., 193, 259
Miller, S. L., 237, 260
Mislow, K., 203ff, 260
Mitscherlich, E., 6, 260
Miyoshi, K., 228, 260
Mjojo, C. C., 161ff, 171, 260
Moeller, T., 191, 260
Moffitt, W., 26, 89ff, 116ff, 182, 260
Mondine, F. A., 29, 260
Morrison, J. D., 203, 249, 260
Moscowitz, A., 54, 67, 139, 152, 254, 259, 260, 262
Mosher, H. S., 203, 249, 254, 260

Nafie, L. A., 49, 252, 253, 260, 261
Nakanishi, K., 79, 83, 255
Nakazaki, M., 200, 260
Neubert, L. A., 64, 260
Newman, M. S., 154, 188, 260
Newman, P., 189, 203ff, 260
Niketić, S. R., 218, 260
Nishikawa, K., 234, 260
Nishikawa, S., 179, 260
Nordén, B., 191, 193, 260
Numan, H., 56, 260
Nyholm, R. S., 249, 260

O'Conner, B. H., 97, 260
Odling, W., 8, 258
Ogston, A. G., 222ff, 260
Oró, J., 238, 261
Osborn, J. A., 219, 261
Overend, J., 139, 259

Palmer, R. A., 106, 251
Parascandola, J., 249, 261
Partington, J. R., 248, 261
Pasteur, L., 1, 6ff, 148ff, 186, 227, 261
Pauling, L., 225, 261

Peacock, R. D., 24, 88, 101, 253, 254, 259
Peart, B. J., 87ff, 106, 259, 261
Peterson, D. L., 92ff, 261
Pfeiffer, P., 202, 261
Pickard, R. H., 191, 261
Piper, T. S., 193, 261
Pirkle, W. H., 193, 261
Polavarapu, P. L., 49, 261
Pope, W. J., 189, 261
Porter, R., 224, 261
Portoghese, P. S., 224, 261
Powles, J. G., 233, 261
Prelog, V., 11, 177, 191, 212ff, 252, 261
Prigogine, I., 156, 261
Prosperi, T., 88, 210, 254, 258, 259

Ramsey, O. B., 248, 261
Raymond, K. N., 110, 254
Rein, D. W., 245ff, 261
Richardson, F. S., 31, 120, 256, 261, 262
Ritchie, P. D., 212, 262
Roberts, D. R., 49, 75, 79, 259
Robin, M. B., 59, 262
Rodberg, L. S., 241, 262
Roozeboom, H. W. B., 152ff, 262
Rosanoff, M. A., 9ff, 177, 262
Rosenfeld, J. S., 246, 262
Rosenfeld, L., 21, 244, 262
Ruch, E., 17, 214ff, 262
Russell, M. F., 27, 262

Saito, Y., 13, 88, 105ff, 172ff, 178ff, 217,
    256, 258, 262
Salam, A., 243, 262
Salvadori, P., 249, 252
Samori, B., 164, 255
Sargeson, A. M., 218, 254, 255, 262
Sarneski, J. E., 112, 262
Scacchi, A., 7, 262
Schellman, J. A., 37, 248, 262
Schnatterly, S. E., 27, 256
Schnepp, O., 27, 248, 262
Schrader, B., 100ff, 257, 262
Schramm, G., 239, 262
Schröder, I., 155ff, 262
Scott, A. I., 61, 262
Seal, R. H., 49, 75, 124, 258, 259
Shemin, D., 222, 262
Shibata, M., 125, 256
Shimura, Y., 158, 265
Simpson, W. T., 92ff, 261
Singh, B. K., 155, 262
Singh, R. D., 49, 262
Smirnoff, A. P., 216, 262
Snatzke, G., 56, 249, 262, 263
Snyder, P. A., 175, 263
Snyder, S. H., 224, 263

Spinat, P., 173, 263
Steinberg, I. Z., 31, 248, 263
Steinstrasser, R., 98, 263
Stephens, P. J., 27, 49, 140, 263
Stevenson, K. L., 198, 263
Stewart, A. W., 220, 263
Stodolsky, L., 243, 255, 263
Su, C. N., 49, 140, 143, 147, 263
Sudmeier, J. L., 109, 263
Sugano, S., 118ff, 263

Taft, R. W. 215, 263
Tanaka, J., 69, 182ff, 263
Theilacker, W., 189, 263
Thomas, W., 25, 263
Thulstrup, E. W., 73, 263
Tiffeneau, M., 212, 263
Timmermans, J., 164, 263
Tinoco, I. Jr., 31, 89, 96, 98, 263
Toftlund, H., 108, 263
Tsuchida, R., 191, 263
Turner, D. H., 248, 263
Turner, E. E., 208ff, 255, 263
Turner, F. E., 177, 263

Ugi, I., 214ff, 262
Ulbricht, T. L. V., 242, 263, 264
Urbach, F. L., 112, 262
Utsuno, S., 134, 254

Vaira, M. Di., 133, 264
Valentine, D., 216, 264
van der Waals, J. D., 153, 228, 264
Vanest, J.-H., 201, 264
van Laar, J. J., 155ff, 264
van't Hoff, J. H., 1, 8, 150, 152ff, 264
Van Vleck, J. H., 35, 116, 254, 264
Velluz, L., 177, 248, 264
Vester, F., 242, 264
Vigneron, J. P., 215, 264

Wagnière, G., 55, 71, 264
Wain, R. L., 223, 264
Walden, P., 9ff, 16, 169, 264
Walker, D. C., 249, 264
Wallach, O., 169, 264
Waser, J., 177, 264
Watanabe, T., 54, 264
Weber, G., 31, 264
Weigang, O. E., Jr., 39, 57, 248, 264, 265
Weinberg, S., 243, 265
Werner, A., 8ff, 103, 150, 163, 187, 265
Weinstein, S., 171, 195ff, 265
Whuler, A., 173, 263
Wick, G. C., 241, 265
Wigner, E. P., 241, 265
Wilczek, F., 241, 265

Wilen, S. H., 189, 249, 265
Wilkinson, G., 219, 261
Winther, C., 162, 189, 265
Woody, R. W., 65, 96, 265
Wood, J. S., 134, 265
Woodward, L. A., 145, 265
Woodward, R. B., 199, 256, 260
Worthing, C. R., 255, 265

Wu, C. S., 241, 265
Wurtz, A., 1, 265
Wynberg, H., 56, 188, 255, 260

Yamagata, Y., 247, 265
Yamamoto, M., 210, 228, 256, 265
Yamamoto, Y., 210, 228, 256, 265
Yamanari, K., 158, 265
Yamasaki, K., 195, 265
Yoneda, H., 173, 195, 228, 257, 260, 265
Yoshikawa, Y., 195, 265
Young, T., 3, 265
Yuki, H., 193, 265

Zaugg, H. E., 225, 265
Zel'dovich, B. Ya., 245, 265
Zorkii, P. M., 165ff, 265

SUBJECT INDEX

absolute stereochemical configuration, 9ff,
  68ff, 175–85
active racemate, 163, 173
alkaloids, 80–3, 140–3, 163, 189, 208, 211
amides, 89–94, 196
amino acids, 89ff, 171, 186–8, 195, 218ff,
  237ff
aniline chromophore, 74, 80–3, 182
anomalous X-ray scattering method for
  absolute configuration, 178–85
asymmetric synthesis, 211–9
asymmetric transformations, 202, 208–10
atrolactate synthesis, 212–6

biaryl series, 49, 72–9, 171–4, 186, 204–8
butan-2-ol, 175

Cahn–Ingold–Prelog convention, 11ff
calycanthine, 80–3, 140–3, 183
carbinols, 139, 147, 175, 203ff, 216
carbonyl chromophore, 37ff, 49, 51ff
charge conservation, 241
chelate coordination compounds, 13, 83–8,
  103–24, 150, 163, 169–73, 178,
  187, 191–6, 209ff, 216–9, 228,
  232–7
chelate ring conformation, 108–15, 217ff
chemical evolution, 237ff
chiral dimers, 42–50, 72–83, 182–4
chirality functions, 17, 215ff
cholesteric mesophase, 98–102, 200, 208
chromatographic optical resolution, 191–6
circular birefringence, 3–6, 19
circular dichroism,
  measurement, 26–32
  relation to ORD, 19ff
  rotational strength equation, 24
circular dichroism spectra,
  dihedral complexes, 87, 107, 232
  dimers, 75, 81, 141
  helical systems, 70, 90, 94
  liquid crystal, 100, 101
  single crystal, 106, 107
  symmetric chromophores, 53, 60

tabulations, 49, 56, 88, 106, 125
tetragonal complexes, 114, 115
trigonal bipyramid complexes, 133
vibrational, 141
circular intensity differential,
  in luminescence, 30ff, 81ff
  in Raman scattering, 143–7
circular photoequilibrium, 197
circular photolysis, 151, 197, 239, 242
circular photosynthesis, 151, 198ff
circularly-polarized radiation, 3–6, 19,
  26–33, 81ff, 144, 150ff, 196–200,
  242
close-pack lattice structures, 166–74
configurational correlations, 9–17, 52ff,
  162–4, 178
configurational nomenclature conventions,
  9–17
Cotton effect, 19–21, 37ff, 51ff
crystal facet growth, 176ff, 186
crystal-field theory, 39, 116ff
crystal-molecule,
  homomorphism, 6ff
  symmetry relations, 165–174

deoxyribonucleic acid (DNA), 94–7
diastereomeric photodiscrimination, 200ff
1,3-diene chromophore, 65ff
dihedral coordination compounds, 13,
  83–8, 103–24, 150, 163, 169–73,
  178, 191–6, 209ff, 216–9, 228,
  232–7
dipole strength, 25ff, 48, 86, 124
dipole–dipole interactions, 43, 74, 86,
  229ff
dipole–hexadecapole interactions, 120ff,
  127ff
dipole-length method, 66ff, 182–5
dipole–quadrupole interactions, 40ff,
  129–37, 231, 236
dipole-velocity method, 66ff, 182–5
dispersion energy, 152, 174, 229–37
dissymmetric forces, 148–52, 240ff
dissymmetry ratio (Kuhn), 21ff, 30, 48ff,

52, 97ff, 126, 129, 139, 145, 146, 194, 197ff, 207
1,2-dithiane chromophore, 64ff
double refraction, 1-6
double solubility rule (Meyerhoffer), 158, 187, 188
dynamic coupling mechanisms, 38ff, 55ff, 120-47

electromagnetic discrimination, 230-3
electron chirality, 240ff
elliptically-polarized radiation, 24, 27
enantiomeric excess, 197ff, 239ff
enantiomer–racemate phase equilibria, 152-62
enantiomorphous crystals, 6, 97, 133, 148, 161, 164-74, 176ff, 186, 201ff
entrainment, 186-8
enzymatic stereoselection, 203, 208, 220ff

Fischer–Rosanoff convention, 9ff, 177
fixed partial charge model, 138ff
fluorescence polarization,
    circular dissymmetry ratio, 30, 81
    linear anisotropy ratio, 31, 81
Friedel's law of X-ray scattering, violation by absorption, 178

Grignard reagent, 212ff

haptophore, 220, 224
helicene series, 69-73, 154, 188, 193, 198-201
helix models, 6, 16, 19, 88-102
high-performance liquid chromatography, 191-6
hydrogen bonding discrimination, 196

independent-systems model, 34ff
ion association, 110-2, 195
isotopic chirality, 56, 138ff, 183

ketones, 37, 49, 51-57, 151, 155, 197, 204-7, 213ff
key-and-lock hypothesis, 220ff
kinetic resolution, 203-8

Laporte selection rule, 22, 35, 241, 243
lattice structures of enantiomer and racemate, 164-74
less-soluble diastereomer method,
    configurational correlation, 162-4
    optical resolution, 189-91, 195
linear birefringence, 5, 19, 26ff, 98, 106
linear dichroism, 19, 73, 91ff, 101, 106
liquid crystals, 98-102, 176

magnetic dipole, 21ff, 35ff, 52, 59, 65, 97, 103ff
mandelates, 140, 154, 171, 203, 212, 227
menthol, 203, 212, 227
molecular rotation, 2, 23
molecule–wavelength ratio (Boltzmann), 19ff, 48, 98ff, 143ff
multipole expansion, 40, 120ff
mutarotation, 208, 212

neutron,
    anomalous diffraction, 183ff
    spin rotation, 243

octahedral coordination complexes, 8-17, 103ff, 125, 163
octant rule, 37ff, 52ff
olefin chromophore, 49, 58ff
one-electron theory, 21, 34-8, 54, 116ff
optical isomerism,
    molecular chirality, 6-17
    nomenclature conventions, 9-17
    wavefunction paradox, 244
optical purity, 193, 202, 247
optical resolution,
    chromatographic methods, 191-6
    diastereomer separation, 189-91
    entrainment, 186-8
optical rotation of quartz,
    infrared region 18ff, 242
    ultraviolet region, 18ff, 242
    visible region, 1ff
    X-ray region, 242
optical rotatory dispersion,
    equations, 2, 18, 21
    measurement, 23ff
    relation to CD, 19ff
oscillator strength, 25

parity non-conservation, 240-7
particle chirality, 240ff
pavine, 193ff
Pfeiffer effect, 202, 208-10
pharmaceuticals, 220ff
phase-equilibria methods for relating configuration, 162-4
photon chirality, 239ff
plane-polarized radiation, 1-6, 31-3, 81ff, 144
plastic crystals, 6, 161, 233
point groups of chiral molecules, 22
polarizability, 16, 20, 39ff, 56, 121-47
polarization direction, transition moment, 22, 31, 35ff, 59, 68, 73ff, 83ff, 89ff, 105ff
polyacrylate chiral adsorbents, 192
polypeptides, 88-94, 239ff

polysaccharides, 88, 191–6, 238ff
pseudoscalar properties, 17, 22, 37,
　117–20, 246

quarter-wave plate, 5, 26
quartz crystal,
　enantiomorphism, 6
　piezoelectric transducer, 27
　optical rotation, 1, 18, 242
quasi-racemate, 163, 173

racemate lattice structures, 164–74
racemisation half-life, 69, 192, 239
radiation field, dipole approximation, 18
Raman optical activity, 143–7
rotational strength, 21ff, 36–48, 70, 86, 90,
　118, 121, 139, 231

sector rules, 37, 52, 57ff, 116ff
selection rules, 22, 35, 38, 241
Sephadex chromatography, 194–6
solid-state photodiscrimination, 201ff
space groups, lattice packing factors,
　165–74
specific rotation, 2
spin–orbit interaction, 245
spontaneous optical resolution on crystal-
　lisation, 152–9, 164, 186–8, 190
sugar series, 211, 237
sum-rule,

oscillator strength, 25
rotational strength, 25, 118, 233

tartrates, 6, 19, 140, 148, 173, 177, 186–9,
　193–6, 209ff, 227
tetragonal coordination complexes, 113ff
tetrahedral coordination complexes, 49,
　125–32
three-point contact model, 221ff
time reversal, 241
transition moment,
　electric dipole, 21ff, 36ff, 67, 104ff
　electric multipole, 38ff, 104ff
　magnetic dipole, 21ff, 36ff, 52, 104ff
trigonal bipyramid coordination
　complexes, 132–7
Tröger's base, 191ff

van't Hoff solubility rule, 153, 187
vitalism and optical activity, 150, 211

Walden constant, 157
Walden inversion, 9–11, 163
Walden rule, 171
Wallach rule, 171
weak nuclear interaction, 240–7

X-ray,
　anomalous diffraction, 177ff
　optical rotation, 242